CHEMICAL AND BIOPROCESS ENGINEERING

Trends and Developments

CHEMICAL AND BIOPROCESS ENGINEERING

Trends and Developments

Edited by
Shirish H. Sonawane, PhD, Y. Pydi Setty, PhD, and
Srinu Naik Sapavatu, PhD

Apple Academic Press Inc.
3333 Mistwell Crescent
Oakville, ON L6L 0A2
Canada

Apple Academic Press Inc.
9 Spinnaker Way
Waretown, NJ 08758
USA

©2015 by Apple Academic Press, Inc.

Exclusive worldwide distribution by CRC Press, a member of Taylor & Francis Group

No claim to original U.S. Government works

Printed in the United States of America on acid-free paper

International Standard Book Number-13: 978-1-77188-077-0 (Hardcover)

All rights reserved. No part of this work may be reprinted or reproduced or utilized in any form or by any electric, mechanical or other means, now known or hereafter invented, including photocopying and recording, or in any information storage or retrieval system, without permission in writing from the publisher or its distributor, except in the case of brief excerpts or quotations for use in reviews or critical articles.

This book contains information obtained from authentic and highly regarded sources. Reprinted material is quoted with permission and sources are indicated. Copyright for individual articles remains with the authors as indicated. A wide variety of references are listed. Reasonable efforts have been made to publish reliable data and information, but the authors, editors, and the publisher cannot assume responsibility for the validity of all materials or the consequences of their use. The authors, editors, and the publisher have attempted to trace the copyright holders of all material reproduced in this publication and apologize to copyright holders if permission to publish in this form has not been obtained. If any copyright material has not been acknowledged, please write and let us know so we may rectify in any future reprint.

Trademark Notice: Registered trademark of products or corporate names are used only for explanation and identification without intent to infringe.

Library and Archives Canada Cataloguing in Publication

Chemical and bioprocess engineering : trends and developments / edited by Shirish Sonawane, PhD, Y. Pydi Setty, PhD, and Srinu Naik Sapavatu, PhD.

Includes bibliographical references and index.
ISBN 978-1-77188-077-0 (bound)

1. Chemical engineering. 2. Biochemical engineering. I. Sonawane, Shirish, author, editor II. Setty, Y. Pydi, author, editor III. Sapavatu, Srinu Naik, editor

TP155.C3285 2015   660   C2015-901594-4

Library of Congress Cataloging-in-Publication Data

Chemical and bioprocess engineering: trends and developments / Shirish Sonawane, PhD, Y. Pydi Setty, PhD, and Srinu Naik Sapavatu, PhD, editors.

pages cm
Includes bibliographical references and index.
ISBN 978-1-77188-077-0 (alk. paper)
1. Chemical engineering. 2. Biochemical engineering. 3. Sustainable development. I. Sonawane, Shirish, editor. II. Setty, Y. Pydi, editor. III. Sapavatu, Srinu Naik, editor.

TP155.C2975 2015   660--dc23   2015009007

Apple Academic Press also publishes its books in a variety of electronic formats. Some content that appears in print may not be available in electronic format. For information about Apple Academic Press products, visit our website at **www.appleacademicpress.com** and the CRC Press website at **www.crcpress.com**

ABOUT THE EDITORS

Shirish H. Sonawane, PhD

Dr. Shirish H. Sonawane is currently working as Associate Professor in Chemical Engineering Department at the National Institute of Technology, Warangal, Telangana, India. His research interests focus on synthesis of hybrid nanomaterials, cavitation-based inorganic particle synthesis, sonochemical synthesis of nanolatex, process intensification, and microreactor for nanoparticles production. Dr. Sonawane is a recipient of fasttrack young scientist project in the year 2007 from the Department of Science and Technology, Government of India. He has industrial experience from reputed chemical industries such as Bayer Polymers (R&D). After M. Tech he carried out research work in University of Dortmund Germany on emulsion polymerization for four months. Dr. Sonawane has published more than 50 research papers in reputed journals, seven book chapters, and six Indian patents applications. He was a recipient of prestigious the BOYSCAST Fellowship from the Department of Science and Technology Government of India in year 2009. He was a visiting academic and worked in the particle Fluid Processing Center, University of Melbourne, Australia. He is also a Heritage Fellow and worked in the chemical engineering department at Instito Superio Technico Lisbon, Portugal in 2013. He was elected young Associate Fellow from Maharashtra Academy of Science. He has completed six consultancies in the chemical industries and four research projects from the Indian Government. Currently ISRO and Ministry of Environment and Forest projects are going on. He has awarded eight MTech and two PhD, and currently two PhD and two MTech students are working with him. He is reviewer of several international journals, notably in chemical engineering science, industrial engineering chemistry research, chemical engineering journal, materials science, chemical engineering processing, and process intensification. He has a good liaison with industries such as Corning India, Calchem, Roha Carbonates, Felix Industries etc. Recently Corning has donated a €30,000 (approximately Rs. 20 Lacs) reactor to Dr. Sonawane's laboratory for working with pharmaceutical and chemical industries in the area of process intensification. Dr. Sonawane received the grant of Rs.80 Lacs for the Minicenter of Excellence in Sonoprocess Engineering for Nanoparticles in 2013 from the Ministry of Human Resources and Development (MHRD).

Y. Pydi Setty, PhD

Professor Y. Pydi Setty is currently working as Professor in the Chemical Engineering Department, National Institute of Technology, Warangal, Telangana, India. He received his MTech (1978) and PhD from IIT Madras (1984), India. His research interests are focused on hydrodynamics and RTD studies in a fluidized bed drying, circulating fluidized bed, biological wastewater treatment using fluidized bed bioreactor, and packed bed bioreactor modeling, simulation, and optimization of chemical engineering processes.

He received the Sir Ganga Ram Memorial Award in January 2000 for the paper titled "Study on heat transfer using limpet coil jacket," published in the year 1998 by IEI (India), Chemical Engineering Division, and for the paper titled "Total liquid holdup in a liquid-liquid-solid fluidized bed" he was given a certificate of merit by IEI (India), Chemical Engineering Division. His paper titled "Heat transfer studies using a limpet coil jacket" was selected as the best paper in the technical session at IEI (India) held at Visakhapatnam, India.

Currently he has ISRO Project for preparation of nano Al particles. He also completed the MHRD sponsored project for Curriculum Design and Development and Consultancy from Pilot Scale Production of Nano Additives Argonate Phase-1. He has 16 research articles in SCI journals and one patent, 16 articles in international and national conferences. Four PhD students have completed their theses, and two are working under his guidance.

Srinu Naik Sapavatu, PhD

Srinu Naik Sapavatu has obtained a PhD in chemical engineering on "Optimization of Parameters for Biological Denitrification of Wastewater using Gas-Liquid-Solid Bioreactor" from the National Institute of Technology, Warangal (2013); an MTech from Chemical Engineering from Osmania University (2003); and a BTech in Chemical Engineering from Nagarjuna University (2001), India. He is an Assistant Professor in the Chemical Engineering

Department at National Institute of Technology, Warangal, Telangana, India. Presently he is on academic loan and working at the University College of Technology, Osmania University, Hyderabad, India. He has 9 years of teaching experience, and his current research includes fluidization, nanofluids, and wastewater treatment. He has published eight articles in reputed journals and six international proceedings, and he guided six postgraduate students. One of his papers was awarded with best conference paper in Malaysia. He has visited the University of Massachusetts, Amherst, USA, and trained in electrospun cellulose nanofiber fabrication. He has conducted an international conference on "Chemical and Bioprocess Engineering-India."

CONTENTS

List of Contributors .... *xiii*

List of Abbreviations .... *xxi*

List of Symbols .... *xxiii*

Preface .... *xxv*

Introduction .... *xxvii*

PART I: CHEMICAL AND BIOPROCESS ENGINEERING

1. **Statistical Optimization of Culture Media Components for Enhanced Production of Sophorolipids from *Starmerella bombicola* MTCC 1910 Using Waste Frying Sun Flower Oil** 3

   Vedaraman Nagarajan, Srinivasa Kannan, and Venkatesh Narayana Murthy

2. **Optimization of Ethanol Production from Mixed Feedstock of Cassava Peel and Cassava Waste by Coculture of *Saccharomycopsis fibuligera* NCIM 3161 and Zymomonas Mobilis MTCC 92** 13

   Selvaraju Sivamani, Shanmugam Anugraka, and Rajoo Baskar

3. **Purification and Concentration of Anthocyanins from Jamun: An Integrated Process** 25

   J. Chandrasekhar and K. S. M. S. Raghavarao

4. **Effect of Pressure Drop and Tangential Velocity on Vane Angle in Uniflow Cyclone** 37

   K. Vighneswara Rao, T. Bala Narsaiah, and B. Pitchumani

5. **Multivariate Statistical Monitoring of Biological Batch Processes Using Corresponding Analysis** 47

   Sumana Chenna, Ankit Mishra, and Abhimanyu Gupta

6. **Extended Kalman Filter-Based Composition Estimation Using Simple Hybrid Model Approximation for a Reactive Batch Distillation Process** 55

   P. Swapna Reddy and K. Yamuna Rani

7. **Estimation of Solubility in Dilute Binary Solutions Using Molar Refraction, Dipole Moment, and Charge Transfer Functions** 67

   N. V. K. Dutt, K. Hemalatha, K. Sivaji, and K. Yamuna Rani

8. **Kinetic Study of Bechamp Process for Nitrobenzene Reduction to Aniline** 75

   Umesh Singh, Ameer Patan, and Nitin Padhiyar

9. **Flow Behavior of Microbubble-Liquid Mixture in Pipe** 91

   Rajeev Parmar and Subrata Kumar Majumder

10. **Methane Production by Butanol Decomposition: Thermodynamic Analysis** 101

    B. Rajesh Kumar, Shashi Kumar, and Surendra Kumar

11. **Novel Method for Increasing the Cleavage Efficiency of Recombinant Bovine Enterokinase Enzyme** 109

    A. Raju, K. Nagaiah, and S.S. Laxman Rao

12. **Application of the Thomas Model for Cesium Ion Exchange on AMP-PAN** 117

    Ch. Mahendra, P. M. Satya Sai, C. Anand Babu, K. Revathy, and K. K. Rajan

13. **Minimizing Post-Harvest Damage in Citrus** 125

    Harbant Singh, Ghassan Al-Samarrai, Muhammad Syarhabil, Sue Yin-Chu, and Boon Beng-Lee

14. **Development of Process Using Ionic Liquid-Based Ultrasound Assisted Extraction of Ursolic Acid from Leaves of *Vitex negundo* Linn** 137

    Anuja Chavhan, Merlin Mathew, Tapasvi Kuchekar, and Suyogkumar Taralkar

15. **Extraction of Chitin and Chitosan from Fishery Scales by Chemical Method** 147

    S. Kumari and P. Rath

16. **Study of Mixing in Shear Thinning Fluid Using CFD Simulation** 157

    Akhilesh Khapre and Basudeb Munshi

17. **Natural Convection Heat Transfer Enhancement in Shell and Coil Heat Exchanger Using Cuo/Water Nanofluid** 171

    T. Srinivas and A. Venu Vinod

18. **Homology Modeling of L-Asparaginase Enzyme from Enterobactor Aerogenes KCTC2190** 185

    Satish Babu Rajulapati and Rajeswara Reddy Erva

19. **Mixing of Binary Mixture in a Spout-Fluid Bed** 195

    B. Sujan Kumar and A. Venu Vinod

Contents

20. Effect of Various Parameters on Continuous Fluidized Bed Drying of Solids 207
G. Srinivas and Y. Pydi Setty

21. Study of Flow Behavior in Microchannels..215
S. Ilaiah, Usha Virendra, N. Anitha, C. E. Alemayehu, and T. Sankarshana

22. Solid Dissolution of Cinnamic and Benzoic Acids in Agitated Vessel with and without Chemical Reaction 227
B. Sarath Babu and P. Sreedhar

PART II: ENVIRONMENTAL ENGINEERING, NANOTECHNOLOGY, AND MATERIAL ENGINEERING

23. Energy Audit, Designing, and Management of Walking Beam Reheating Furnace in Steel Industry...241
Shabana Shaik, K. Sudhakar, and Deepa Meghavatu

24. Synthesis and Characterization of PVDF/PAN Hollow Fiber Blend Membrane for Surface Water Treatment 257
K. Praneeth, James Tardio, Suresh K. Bhargava, and S. Sridhar

25. Development of Epoxide Material from Vegetable Oil.............................269
Srikanta Dinda, Nihit Bandaru, and Radhev Paleti

26. Biosorption of Copper from Aqueous Solution Using *Oscillatoria splendida*...281
G. Baburao, P. Satyasagar, and M. Krishna Prasad

27. Fermentation of Starch and Starch-Based Packing Peanuts for ABE Production: Kinetic Study 293
Ayushi Verma, Shashi Kumar, and Surendra Kumar

28. Production of Holocellulolytic Enzymes by *Cladosporium cladosporioides* Under Submerged and Solid State Conditions Using Vegetable Waste as Carbon Source..299
Chiranjeevi Thulluri, Uma Addepally, and Baby Rani Goluguri

29. Removal of Ammonia from Wastewater Using Biological Nitrification...309
P. B. N. Lakshmi Devi and Y. Pydi Setty

30. Removal of Chromium from Aqueous Solution Using a Low Cost Adsorbents 321
P. Akkila Swathanthra and V. V. Basava Rao

31. **Preparation of PANI/Calcium Zinc Phosphate Nanocomposite Using Ultrasound Assisted *In Situ* Emulsion Polymerization and Its Application in Anticorrosion Coatings** 331

    B. A. Bhanvase, Shirish H. Sonawane, and M. P. Deosarkar

32. **Synthesis of 2K Polyurethane Coating Containing Combination of Nano Bentonite Clay, Nano CaCO3, and Polyester Polyol and Its Performance Evaluation on ABS Substrate** 343

    S. A. Kapole, B. A. Bhanvase, R. D. Kulkarni, and Shirish H. Sonawane

33. **Size-Controlled Biosynthesis of Ag Metal Nanoparticles Using Carrot Extract** 357

    Shrikaant Kulkarni

34. **A Combined Effect of Ultrasound Cavitation on Adsorption Kinetics in Removal of 4-[(4-Dimethyl Amino Phenyl) Phenyl-Methyl]-N, N-Dimethyl Aniline Along With Bio-Adsorbent** 369

    G. H. Sonawane, B. S. Bhadane, A. D. Mudawadkar, and A. M. Patil

35. **Formation of Iron Oxide Nanoparticles in Continuous Flow Microreactor System** 391

    Mahendra L. Bari, Shirish H. Sonawane, and S. Mishra

36. **Comparative Study of Production of Stable Colloidal Copper Nanoparticles Using Microreactor and Advanced-Flow Reactors®** 399

    M. Suresh Kumar, S. Niraj, Anshul Jain, Sonal Gupta, Vikash Ranjan, Makarand Pimplapure, and Shirish H. Sonawane

Index 417

LIST OF CONTRIBUTORS

Uma Addepally
Centre for Innovative Research, CBT, IST, Jawaharlal Nehru Technological University Hyderabad (JNTUH), Kukatpally, Hyderabad, Telangana – 500085, India, E-mail: vedavathi1@jntuh.ac.in

C. E. Alemayehu
University College of Technology, Osmania University, Hyderabad, Telangana, India.

Ghassan Al-Samarrai
School of Bioprocess Engineering, University Malaysia Perlis (UniMAP), Kompleks Pusat Pengajian Jejawi 3, 02600 Arau, Perlis, Malaysia.

N. Anitha
University College of Technology, Osmania University, Hyderabad, Telangana, India

Shanmugam Anugraka
Department of Biotechnology, Kumaraguru College of Technology, Coimbatore, Tamil Nadu, India.

C. Anand Babu
PSG College of Technology, Coimbatore, Tamil Nadu, India.

B. Sarath Babu
Department of Chemical Engineering, S. V. University College of Engineering, Tirupati, Andra Pradesh, India.

G. Baburao
Department of Chemical Engineering, GMRIT, Rajam, Srikakulam, E-mail: baburao01803@gmail.com

Nihit Bandaru
Department of Chemical Engineering, BITS Pilani Hyderabad Campus, Hyderabad, Telangana – 500078, India.

Mahendra L. Bari
University Institute of Chemical Technology North Maharashtra University Jalgaon MS India.

Rajoo Baskar
Department of Chemical Engineering, Kongu Engineering College, Perundurai, Erode, Tamil Nadu, India. E-mail: erbaskar@kongu.ac.in

Boon Beng-Lee
School of Bioprocess Engineering, University Malaysia Perlis (UniMAP), Kompleks Pusat Pengajian Jejawi 3, 02600 Arau, Perlis, Malaysia.

B. S. Bhadane
Department of Chemistry, Kisan Arts Commerce and Science College, Parola, Jalgaon Dist., Maharashtra – 425111, India.

B. A. Bhanvase
Department of Chemical Engineering Laxminarayan Institute of Technology Nagpur MS India.

Suresh K. Bhargava
Royal Melbourne Institute of Technology (RMIT), School of Applied Sciences, Melbourne, VIC-3001, Australia.

J. Chandrasekhar
CSIR-Central Food Technological Research Institute, Mysore, Karnataka – 570020, India.

Anuja Chavhan
Chemical Engineering Department, MIT Academy of Engineering, Alandi(D), Pune, Maharashtra – 412105, India.

Sumana Chenna
Process Dynamics and Controls Group, Chemical Engineering Sciences, Indian Institute of Chemical Technology, Hyderabad, Telangana, India. Email: sumana@iict.res.in, sumana.chenna@gmail.com

M. P. Deosarkar
Vishwakarma Institute of Technology, Pune, Maharashtra, India.

Srikanta Dinda
Department of Chemical Engineering, BITS Pilani Hyderabad Campus, Hyderabad, Telangana – 500078, India, E-mail: srikantadinda@gmail.com

N. V. K. Dutt
Process Dynamics and Control Group, Chemical Engineering Sciences, Indian Institute of Chemical Technology, Hyderabad, Telangana – 500607, India.

Rajeswara Reddy Erva
Department of Biotechnology, National Institute of Technology, Warangal, Telangana – 506004, India, E-mail: rajeshreddy.bio@gmail.com

Baby Rani Goluguri
Centre for Innovative Research, CBT, IST, Jawaharlal Nehru Technological University Hyderabad (JNTUH), Kukatpally, Hyderabad, Telangana – 500085, India.

Abhimanyu Gupta
Process Dynamics and Controls Group, Chemical Engineering Sciences, Indian Institute of Chemical Technology, Hyderabad, Telangana, India.

Sonal Gupta
Chemical Engineering Department, National Institute of Technology, Warangal, Telangana, India.

K. Hemalatha
Process Dynamics and Control Group, Chemical Engineering Sciences, Indian Institute of Chemical Technology, Hyderabad, Telangana – 500607, India.

S. Ilaiah
University College of Technology, Osmania University, Hyderabad, Telangana, India

Anshul Jain
Chemical Engineering Department, National Institute of Technology, Warangal, Telangana, India.

Srinivasa Kannan
Chemical Engineering Program, The Petroleum Institute, Abu Dhabi, UAE.

S. A. Kapole
Department of Chemical Engineering, Vishwakarma Institute of Technology, 666, Upper Indiranagar, Pune, Maharashtra – 411037, India.

List of Contributors

Akhilesh Khapre
Department of Chemical Engineering, National Institute of Technology Rourkela, Odisha, India.

Tapasvi Kuchekar
Chemical Engineering Department, MIT Academy of Engineering, Alandi(D), Pune, Maharashtra – 412105, India.

R. D. Kulkarni
University Institute of Chemical Technology North Maharashtra University Jalgaon, Maharashtra – 402103, India.

Shrikaant Kulkarni
Vishwakarma Institute of Technology, Pune, Maharashtra, India, E-mail: srkulkarni21@gmail.com

Brajesh Kumar
Department of Chemical Engineering, Indian Institute of Technology Roorkee, Roorkee, Uttarakhand – 247667, India.

Shashi Kumar
Department of Chemical Engineering, Indian Institute of Technology Roorkee, Roorkee, Uttarakhand – 247667, India.

B. Sujan Kumar
Department of Chemical Engineering, National Institute of Technology, Warangal, Telangana, India.

Surendra Kumar
Department of Chemical Engineering, Indian Institute of Technology Roorkee, Roorkee, Uttarakhand – 247667, India, E-mail: skumar.iitroorkee@gmail.com

M. Suresh Kumar
Chemical Engineering Department, National Institute of Technology, Warangal, Telangana – 506004, India.

S. Kumari
Department of Chemical Engineering, Research Scholar, National Institute of technology, Rourkela, Odisha, India, E-mail: Suneetak7@gmail.com

P. B. N. Lakshmi
Department of Chemical Engineering, National Institute of Technology, Warangal, Telangana – 506004, India.

Ch. Mahendra
Fast Reactor Technology Group, Indira Gandhi Centre for Atomic Research, Kalpakkam, Tamil Nadu, India.

Subrata Kumar Majumder
Department of Chemical Engineering, Indian Institute of Technology Guwahati, Guwahati, Assam – 781039, India, E-mail: skmaju@iitg.ernet.in

Merlin Mathew
Chemical Engineering Department, MIT Academy of Engineering, Alandi(D), Pune, Maharashtra – 412105, India.

Deepa Meghavatu
Depertment of Chemical Engineering, Andhra University College of Engineernig, Andhra University (A), Vishakapatnam, Andra Pradesh, India.

Ankit Mishra
Process Dynamics and Controls Group, Chemical Engineering Sciences, Indian Institute of Chemical Technology, Hyderabad, Telangana, India.

S. Mishra
University Institute of Chemical Technology, North Maharashtra University, Jalgaon, Maharashtra – 425001, India.

A. D. Mudawadkar
Department of Chemistry, Kisan Arts Commerce and Science College, Parola, Jalgaon Dist., Maharashtra – 425111, India.

Basudeb Munshi
Department of Chemical Engineering, National Institute of Technology, Rourkela, Odisha, India.

Venkatesh Narayana Murthy
Department of Chemical Engineering, St. Joseph's College of Engineering, Chennai, Tamil Nadu, India, E-mail: chemvenkatesh@gmail.com

K. Nagaiah
Gland Pharma Limited, D. P. Pally (Narsapur Road), Near Gandimaisamma 'X' Road, Hyderabad, Telangana – 500043, India.

Vedaraman Nagarajan
Chemical Engineering Division, Central Leather Research Institute, Chennai, Tamil Nadu, India.

T. Bala Narsaiah
Centre for Chemical Sciences & Technology, I.S.T, J.N.T.U.H, Kukatpally, Hyderabad, Telangana, India.

S. Niraj
Chemical Engineering Department, National Institute of Technology, Warangal, Telangana, India.

Nitin Padhiyar
Department of Chemical Engineering, Indian Institute of Technology Gandhinagar, Gujarat, India, E-mail: nitin@iitgn.ac.in

Radhev Paleti
Department of Chemical Engineering, BITS Pilani Hyderabad Campus, Hyderabad, Telangana – 500078, India.

Rajeev Parmar
Department of Chemical Engineering, Indian Institute of Technology Guwahati, Guwahati, Assam – 781039, India, Email: r.parmar@iitg.ernet.in

Ameer Patan
Department of Chemical Engineering, Indian Institute of Technology Gandhinagar, Gujarat, India.

A. M. Patil
Department of Chemistry, Kisan Arts Commerce and Science College, Parola, Jalgaon Dist., Maharashtra – 425111, India.

Makarand Pimplapure
Corning AFR Technologies Pune, India.

B. Pitchumani
Department of Chemical Engineering, Indian Institute of Technology, New Delhi, India.

List of Contributors

K. Praneeth
Royal Melbourne Institute of Technology (RMIT), School of Applied Sciences, Melbourne, VIC-3001, Australia; Membrane Separations Group, Chemical Engineering Division, Indian Institute of Chemical Technology (CSIR-IICT), Hyderabad, Telangana – 500007, India.

M. Krishna Prasad
Department of Chemical Engineering, GMRIT, Rajam, Srikakulam, Andra Pradesh, India.

K. S. M. S. Raghavarao
Department of Food Engineering, CSIR-Central Food Technological Research Institute, Mysore, Karnataka – 570020, India, E-mail: raghavarao@cftri.res.in

K. K. Rajan
Fast Reactor Technology Group, Indira Gandhi Centre for Atomic Research, Kalpakkam, Tamil Nadu, India.

A. Raju
Jawaharlal Nehru Technological University Hyderabad, Kukatpally, Hyderabad, Telangana – 500085, India, E-mail: arajubio@gmail.com

Satish Babu Rajulapati
Department of Biotechnology, National Institute of Technology, Warangal, Telangana – 506004, India.

K. Yamuna Rani
Process Dynamics and Control Group, Chemical Engineering Division, Indian Institute of Chemical Technology, Hyderabad, Telangana – 500607, India, E-mail: kyrani@iict.res.in

Vikash Ranjan
Chemical Engineering Department, National Institute of Technology, Warangal, Telangana, India.

V. V. Basava Rao
College of Technology Osmania University, Hyderabad, Telangana, India, E-mail: akhilasweety9@gmail.com

S. S. Laxman Rao
Sreenidhi Institute of Science and Technology, Hyderabad, Telangana, India.

K. Vighneswara Rao
Padmasri Dr. B.V. Raju Institute of Technology, Narsapur, Andra Pradesh, India, E-mail: vignesh.che@gmail.com

P. Rath
Department of Chemical Engineering, National Institute of technology, Rourkela, Odisha, India.

P. Swapna Reddy
Process Dynamics and Control Group, Chemical Engineering Division, Indian Institute of Chemical Technology, Hyderabad, Telangana – 500607, India.

K. Revathy
Fast Reactor Technology Group, Indira Gandhi Centre for Atomic Research, Kalpakkam, Tamil Nadu, India.

T. Sankarshana
University College of Technology, Osmania University, Hyderabad, Telangana, India, E-mail: tsankarshana@yahoo.com

P. Satyasagar
Department of Chemical Engineering, GMRIT, Rajam, Srikakulam, Andra Pradesh, India.

P. M. SatyaSai
Waste Immobilization Plant, Bhaba Atomic Research Centre Facilities, Kalpakkam, Tamil Nadu, India.

Y. Pydi Setty
Department of Chemical Engineering, National Institute of Technology Warangal, Warangal, Telangana – 506004, India.

Shabana Shaik
Depertment of Chemical Engineering, Andhra University College of Engineernig, Andhra University (A), Vishakapatnam, Andra Pradesh, India, E-mail: shabana89.chem@gmail.com

Harbant Singh
School of Bioprocess Engineering, University Malaysia Perlis (UniMAP), Kompleks Pusat Pengajian Jejawi 3, 02600 Arau, Perlis, Malaysia, E-mail: harbant@unimap.edu.my

Umesh Singh
Department of Chemical Engineering, Indian Institute of Technology Gandhinagar, Gujarat, India.

K. Sivaji
Process Dynamics and Control Group, Chemical Engineering Sciences, Indian Institute of Chemical Technology, Hyderabad, Telangana – 500607, India.

Selvaraju Sivamani
Department of Biotechnology, Kumaraguru College of Technology, Coimbatore, Tamil Nadu, India.

G. H. Sonawane
Department of Chemistry, Kisan Arts Commerce and Science College Parola, Jalgaon Dist., Maharashtra – 425111, India, E-mail: drgunvantsonawane@gmail.com

Shirish H. Sonawane
Department of Chemical Engineering, National Institute of Technology, Warangal, Telangana – 506004, India.

M. P. Sreedhar
Department of Chemical Engineering, S. V. University College of Engineering, Tirupati, Andra Pradesh, India.

S. Sridhar
Membrane Separations Group, Chemical Engineering Division, Indian Institute of Chemical Technology (CSIR-IICT), Hyderabad, Telangana – 500007, India, E-mail: sridhar11in@yahoo.com

G. Srinivas
Department of Chemical Engineering, National Institute of Technology, Warangal, Telangana – 506004, India.

T. Srinivas
Department of Chemical Engineering, National Institute of Technology, Warangal, Telangana – 506004, India.

K. Sudhakar
EMD Department, Vizag Steel Plant, Visakhapatnam, Andra Pradesh, India.

P. Akkila Swathanthra
Department of Chemical Engineering, S.V. University, Tirupati, Andra Pradesh, India.

Muhammad Syarhabil
School of Bioprocess Engineering, University Malaysia Perlis (UniMAP), Kompleks Pusat Pengajian Jejawi 3, 02600 Arau, Perlis, Malaysia.

Suyogkumar Taralkar
Chemical Engineering Department, MIT Academy of Engineering, Alandi(D), Pune, Maharashtra – 412105, India.

James Tardio
Royal Melbourne Institute of Technology (RMIT), School of Applied Sciences, Melbourne, VIC-3001, Australia.

Chiranjeevi Thulluri
Centre for Innovative Research, CBT, IST, Jawaharlal Nehru Technological University Hyderabad (JNTUH), Kukatpally, Hyderabad, Telangana – 500085, India.

Ayushi Verma
Department of Chemical Engineering, Indian Institute of Technology Roorkee, Roorkee, Uttarakhand – 247667, India.

A. Venu Vinod
Department of Chemical Engineering, National Institute of Technology, Warangal, Telangana – 506004, India. E-mail: avv122@yahoo.com

Usha Virendra
Indian Institute of Chemical Technology, Hyderabad, Telangana, India.

Sue Yin-Chu
School of Bioprocess Engineering, University Malaysia Perlis (UniMAP), Kompleks Pusat Pengajian Jejawi 3, 02600 Arau, Perlis, Malaysia.

LIST OF ABBREVIATIONS

AA	acetic acid
AAS	atomic absorption spectrophotometer
ABE	acetone, butanol, ethanol
ABS	acrylonitrile butadiene styrene
AMP	ammonium molybdophosphate
BSA	bovine serum albumin
CA	correspondence analysis
CBD	chitin binding domain
CCD	central composite design
CMC	carboxy methyl cellulose
CTAB	cetyl trimethyl ammonium bromide
CTF	charge transfer functions
DA	degree of deacetylation
DM	demineralized
DNS	di-nitro salicylic acid
EKF	extended Kalman Filter
EU	ethylenic unsaturation
FESEM	field emission scanning electron microscopy
FO	forward osmosis
FTIR	fourier transform infrared
FT-NMR	Fourier transforms nuclear magnetic resonance
GC	gas chromatograph
HPLC	high performance liquid chromatography
IARI	Indian Agricultural Research Institute
ITCC	Indian Type Culture Collection
KF	Kalman filter
LLDPE	low density polyethylene tube
MAP	modified atmosphere packaging
MCA	multi-way correspondence analysis
MCFC	molten-carbonate fuel cell
MG	malachite green
MP	membrane pertraction
MSM	multivariate statistical monitoring
MWCO	molecular weight cut off

NB	nitrobenzene
NLE	non- local environment
NOC	normal operating conditions
NPG	neopentyl glycol
OFAT	one-factor-at-a-time
PAA	peroxyacetic acid
PAN	polyacrylonitrile
PAT	process analytical technology
PCA	principal component analysis
PCD	pitch circle diameter
PDA	potato dextrose agar
PIV	particle image velocimetry
PU	polyurethanes
PVC	poly(vinylchloride)
PVP	polyvinylpyrrolidine
RBD	reactive batch distillation
SCF	super computing facility for bioinformatic
SEM	scanning electron microscopy
SLE	solid liquid equilibrium
SLs	sophorolipids
SOFC	solid-oxide fuel cell
SPR	surface plasmon resonance
SSF	solid state fermentation
TEM	transmission electron microscopy
TMP	trans-membrane pressure
TMP	trimethylol propane
UF	ultrafiltration
VLLE	vapour-liquid-liquid equilibria
WFSFO	waste fried sunflower oil
XRD	X-ray diffraction
YEGS	yeast extracts glucose salt

LIST OF SYMBOLS

A_j	molar holdup on jth plate (mol)
D	distillate flow rate (mol/s)
$BuAc$	butyl acetate
$BuOH$	butanol
HAc	acetic acid
H_2O	water
L_j	liquid flow rate from jth plate (mol/s)
N	no. of stages excluding reboiler and condenser = 31
Q_0	heat duty on condenser-reflux drums (J/s)
Q_{n+1}	heat duty on reboiler (J/s)
R_{ij}	rate of reaction of ith component on jth plate (s^{-1})
t	time (s)
V	vapour flow rate (mol/s)
x_{ji}	mole fraction of ith component in liquid phase on jth plate
y_{ji}	mole fraction of ith component in vapour phase on jth plate
C_s	concentration of surfactant (ppm)
D	pipe diameter (m)
K	consistency of fluid (Pa.s^n)
L	length (m)
n	flow behavior index (-)
ΔP	pressure drop (N/m^2)
U_f	fluid velocity (m/s)
ρ_f	density of fluid (kg/m^3)
τ_w	wall shear stress (Pa)
γ_w	wall shear rate (s^{-1})
γ_a	apparent shear rate (s^{-1})

NOMENCLATURE

A	area of heat transfer (m^2)
C_p	specific heat (kJ kg^{-1} K^{-1})
d	diameter (m)
De	Dean number

g	gravity (m/s^2)
k	thermal conductivity (W m^{-1} K^{-1})
L	length of coil tube
Nu	Nusselt number
Ra	Rayleigh number
R_c	curvature radius of the coil (m)
Re	Reynolds number
T	temperature (K)
v	velocity (m/s)

GREEK LETTERS

ΔT	temperature difference (K)
μ	viscosity (kg m^{-1} s^{-1})
ρ	density (kg m^{-3})
ϕ	volume fraction
β	thermal expansion coefficient (K^{-1})
α	thermal diffusivity (m^2/s)
η	kinematic viscosity

SUBSCRIPT

BF	basefluid
crit	critical
i	inside
NF	nanofluid
NP	nanoparticle
o	outside
s	shell

PREFACE

This book, **Chemical and Bioprocess Engineering: Trends and Development**, is the result of 2013 international conference organized by the Department of Chemical Engineering at the National Institute of Technology, Warangal, India. Out of 165 articles, we have selected 36 articles to publish in this book. We received our inspiration from Professor T. Srinivasa Rao, Director of the National Institute of Technology, Warangal, and the Eminent Scientist Dr. B. D. Kulkarni of National Chemical Laboratory, Pune. They have inspired us to take up the task for writing this book.

This book covers different areas of chemical and biochemical engineering. Some of the important and recent topics in this field include CFD simulation, statistical optimization, process control, wastewater treatment, microreactors, fluidized bed drying, and hydrodynamic studies of gas liquid mixture in pipe. Some additional important and special topics include ultrasound-assisted extraction, process intensification, polymers and coatings, etc. Modeling of bioreactor and enzyme systems and biological nitrification are notable topics from biochemical engineering. We, the editors, are proud to bring out this special type of book that reports on these areas.

All the chapters went through a crucial process of peer review and were improved through the addition of the experts' comments and suggestions.

We would also like to acknowledge the team of Apple Academic Press, Ms. Sandra Jones Sickels, Vice President (Editorial and Marketing), Mr. Ashish Kumar, President and Publisher, and Mr. Rakesh Kumar from Apple Academic Press. Inc., USA, for their prompt and supportive attention to all our queries related to editorial assistance.

With all humility, we acknowledge the initial strength derived for this book from Professor A. B. Pandit, ICT Mumbai, and Professor S. Mishra and Professor R. D. Kulkarni from UICT Jalgaon for their unwavering encouragement. Dr. Shirish Sonawane wishes to acknowledge his wife Kiran, children, Satwik and Vedika, who suffered seclusion and neglect due to his involvement with this book for last six months. Dr. S. Srinu Naik acknowledges the support of his wife, Dr. Nagaveni, and son, Snithik Chauhan. Prof. Pydi Setty also acknowledges the support of his wife, Manikyam.

Dr. Shirish Sonawane
Professor Y. Pydi Setty
Dr. S. Srinu Naik

INTRODUCTION

This book is divided into two parts. The first part contains 22 chapters covering different areas of chemical and bioprocess engineering. Selected articles from the 2013 International Conference on Chemical and Biochemical Engineering in India, held at the Chemical Engineering Department of the National Institute of Technology, Warangal, are organized in two parts. The two parts cover a wide variety of interdisciplinary topics from nanoscience with emphasis on particle production to applications in wastewater treatment, energy materials, etc.

In Part I, we attempted to highlight recent trends in chemical and biochemical engineering, such as integration of processes syrup and concentrate, uniflow cyclone design and its optimization, batch fermentation process and para multivariate statistical MCA model analysis, extended Kalman filter (EKF) based state and parameter estimation for reactive batch distillation process, statistical optimization, newer feedstock for ethanol production, and optimization of these processes. Some of the reaction kinetics problems of aniline nitration reaction are also addressed in Chapter 8. Detailed experimental rheological studies of microbubbles and suspension are also reported as are several thermodynamics studies of petroleum feed-stock butanol decomposition. A novel method developed to increase the cleavage efficiency of recombinant bovine enterokinase using magnesium ion as cofactor is also included as part of the coverage of biochemical engineering. A detailed study of the development of an optimum condition for production of chitosan from fish scales is also reported in Chapter 15. CFD study of mixing in shear thinning fluid in stirred tanks is also reported in this book. Also included is heat transfer analysis, CuO in helical coil. Homology modeling of a 3-D structure of L-asparaginase from *Enterobacter aerogenes* KCTC2190 using Swiss-Model is also discussed. The hydrodynamic behavior of mixing of binary mixture in a spout-fluid bed is strongly influenced by the difference in properties of the respective particles and is also covered. The effect of various parameters of continuous fluidized bed drying is reported in Chapter 20. Flow behavior in microchannels by using CFD simulation and velocity profile at outlet of the microchannels is also reported. Also in this book the experimental study and work that involves mass transfer without chemical reaction and mass transfer with chemical reaction is examined.

The second section of this book consists of chapters related to energy, the environment, and nanotechnology and allied chemical engineering themes.

Energy generationand storage is one of the important challenges today, and chemical engineering will play an important role in this area; hence we felt it was important to include some coverage of subject area of materials engineering and energy. Chapter 23 reports on the finding of the use of proper equipment and optimization of air fuel mixture that can save significant energy and can be helpful for the environment also. Membrane technology is a very important area in terms of clean drinking water, and separation technology is a concern, which is explained in Chapter 24. Another chapter reports on the performance of the hollow fiber (HF) membrane synthesized by using PVDF as a semicrystalline polymer. Chapter 25 recounts a study on epoxidation of vegetable oil that can be carried out *in situ* with formed or preformed peroxyacid in the presence of an acidic catalyst. Chapter 30 deals with the removal of toxic heavy ions metals, mainly copper and their ions from pollutant industrial wastewater.

The development of a kinetic model for the production of acetone, butanol, and ethanol (ABE) by using fermentation of starch is also related in Chapter 27. In Chapter 28 vegetable biomass used for production of holocellulolytic enzymes is examined. The removal of ammonium from wastewater by using method of biological nitrification is reported in Chapter 29, and in Chapter 30, the removal of chromium from aqueous solution by using sawdust, fly ash, and bagasse as an adsorbent is reported. In Chapter 31 the preparation of PANI-calcium zinc phosphate nanocomposite by ultrasound assisted *in situ* emulsion polymerization of aniline is covered. The performance of 2K epoxy coatings in presence of nanoclay filler is related in Chapter 32. In Chapter 33 the biosynthesis of silver nanoparticles using natural reducing agents is described. Degradation and adsorption of azo dye Malachite Green (MG) that is widely used in textile dyeing by means of ultrasonic cavitation is also reported in Chapter 34. Synthesis of nanoparticles of iron oxide is examined in Chapter 35. In addition, a comparative study of the production of nanoparticles in microreactor and advanced flow reactor is presented in Chapter 36.

We trust that the varied chapters in this book will stimulate new ideas, methods, and applications in ongoing advances in this growing area of chemical and biochemical engineering.Through this book we have provided an exploration of the emerging areas in this field, such as water treatment, CFD simulation, nanotechnologies, polymer engineering, drying,fluidization, mass transfer, biochemical engineering, fermentation technologies, reactive distillation, microreactor, and more, that are poised to make the new revolution become a reality.

PART I
CHEMICAL AND BIOPROCESS ENGINEERING

CHAPTER 1

STATISTICAL OPTIMIZATION OF CULTURE MEDIA COMPONENTS FOR ENHANCED PRODUCTION OF SOPHOROLIPIDS FROM *STARMERELLA BOMBICOLA* MTCC 1910 USING WASTE FRYING SUN FLOWER OIL

VEDARAMAN NAGARAJAN[1], SRINIVASA KANNAN[2], and VENKATESH NARAYANA MURTHY[3*]

[1]Chemical Engineering Division, Central Leather Research Institute, Chennai, India;
[2]Chemical Engineering Program, The Petroleum Institute, Abu dhabi, UAE;
[3]Department of Chemical Engineering, St. Joseph's College of Engineering, Chennai, India; *E-mail: chemvenkatesh@gmail.com

CONTENTS

1.1	Introduction	4
1.2	Materials and Methods	5
1.3	Results and Discussion	8
1.4	Conclusions	11
	Keywords	11
	References	12

1.1 INTRODUCTION

Surfactants constitute an important class of industrial chemicals used widely in almost every sector of modern industry. Present worldwide production of surfactants is around 12.5 million tons per year worth approximately US$28 billion, growing at the rate of about 0.5 million tons per year. Around 60 percent of surfactant production is used in household detergents, 30 percent in industrial and technical applications, 7 percent in industrial and institutional cleaning, and 3 percent in personal care. Asia accounts for largest share of global detergents consumption at 33 percent, whereas Eastern Europe's soaps and detergents market is reported to grow at a faster rate than Asia [1]. Almost all of these surfactants are petroleum based and are produced by chemical processes, which often damaging to environment are leading to significant ecological problems. Hence interest in biosurfactants has been steadily increasing in recent years due to their eco-friendly nature [2].

Microbially produced surfactants which are referred to as biosurfactants must compete with surfactants in three aspects—cost, functionality, and production capacity to meet the needs of intended applications [3]. Biosurfactants like glycolipids, rhamnolipids that are environmental friendly, nontoxic, and active over wide range of temperature and pH, can be produced with reasonable yields by microorganisms. One class of biosurfactants that is being currently produced industrially is glycolipids [4]. They are made up of disaccharide sophorose linked to a hydroxy fatty acyl moiety. The hydroxy fatty acid itself counts in general 16 or 18 carbon atoms and can have one or more unsaturated bonds [5]. They possess good surface active properties and show excellent skin compatibility, a property that is very important for cosmetic and personal care applications [6, 7].

The cost of raw materials makes up approximately 40–50 percent of the whole expense for biosurfactants production [8]. To overcome this problem, processes should be coupled to utilization of waste substrates combating at the same time their polluting effect which balances the overall costs [9]. Frying oil is produced in large quantities for use in both food industry and at the domestic scale. After usage, cooking oil changes its composition and contains more than 30 percent of polar compounds depending on the variety of food, the type of frying, and the number of times it has been used [10]. In the present work along with glucose, waste fried sunflower oil (WFSFO) has been used as carbons substrate, to synthesize sophorolipids from *Starmerella bombicola*, with two objectives. Earlier researchers have already studied the synthesis of sophorolipids using glucose and sunflower oil from *Starmerella bombicola* and reported and yield of 38.6 g/l [11].

Starmerella bombicola a teleomorph of *Candida bombicola* is a less reported species capable of producing sophorolipids (SLs). Sophorolipids fermentation is a two-step process in which production occurs after growth when one of the nutrients, such as the nitrogen source, has been exhausted [12, 13]. Any attempt to increase the yield of a biosurfactants demands optimal addition of media components and selection of the optimal culture conditions that will induce the maximum or the optimal production [14]. Hence for maximizing the production of sophorolipids, the concentrations of different media components are critical. A very efficient way to enhance the value and quality of research and to cut the process development time and cost is through statistically designed experiments. The full factorial central composite design (CCD) consists of a complete 2^K factorial design, where K is the number of test variables, n_0 center points ($n_0 \geq 1$) and two axial points on the axis of each design variable at a distance α ($=2^{K/4}$) from the design center [15]. A full factorial composite design is used to acquire data to fit an empirical second-order polynomial model. For four factors, the quadratic model takes the following form,

$$Y = \beta_0 + \beta_1 x_1 + \beta_2 x_2 + \beta_3 x_3 + \beta_4 x_4 + \beta_{11} x_1^2 + \beta_{22} x_2^2 + \beta_{33} x_3^2 + \beta_{44} x_4^2 + \beta_{12} x_1 x_2 + \beta_{13} x_1 x_3 \\ \beta_{14} x_1 x_4 + \beta_{23} x_2 x_3 + \beta_{24} x_2 x_4 + \beta_{34} x_3 x_4 \quad (1.1)$$

where x_1, x_2, x_3, and x_4 represent four selected factors, concentration of glucose, WFSFO, yeast extract, and urea. Y is the response value, concentration of sophorolipids produced at the end of incubation period. β_i and β_{ij} are the model coefficient parameters estimated from regression results.

1.2 MATERIALS AND METHODS

1.2.1 MICROORGANISM

Starmerella bombicola MTCC 1910 was procured from Institute of Microbial Technology, Chandigarh, India. The organism was maintained on dextrose—peptone—yeast extract agar slants and subcultures every 4–6 weeks. Seed culture was developed from frozen *Starmerella bombicola* in 250 ml Erlenmeyer flasks containing sterile 50 ml medium. The composition of the inoculum medium was (g/l): glucose 100 and yeast extract 10. This was inoculated with one loop of the strain from agar slant and incubated for 48 hr at 30°C on a rotary shaker at 200 rpm [15, 16].

1.2.2 EXPERIMENTAL DESIGN

A 2^4 full factorial CCD with 7 replicates at the center point, 16 cube points and 8 axial points was used to optimize the production of sophorolipids. The upper and lower limits of range of variables covered in the present study are shown in Table 1.1.

TABLE 1.1 Experimental range and levels of independent variables

	−2	−1	0	1	2
Glucose (g/l)	0	50	100	150	200
WFSFO (g/l)	0	50	100	150	200
Yeast extract (g/l)	0	5	10	15	20
Urea (g/l)	0	0.5	1.0	1.5	2.0

The details of the experimental conditions generated by the software are shown in Table 1.2.

TABLE 1.2 F.F.C.composite design matrix of four variables in coded and natural units with the observed responses

Obs No	x_1	x_2	x_3	x_4	Glucose (g/l)	WFSFO (g/l)	Y. Ext (g/l)	Urea (g/l)	SLs Prod (g/l)
1	−1	−1	−1	−1	50	50	5	0.5	30.06
2	+1	−1	−1	−1	150	50	5	0.5	33.21
3	−1	+1	−1	−1	50	150	5	0.5	32.12
4	+1	+1	−1	−1	150	150	5	0.5	33.02
5	−1	−1	+1	−1	50	50	15	0.5	31.19
6	+1	−1	+1	−1	150	50	15	0.5	28.41
7	−1	+1	+1	−1	50	150	15	0.5	26.10
8	+1	+1	+1	−1	150	150	15	0.5	19.24
9	−1	−1	−1	+1	50	50	05	1.5	23.66
10	+1	−1	−1	+1	150	50	05	1.5	21.52
11	−1	+1	−1	+1	50	150	05	1.5	21.12
12	+1	+1	−1	+1	150	150	05	1.5	17.12
13	−1	−1	+1	+1	50	50	15	1.5	15.23
14	+1	−1	+1	+1	150	50	15	1.5	10.12
15	−1	+1	+1	+1	50	150	15	1.5	9.52
16	+1	+1	+1	+1	150	150	15	1.5	12.62
17	−2	0	0	0	0	100	10	1	13.24

Statistical Optimization of Culture Media Components

TABLE 1.2 *(Continued)*

Obs No	x_1	x_2	x_3	x_4	Glucose (g/l)	WFSFO (g/l)	Y. Ext (g/l)	Urea (g/l)	SLs Prod (g/l)
18	+2	0	0	0	200	100	10	1	17.25
19	0	−2	0	0	100	0	10	1	25.21
20	0	+2	0	0	100	200	10	1	18.25
21	0	0	−2	0	100	100	0	1	30.12
22	0	0	+2	0	100	100	20	1	24.85
23	0	0	0	−2	100	100	10	0	33.31
24	0	0	0	+2	100	100	10	2	15.92
25	0	0	0	0	100	100	10	1	33.12
26	0	0	0	0	100	100	10	1	33.52
27	0	0	0	0	100	100	10	1	32.14
28	0	0	0	0	100	100	10	1	32.56
29	0	0	0	0	100	100	10	1	33.29
30	0	0	0	0	100	100	10	1	32.92
31	0	0	0	0	100	100	10	1	33.10

All the experiments were carried out in duplicate and the amount of sophorolipids produced was estimated based on the average value of two experiments, which are also listed in Table 1.2. The optimal concentrations of the media components were obtained by solving the regression equation which was developed by fitting the process variables against the concentration of the sophorolipids produced.

1.2.3 RECOVERY, QUANTIFICATION, AND CHARACTERIZATION OF SOPHOROLIPIDS

The volume of entire culture medium after fermentation was measured and centrifuged at 10,000 rpm at 4°C for 20 min. The supernatant was extracted twice with equal volumes of ethyl acetate. Since the recovery and concentration of biosurfactants from fermentation broth largely determines the production cost, ethyl acetate was identified to be a better choice than the highly toxic chloro-organic compounds [17]. The pure sophorolipids were recovered from the hexane by vacuum evaporation and then accurately weighed. Sophorolipids synthesized using glucose and WFSFO was characterized by Fourier transforms infrared spectroscopy (FT-IR) and Fourier transforms nuclear magnetic resonance (FT-NMR) techniques.

1.3 RESULTS AND DISCUSSION

1.3.1 RESPONSE SURFACE METHODOLOGY

The amount of sophorolipids (g/l) produced, under each of the experimental conditions is listed in Table 1.2 and are related to the process variables using the regression equation by surface response methodology as,

$$Y = 3.65 + 0.37x_1 + 0.21x_2 + 1.3x_3 + 7.77x_4 - 0.002x_1^2 - 0.001x_2^2 - 0.051x_3^2 - 8.00x_4^2 \\ - 0.0005x_1x_2 - 0.002x_1x_3 - 0.06x_1x_4 - 0.003x_2x_3 - 0.006x_2x_4 - 0.31x_3x_4 \quad (1.2)$$

Where Y is the amount of sophorolipids and x_1, x_2, x_3, and x_4 are the coded values of the test variables glucose, WFSFO, yeast extract, and urea concentrations respectively. The significance of each coefficient was determined by student's t test and p values which are listed in Table 1.3.

TABLE 1.3 Analysis of variance for sophorolipids production

Source	Degrees of Freedom	Sum of Squares	Adjusted Sum of Squares	Adjusted Mean of Squares	t Value	p Value
Regression	14	1,816.04	1,816.036	129.717	14.57	0.000
Linear	4	1,045.12	292.427	73.107	8.21	0.001
Square	4	745.16	745.163	186.291	20.93	0.000
Interaction	6	25.75	25.749	4.291	0.48	0.812
Residual error	16	142.44	142.439	8.902		
Lack of fit	10	141.14	141.138	14.114	65.09	0.000
Pure error	6	1.30	1.301	0.217		
Total	30	1,958.47				

$R^2 = 0.927$; adj $R^2 = 0.864$; and $R = 0.96$.

The p values were used as a tool to determine the significance of each of the coefficients, which in turn, are necessary to understand the pattern of mutual interactions between the test variables [18]. The larger of the magnitude of the t-value and smaller of the value of p-value, the more significant is the corresponding coefficient [15]. Based on the p-values, interactions of glucose-

WFSFO, glucose-urea, and WFSFO-urea are insignificant. After removing the insignificant parameters, Eq. (1.2) can be simplified as,

$$Y = 3.65 + 0.37x_2 + 0.21x_3 + 1.3x_3 + 7.77x_4 - 0.002x_1^2 - 0.001x_2^2 - 0.051x_3^2 - 8.00x_4^2 \\ - 0.002x_1x_3 - 0.003x_2x_3 - 0.31x_3x_4 \quad (1.3)$$

A high R^2 value of 0.927 and R of 0.96 indicate the close proximity of the model equation with the experimental data. Among the first order terms, glucose has the lowest p value, which implies the first order effects of glucose is very significant, and it has to be maintained at zero level for the maximum production of sophorolipids. WFSFO has also got low p value (0.007). Increasing the concentration of oil resists the growth of microorganism as it creates stress to the microorganism and retards the production of sophorolipids.

The quadratic main effects of glucose and sunflower oil are also more pronounced than the others. This suggests that any minor change in these variables from their zero level values would cause a second-order positive or negative shift in the production of sophorolipids. The initial concentration of glucose is very important as *Starmerella bombicola* grows by consuming glucose, while the stress created by the presence of WFSFO after the depletion of glucose, produces extracellular product sophorolipids. Sophorolipids produced acts as emulsifier which facilitates the consumption of lipidic carbon source by microorganism. As concentration of the sunflower oil has to be maintained around zero level, concentration of the yeast extract can be minimized which increases the production of sophorolipids. Process economics favors such a reduction as yeast extract is costlier than sunflower oil. It is also well known that sophorolipids production occurs under nitrogen limiting conditions [12–16]. The summary of the results of the ANOVA analysis of the quadratic response surface model is presented in Table 1.4.

TABLE 1.4 The least squares fit and parameter estimates(significance of regression coefficients)

Term	Factor Errors	Standard Errors	t	p
Intercept	3.650	8.760	0.417	0.682
x_1	0.373	0.069	5.377	0.000
x_2	0.213	0.069	3.067	0.007
x_3	1.304	0.069	1.880	0.078
x_4	7.767	6.94	1.120	0.279
$x_1 \times x_1$	−0.001	0.000	−7.784	0.000
$x_2 \times x_2$	−0.001	0.000	−4.879	0.000

TABLE 1.4 *(Continued)*

Term	Factor Errors	Standard Errors	t	p
$x_3 \times x_3$	−0.051	0.022	−2.300	0.035
$x_4 \times x_4$	−8.003	−8.003	−3.586	0.002
$x_1 \times x_2$	0.000	0.000	0.002	0.999
$x_1 \times x_3$	−0.002	0.003	−0.801	0.435
$x_1 + x_4$	−0.006	0.030	−0.214	0.833
$x_2 \times x_3$	−0.003	0.003	−1.039	0.314
$x_2 \times x_4$	0.006	0.030	0.188	0.853
$x_3 \times x_4$	−0.312	0.298	−1.044	0.312

Regression Eq. (1.3) was solved to get optimal design conditions. The optimum levels of glucose, WFSFO, yeast extract, and urea were found to be 100 g/l, 87.6 g/l, 6.9 g/l, and 0.3 g/l respectively. Under the optimized experimental conditions the model predicts a highest concentration sophorolipids corresponding to 38.67 g/l. To ensure the repeatability of the optimum conditions, experiments were conducted at the optimized conditions, which resulted in a sophorolipids concentration of 37.44 g/l in close agreement with the model prediction.

^{1}H-NMR spectrum of the sophorolipid taken in chloroform is as shown in the Figure 1.1.

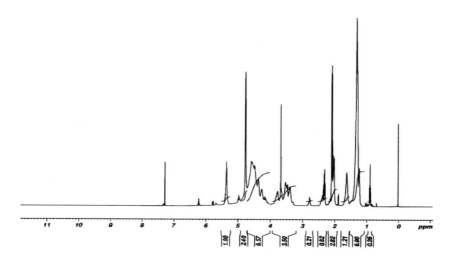

FIGURE 1.1 ^{1}H NMR spectra of sophorolipids synthesized from glucose and WFSFO.

It was assigned to a typical glycolipid type structure. The protons of glucose-H-1' and glucose-H-2" were resonated at 4.46 and 4.48, and 4.56 and 4.58 ppm respectively. The other protons of two glucoses were resonated at 3.50–4.40 ppm. Multiple signals of protons at 1.23–1.32 revealed the existence of a fatty acid chain moiety and signals at 5.35–5.36 revealed –CH=CH– group in the fatty acid chain. A signal at 2.09 ppm revealed the presence of (–COCH$_3$) group in the sophorolipid.

1.4 CONCLUSIONS

In the present work mixed carbon source glucose and WFSFO was employed for the synthesis of sophorolipids from *Starmerella bombicola*. It can be concluded that sunflower oil can be replaced by WFSFO without any major reduction in the amount of sophorolipids synthesized. Gas chromatographic analysis of WFSFO revealed that, there was an increase in the oleic acid level from 31.38 to 41.28 percent, which is desirable for the synthesis of sophorolipids. In the present work culture media components were optimized for the maximum production sophorolipids by using 2^4—full factorial response surface model based on central composite design. The regression model developed relates sophorolipids produced to the process variables which enable to analyze the individual, cumulative, and interactive effects of the medium components. The optimum condition for maximizing the sophorolipids production was obtained by solving the regression equation. Based on the model the optimum concentrations of glucose, WFSFO, yeast extract, and urea were found to be 100 g/l, 87.6 g/l, 6.9 g/l, and 0.3 g/l respectively, with the amount of sophorolipids produced being 38.67 g/l. The study identified the interaction between sunflower oil and yeast extract to be critical for the maximum production of sophorolipids. FT-IR and ^{1}H-NMR analysis confirmed the structure of glycolipid for the product formed.

KEYWORDS

- **Biosurfactant**
- **Mixed carbon substrate**
- **Response surface methodology**
- **Sophorolipids**
- ***Starmerella bombicola***

REFERENCES

1. Edser, C.; *Focus Surfactants*. **2006**, *5*, 1–2.
2. Banat, I. M.; Makkar, R. S.; and Cameotra, S. S.; *Appl. Microbiol. Biotechnol*. **2000**, *53*, 495–508.
3. Kosaric, N.; Cairns, W. L.; Gray, N. C. C.; Stechey, D.; and Wood, J.; *J. Am. Oil Chem. Soc*. **1984**, *61*, 1735–1743.
4. Ashyby, R. D.; Solaiman, D. K. Y.; and Foglia, T. A.; *Biotechnol. Lett*. **2008**, *30*, 1093–1100.
5. Asmer, H. J.; Lang, S.; Wagner, F.; and Wray, V.; *J. Am. Oil Chem. Soc*. **1998**, *65*, 1460–1466.
6. Inge, N. A.; Bogaert, V.; Saerens, K.; Muynck, C. D.; Develter, D.; Soetaert, W.; and Vandamme, W. J.; *Appl. Microbiol. Biotechnol*. **2007**, *76*, 23–34.
7. Tulloch, A. P.; Spenser, J. F. T.; and Gorin, P. A. J.; *Can. J. Chem*. **1962**, *46*, 1326–1338.
8. Kim, H. S.; Kim, Y.B.; Lee, B.S.; and Kim, E.K.; *J. Microbiol. Biotechnol*. **2005**, *15*, 55–58.
9. Kosaric, N.; *Pure. Appl. Chem*. **1992**, *64*, 1731–1737.
10. Kock, J. L. F.; Botha, A.; Blerh, J.; and Nigam, S.; *S. Afr. J. Sci*. **1996**, *50*, 513–514.
11. Vedaraman, N.;a nd Venkatesh, N.; *Pol. J. Chem. Technol*. **2010**, *12*, 9–13.
12. Cooper, D. G.; and Paddock, D. A.; *Appl. Environ. Microbiol*. **1984**, *47*, 173–176.
13. Davila, A. M.; Marchal, R.; and Vandecasteele, J. P.; *Appl. Microbiol. Biotechnol*. **1992**, *38*, 6–11.
14. Mukherjee, S.; Das, P.; and Sen, R.; *Trends Biotechnol*. **2006**, *24*, 509–515.
15. Khuri, A. L.; and Cornell, J. A.; Response Surfaces: Design and Analysis. New York: Marcel Dekker, Inc.;**1987**.
16. Daverey, A.; and Pakshirajan, K.; *Colloids Surf B Biointerfaces*. **2010**, *79*, 246–253.
17. Surubbo, L. A.; Farias, B. B.; and Campos-Takaki, G. M.; *Curr. Microbiol*. **2007**, *54*, 68–73.
18. Pal, M. A.; Vaidya, K. B.; Desai, K. M.; Joshi, R. M.; Nene, S. N.; and Kulkarni, B. B.; *J. Ind. Microbiol. Biotechnol*. **2009**, *36*, 747–756.

CHAPTER 2

OPTIMIZATION OF ETHANOL PRODUCTION FROM MIXED FEEDSTOCK OF CASSAVA PEEL AND CASSAVA WASTE BY COCULTURE OF *SACCHAROMYCOPSIS FIBULIGERA* NCIM 3161 AND *ZYMOMONAS MOBILIS* MTCC 92

SELVARAJU SIVAMANI[1], SHANMUGAM ANUGRAKA[1], and RAJOO BASKAR[2*]

[1]Department of Biotechnology, Kumaraguru College of Technology, Coimbatore, Tamil Nadu, India;

[2]Professor, Department of Chemical Engineering, Kongu Engineering College, Perundurai, Erode, Tamil Nadu, India, *E-mail: erbaskar@kongu.ac.in

CONTENTS

2.1	Introduction	14
2.2	Materials and Methods	14
2.3	Results and Discussion	17
2.4	Conclusion	23
	Keywords	24
	References	24

2.1 INTRODUCTION

The overall decline in the conventional oil production, emission of carbon monoxide due to burning of fossil fuels and threatening global climatic change highlights the need for an alternative energy fuel [1]. Ethanol serves this purpose well as it possesses the following advantages over conventional fuel: it reduces CO emissions, it can easily blend with petrol, ensures cleaner environment, and also renewable. Bioethanol can be produced from domestically abundant sources of biomass including agricultural and forestry residues, waste papers and other sizeable portions of municipal solid wastes [2].

Cassava is cultivated in tropical and subtropical regions of Africa, Asia, Latin America, and the Caribbean by poor farmers, many of them women, often on marginal land. It is the third largest source of food carbohydrates in the tropics after rice and maize—providing a basic diet for around 500 million people, 200 million of them in sub-Saharan Africa. The cassava plant gives the highest yield of carbohydrates per cultivated area among crop plants, only surpassed by sugarcane and sugar beets [3]. Cassava peel and waste are the wastes generated during the processing of their tuber to starch. They pose serious threat to environment because they are highly organic in nature [4]. Starch and total sugars present in cassava peel and waste are hydrolyzed to fermentable sugars by amylolytic yeast and then fermented to ethanol by yeasts or bacteria.

The objectives of the present study was to demonstrate the fitness of mixed feedstock of cassava peel and cassava waste for ethanol production through one step fermentation by employing one-factor-at-a-time (OFAT) and to optimize the process parameters for maximum ethanol production.

2.2 MATERIALS AND METHODS

2.2.1 MATERIALS

Cassava peel and waste were collected from starch processing industry in Namakkal district, Tamil Nadu. *Saccharomycopsis fibuligera* NCIM 3161 was received from National Collection of Industrial Microorganisms, National Chemical Laboratory, Pune, Maharashtra and *Zymomonas mobilis* MTCC 92 was received from Microbial Type Culture Collection, Institute of Microbial Technology, Chandigarh, Punjab. *Saccharomycopsis fibuligera* NCIM 3161 was grown on 100 ml sterile broth in 250 ml Erlenmeyer flask containing 3% (w/v) cassava starch, 0.05% (w/v) yeast extract and 0.05% (w/v) peptone by incubating at 28°C. *Zymomonas mobilis* MTCC 92 was grown separately on 100 ml serile broth in 250 ml Erlenmeyer flask containing yeast extract glucose salt (YEGS) medium (20% (w/v) glucose, 1.11% (w/v) yeast extract, 1 g/10 ml

$MgCl_2$, 1 g/10 ml $(NH_4)_2SO_4$, 1 g/10 ml KH_2PO_4) by incubating at 30°C. Both cultures were subcultured every fourth week.

2.2.2 BATCH FERMENTATION STUDIES

The batch fermentation studies were carried out in 250 ml Erlenmeyer flasks by mixing substrate and inoculum providing optimum environmental conditions for bioconversion. The flasks were kept on a thermostat incubated shaker for different periods of time. The supernatant was used for measurement of ethanol concentration.

2.2.2.1 EFFECT OF CASSAVA PEEL-TO-CASSAVA WASTE RATIO ON ETHANOL PRODUCTION

In single-step fermentation, the effect of cassava peel-to-cassava waste ratio was studied with different ratios of mixed feedstock of cassava peel-to-cassava waste (1:1, 2:1, and 3:1) maintaining constant substrate concentration of 50 g/L, 4.5 pH, temperature of 37°C, inoculum size of 10% (v/v), and agitation of 100 rpm for 72 h. The quantity of ethanol produced was quantified using dichromate method [5].

2.2.2.2 EFFECT OF SUBSTRATE CONCENTRATION ON ETHANOL PRODUCTION

The effect of substrate concentration was studied by varying from 10 to 90 g/L maintaining constant cassava peel-to-cassava waste ratio of 2:1, 4.5 pH, and temperature of 37°C, inoculum size of 10% (v/v) and agitation of 100 rpm for 72 h in single-step fermentation.

2.2.2.3 EFFECT OF REACTION TIME ON ETHANOL PRODUCTION

Single-step fermentation was studied by varying reaction time from 72 to 168 h to study the effect of reaction time on ethanol concentration maintaining constant cassava peel-to-cassava waste ratio of 2:1, substrate concentration of 50 g/L, 4.5 pH, temperature of 37°C, inoculum size of 10% (v/v), and agitation of 100 rpm.

2.2.2.4 EFFECT OF TEMPERATURE ON ETHANOL PRODUCTION

In single-step fermentation, the effect of temperature was studied by varying from 27 to 47°C maintaining constant cassava peel-to-cassava waste ratio of 2:1, substrate concentration of 50 g/L, 4.5 pH, inoculum size of 10% (v/v) and agitation of 100 rpm for 120 h.

2.2.2.5 EFFECT OF PH ON ETHANOL PRODUCTION

The effect of pH was studied by varying from 3.5 to 5.5 using 0.1 N HCl or 0.1 N NaOH. The cassava peel-to-cassava waste ratio of 2:1, substrate concentration of 50 g/L, temperature of 37°C, inoculum size of 10% (v/v) and agitation of 100 rpm for 120 h were maintained constant in single-step fermentation.

2.2.2.6 EFFECT OF INOCULUM SIZE ON ETHANOL PRODUCTION

Single-step fermentation was studied by varying inoculum size from 5 to 15% (v/v) with *Saccharomycopsis fibuligera* to *Zymomonas mobilis* ratio of 1:1 to study the effect of inoculum size on ethanol concentration maintaining constant cassava peel-to-cassava waste ratio of 2:1, substrate concentration of 50 g/L, 4.5 pH, temperature of 37°C, and agitation of 100 rpm for 120 h.

2.2.2.7 EFFECT OF AGITATION SPEED ON ETHANOL PRODUCTION

A mixed feedstock of cassava peel-to-cassava waste in the ratio of 2:1 with concentration of 50 g/L was inoculated with 10% (v/v) of inoculum for 120 h at a constant temperature of 37°C on shaker varying speed from 50 to 150 rpm for bioconversion studies for 120 h with pH of 4.5.

2.2.3 ANALYTICAL METHODS

Cell growth was measured by measuring optical density at 660 nm [6]. The total reducing sugars were determined in the broth after centrifuging at 5,000 rpm for 10 min by 3,5-dinitrosalicylic acid method [7]. Starch concentration was estimated colorimetrically by anthrone method [8].

2.3 RESULTS AND DISCUSSION

2.3.1 EFFECT OF CASSAVA PEEL-TO-WASTE RATIO ON ETHANOL PRODUCTION

Effect of various cassava peel-to-waste ratios on concentrations of biomass, starch, TRS, and ethanol was studied with 50 g/L of substrate. The results are shown in Fig. 2.1. The feedstock with cassava peel-to-waste ratio of 2:1 showed the highest ethanol concentration. There was less amount of reducing sugar in the feedstock with more cassava waste. So the mixed feedstock with more cassava peel was suitable for maximum ethanol production. Hence the cassava peel-to-waste ratio of 2 was used for further study.

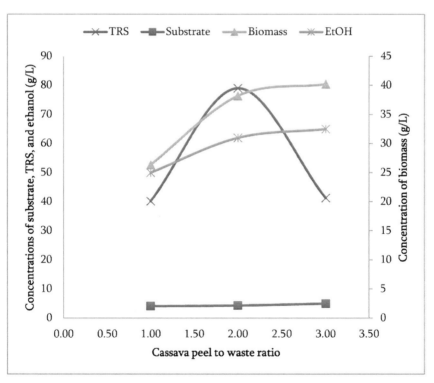

FIGURE 2.1 Effect of cassava peel-to-waste ratio on concentrations of biomass, substrate, TRS, and ethanol.

2.3.2 EFFECT OF SUBSTRATE CONCENTRATION ON ETHANOL PRODUCTION

Effect of different concentrations of substrate on ethanol co-fermentation by *Saccharomycopsis fibuligera* NCIM 3161 and *Zymomonas mobilis* MTCC 92 was investigated when the culture was grown at 37°C and shaking at 100 rpm. Fig. 2.2 shows that the ethanol concentration increased with increase in the substrate concentration from 10 to 50 g/L and then remains constant. From this study, the optimum concentration of substrate for ethanol fermentation was 50 g/L.

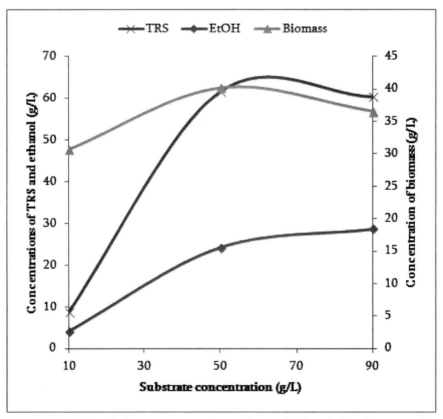

FIGURE 2.2 Effect of substrate concentration on biomass, TRS, and ethanol concentrations.

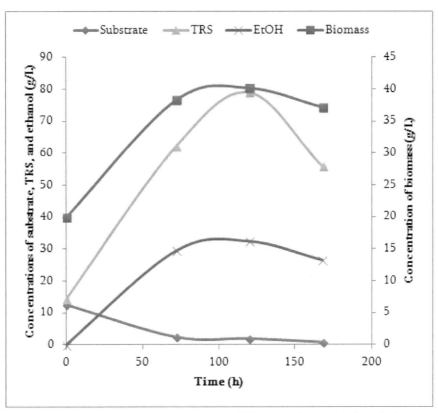

FIGURE 2.3 Effect of time on concentrations of biomass, substrate, TRS, and ethanol.

2.3.3 EFFECT OF TIME ON ETHANOL PRODUCTION

The results showed that increasing contact time from 72 to 168 h significantly varied the amount of ethanol produced during seven days of cultivation (Fig. 2.3). Therefore, five days was selected for the further experiments.

2.3.4 EFFECT OF TEMPERATURE ON ETHANOL PRODUCTION

The impact of the temperature on the ethanol fermentation was conducted in 50 g/L of substrate by controlling the growth temperatures at 27, 37, and 47°C. It was observed that the ethanol concentration increased as the temperature increased from 27 to 37°C and then decreased at 47°C (Fig. 2.4). Hence, the optimal temperature for this process is 37°C [9].

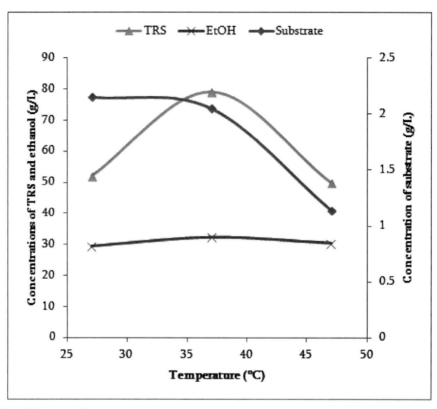

FIGURE 2.4 Effect of temperature on concentrations of substrate, TRS, and ethanol.

2.3.5 EFFECT OF PH ON ETHANOL PRODUCTION

The effect of pH on the fermentation of ethanol was carried out in 50 g/L of substrate at 37°C. Figure 2.5 shows that the highest ethanol production was obtained with pH of 4.5. The optimal pH for ethanol fermentation was 4.5. At pH of 3.5 and 5.5, the co-culture produced a lesser ethanol concentration [10]. Current results revealed that feedstock ratio, substrate concentration, pH, and temperature affected concentrations of biomass, starch, TRS, and ethanol.

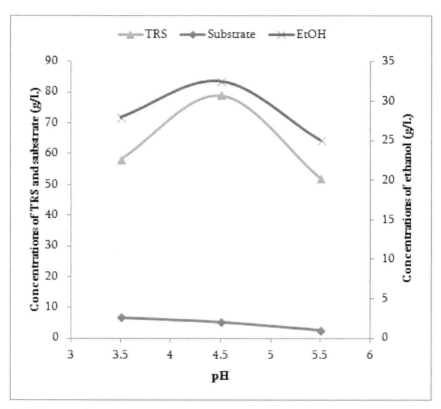

FIGURE 2.5 Effect of pH on concentrations of substrate, TRS, and ethanol.

2.3.6 EFFECT OF INOCULUM SIZE ON ETHANOL PRODUCTION

The results showed that increasing of the inoculum size from 5 to 15% (v/v) significantly altered the amount of ethanol produced after five days of cultivation (Fig. 2.6). Therefore, the 10% (v/v) inoculum size was selected for the next experiment.

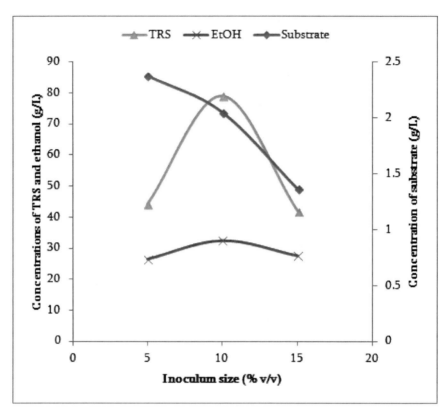

FIGURE 2.6 Effect of inoculum size on concentrations of substrate, TRS, and ethanol.

2.3.7 EFFECT OF AGITATION SPEED ON ETHANOL PRODUCTION

Effect of oxygen transfer rate was studied by varying shaking speed from 50 to 150 rpm. The result shows that the co-culture of *Saccharomycopsis fibuligera* NCIM 3161 and *Zymomonas mobilis* MTCC 92 with the shaking speeds of 100 rpm produced more ethanol than that cultivated at 50 and 150 rpm (Fig. 2.7). During oxygen limitation, the bacteria might consume less sugar for growth. Hence, the co-culture produced maximum ethanol at 100 rpm [11].

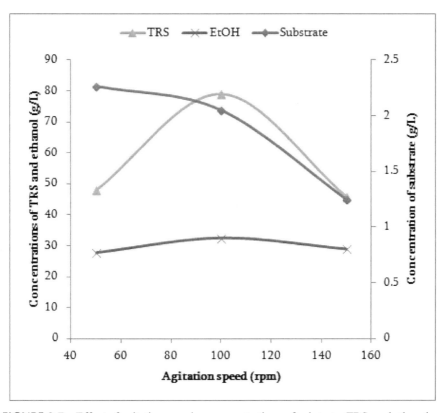

FIGURE 2.7 Effect of agitation speed on concentrations of substrate, TRS, and ethanol.

2.3.8 EXPERIMENTS UNDER OPTIMUM CONDITIONS

The optimal process conditions were cassava peel-to-waste ratio of 2, substrate concentration of 50 g/L, pH of 4.5, temperature of 37°C, reaction time of 120 h, inoculum size of 10% (v/v), and agitation speed of 100 rpm for mixed feedstock of cassava peel and waste. Under these conditions, the highest ethanol concentration achieved was 26.46 g/L for mixed feedstock (93.75% of the theoretical yield).

2.4 CONCLUSION

The co-culture of *Saccharomycopsis fibuligera* NCIM 3161 and *Zymomonas mobilis* MTCC 92 was demonstrated to be a potential feedstock for ethanol production through one step fermentation. The optimized parameters of cassava peel-to-waste ratio of 2, substrate concentration of 50 g/L, pH of 4.5, temperature of 37°C, reaction time of 120 h, inoculum size of 10% (v/v), and agitation speed of 100 rpm was obtained to produce 93.75% yield of ethanol.

KEYWORDS

- **Cassava peel**
- **Cassava waste**
- **Ethanol**
- ***Saccharomycopsis fibuligera***
- ***Zymomonas mobilis***

REFERENCES

1. Singh, A.; and Bishnoi, N. R.; *Ind. Crop. Prod.* **2012**, *37*, 334–341.
2. Mosier, N.; Wyman, C.; Dale, B.; Elander, R.; Lee, Y. Y.; Holtzapple, M.; and Ladisch, M.; *Bioresour. Technol.* **2005**, *96*, 673–686.
3. Ekundayo, J. A.; Biotransformation of cassava peel in the Niger Delta area of Nigeria using enhanced natural attenuation. In: J. E. Berry; and D. R. Kristiansen (Eds.), Fungal Biotechnology. London: Academic Press; **1980**, 244–270.
4. Jyothi, A. N.; Sasikiran, K.; Nambisan, B.; and Balagopalan, C.; *Process Biochem.* **2005**, *40*, 3576–3579.
5. Caputi, A. J.; Ueda, M.; and Brown, T.; *Am. J. Enol. Vitic.* **1968**, *19*, 160–165.
6. Maynard, A. J.; Methods in Food Analysis. New York: Academic Press; **1970**.
7. Miller, G. L.; *Anal. Chem.* **1959**, *31*, 426–428.
8. Hodge, J. E.; and Hofreiter, B. T.; Methods in Carbohydrate Chemistry. New York: Academic Press; **1962**.
9. Hostinova, E.; *Biol. Bratislava.* **2002**, *11*, 247–251.
10. Garg, S. K.; and Doelle, H. W.; *MIRCEN J.* **1989**, *5*, 297–305.
11. Sandhu, D. K.; Vilkhu, K. S.; and Soni, S. K.; *J. Ferm. Technol.* **1987**, *65*, 387–394.

CHAPTER 3

PURIFICATION AND CONCENTRATION OF ANTHOCYANINS FROM JAMUN: AN INTEGRATED PROCESS

J. CHANDRASEKHAR and K. S. M. S. RAGHAVARAO*

Department of Food Engineering, CSIR-Central Food Technological Research Institute, Mysore – 570 020, India; *E-mail: raghavarao@cftri.res.in

CONTENTS

3.1	Introduction	26
3.2	Materials and Methods	26
3.3	Results and Discussion	29
3.4	Conclusions	34
	Keywords	34
	References	34

3.1 INTRODUCTION

Demand of consumers for natural food colorants is driving manufacturers to find natural sources of colors and flavors. Anthocyanins is a group of water soluble natural pigments found primarily in red-toned fruits and vegetables. Anthocyanins have a limitation due to their relative instability, which causes loss of color, particularly under processing conditions. Extraction methods of anthocyanins from plant material are nonselective and yield pigment solutions with large amounts of other compounds such as sugars, sugar alcohols, organic acids, and amino acids which are detrimental to the stability of pigments. Stability of anthocyanins is strongly influenced by sugars, oxygen, pH [1], temperature, light [2], UV-light [3] and concentration of colorant itself [4]. In this regard, attention on the purification of these pigments has been on the rise. There are a good number of sources rich in the anthocyanins content, jamun (*Jambolao*) being one among them. Various medicinal properties such as stomachic, astringent, antiscorbutic, diuretic, antidiabetic, antioxidant, and antiproliferative have been reported for jamun anthocyanins [5–8]. Inspite of having enormous medicinal and food applications, there is a paucity of information regarding the purification and concentration of jamun anthocyanins. Hence, the objective of the present work is to develop an integrated process involving purification of anthocyanins from jamun employing adsorption followed by membrane pertraction (MP) and forward osmosis (FO) for the concentration of purified anthocyanins.

3.2 MATERIALS AND METHODS

3.2.1 CHEMICALS

Amberlite XAD7HP was procured from Sigma Aldrich, St. Louis, USA. Calcium chloride dihydrate ($CaCl_2 \cdot 2H_2O$) and sodium chloride (NaCl) were obtained from Ranbaxy chemicals, Mumbai, India. Jamun fruits were purchased from local market. All the chemicals used were of analytical grade.

3.2.2 MEMBRANES

Hydrophobic polypropylene (PP) membrane (Accurel, Enka, Germany) of pore size 0.05 µm was used for membrane pertraction. Hydrophilic membrane (Osmotek, Inc., Corvallis, OR, USA), used for forward osmosis, is asymmetric, which has a very thin semipermeable nonporous active skin layer of cellulose triacetate and porous support layer.

3.2.3 METHODS

3.2.3.1 PREPARATION OF CRUDE EXTRACT OF ANTHOCYANINS

Crude extract of anthocyanins was prepared according to the procedure mentioned briefly and details are given elsewhere [9].

3.2.3.2 ADSORPTION AND DESORPTION

Process parameters such as volume of feed and eluent, flow rate of feed, and eluent were standardized for the purification of jamun anthocyanins employing adsorption and the details are given elsewhere [10]. Purification of anthocyanins from the crude extract was carried out using nonionic acrylic ester adsorbent Amberlite XAD-7HP based on our earlier report [9]. Adsorption was carried out using 20 ml of adsorbent in a glass column (2 cm × 12 cm length) at room temperature (27 ± 1°C). Anthocyanin solution was passed through the glass column at a flow rate of 1 ml/min. Fractions of 10 ml were collected and concentration of anthocyanins was determined. After reaching adsorptive saturation, the adsorbent in the column was washed thoroughly using deionized water and desorption was carried out using acidified aqueous ethanol (50%, v/v) at a flow rate of 1 ml/min.

3.2.3.3 EVAPORATIVE PERTRACTION AND FORWARD OSMOSIS

A polyester mesh (0.25 mm, on osmotic agent side), membrane, and a Viton gasket (3.0 mm, on the feed side) were supported between two stainless steel (SS 316) frames of a flat module having an area of 0.0116 m^2 as shown in Figure 3.1.

Feed (anthocyanins solution) and stripping solutions were circulated on either side of the membrane in co-current mode using peristaltic pumps (Model 72-315-230, Barnant Company, IL, and USA). The ratio of feed to stripping solution was maintained as 1:3 for all the experiments. The transmembrane flux was calculated by measuring the increase in weight of stripping solution at regular intervals of 1 hr. Double distilled water and sodium chloride solutions were used as stripping solutions in MP and FO, respectively. All the experiments were carried out at room temperature (27 ± 1°C).

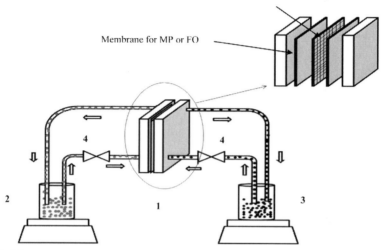

FIGURE 3.1 Flat membrane module for membrane pertraction/forward osmosis. (1) flat membrane module (2) feed reservoir (3) osmotic agent reservoir (4) peristaltic pump. FO—forward osmosis, MP—membrane pertraction.

3.2.3.4 THERMAL EVAPORATION

Crude extract of anthocyanins was concentrated by conventional process of thermal evaporation in agitated falling film vacuum evaporator (Model no. 04-012, Chemetron Corporation, USA) and used as control.

3.2.3.5 QUANTIFICATION OF ANTHOCYANINS

Concentration of anthocyanins (mg/L) was estimated according to the pH differential method [11] using the following equation

$$\text{Concentration of anthocyanins} = \frac{A \times \text{MW} \times \text{DF} \times 1000}{\varepsilon \times L} \quad (3.1)$$

where $A = [(A530 - A700) \text{ pH } 1.0 - (A530 - A700) \text{ pH } 4.5]$, MW is the molecular weight of anthocyanins (cyanidin-3-glycoside – 449.2 g/mol), DF the dilution factor, ε the extinction coefficient (26,900 L/cm mol) and L is the path length (1 cm).

3.2.3.6 ESTIMATION OF ALCOHOL, TOTAL SUGARS, SOLUBLE SOLIDS, AND DEGRADATION CONSTANT

The percentage of ethanol present in the feed was estimated according to the specific gravity bottle method, using water as a standard, at 25 ± 2°C. The measured density values were compared with standard values obtained according to AOAC method [12]. Soluble solid content, expressed as refractive index, was measured using refractometer (ERMA, Japan). Estimation of total sugars present in the solution was carried out using the Dubois method [13].

Degradation constant (K_d) of the anthocyanins was determined considering first-order degradation kinetics as per the following equation [14].

$$\ln\left(\frac{C_t}{C_0}\right) = K_d t \quad (3.2)$$

Where, C_0 is the initial concentration of anthocyanins (mg/L), C_t is concentration of anthocyanins (mg/L) at a specified time "t." The degradation studies were carried out by measuring the concentration of anthocyanins at regular intervals of one day, up to 15 days at room temperature (27 ± 1°C).

3.3 RESULTS AND DISCUSSION

3.3.1 ADSORPTION AND DESORPTION OF ANTHOCYANINS (DYNAMIC BREAK-THROUGH)

The dynamic break-through curve of anthocyanins on Amberlite XAD-7 resin was obtained based on the volume of effluent liquid and the concentration of solute therein and the results are presented in Figure 3.2.

It can be observed from the figure that anthocyanins were completely adsorbed before the cumulative volume of the fractions reaching 130 ml (break-through point). Above this point, the concentration of anthocyanins in the solution (fractions coming out of the adsorption column) increased until it reached a steady plateau when the cumulative volume of 260 ml (adsorption break-through) is reached.

Dynamic desorption curve was obtained based on the volume of the desorption solution required for complete recovery of anthocyanins and concentration of anthocyanins therein and is given in Figure 3.2. It can be seen from the figure that anthocyanins could be desorbed completely by approximately 80 ml of desorption solution.

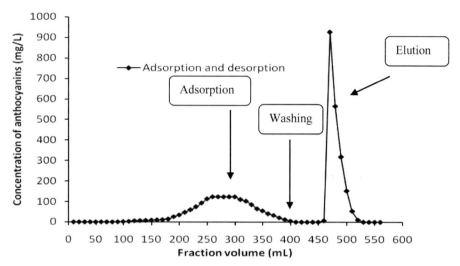

FIGURE 3.2 Dynamic breakthrough curve of anthocyanins.

3.3.2 MEMBRANE PERTRACTION

In order to use the anthocyanins for the intended applications, removal of alcohol is necessary. Besides this, increase in concentration of anthocyanins improves their stability through self-association [15]. Dealcoholisation of purified anthocyanins was carried out using hydrophobic polypropylene membrane of 0.05 μm pore size [16]. Feed (anthocyanins, 600 ml) and stripping (water, 1,600 ml) solutions were circulated at a constant flow rate of 50 ml/min. Transmembrane flux was calculated at regular interval of 1 hr and the results are presented in Figure 3.3.

The transmembrane flux decreased with an increase in process time, which can be attributed to the decrease in driving force, vapor pressure difference (gradient), for alcohol transport through the membrane. The transport of alcohol to the stripping side leads to an increase in vapor pressure of stripping solution, which in turn decreases the driving force. In order to maintain the maximum possible vapor pressure difference between the feed and stripping solutions, fresh water was circulated once in every hour on stripping side. Transmembrane flux was calculated at regular interval of 1 hr and the results are presented in Figure 3.3. It can be observed from the figure that the transmembrane flux was high in case of fresh water circulation on stripping side. However, decrease in transmembrane flux was observed with an increase in process time. This can be attributed to the decrease in alcohol content with respect to process time. The concentration of anthocyanins was found to increase from 397.78 (8° Brix) to

768.89 mg/L (13° Brix) with an increase in time during the dealcoholization. Nearly complete removal of alcohol (from 50 to 0.25%) from the alcoholic extract of the anthocyanin was achieved in 8 hr using this process.

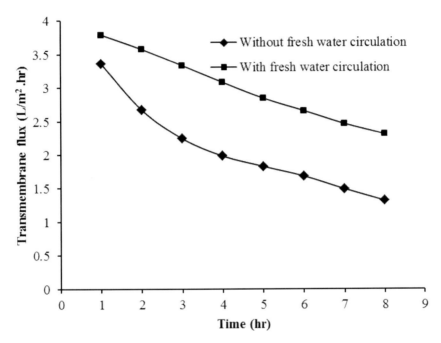

FIGURE 3.3 Effect of process time on transmembrane flux during membrane pertraction.

3.3.3 FORWARD OSMOSIS

After the removal of alcohol employing membrane pertraction, the anthocyanins solution was subjected to forward osmosis for further concentration. The feed (nearly alcohol free anthocyanin solution) and stripping solutions (NaCl) were circulated on either side of the membrane at a flow rate of 100 ml/min. Transmembrane flux was calculated at regular intervals of 1 hr and the results are presented in Figure 3.4.

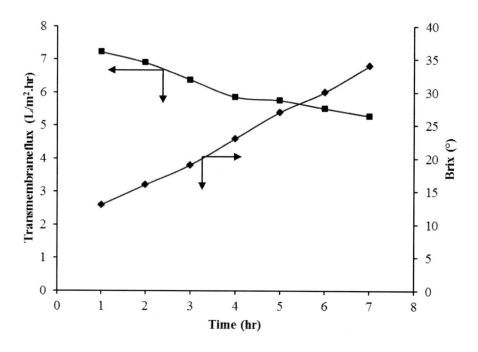

FIGURE 3.4 Change in transmembrane flux and brix during forward osmosis.

It can be observed from the figure that the transmembrane flux decreased with an increase in time. This decrease is mainly due to the decrease in driving force (concentration gradient) associated with the dilution of stripping solution and concentration of feed solution. Dilution of osmotic agent solution leads to the decrease in osmotic pressure difference, which in turn reduces the driving force for water transport from feed to osmotic agent side trough the membrane. The concentration of anthocyanins increased with an increase in process time, reaching a value of 2,321.86 mg/L (34° Brix) in 7 hr.

3.3.4 PHYSIOCHEMICAL CHARACTERISTICS OF ANTHOCYANINS

The anthocyanin concentrate produced by the integrated process (adsorption + MP + FO) was compared with the conventional process (thermal evaporation) and crude extract by evaluating the physiochemical characteristics of anthocyanins and the results are presented in Table 3.1.

TABLE 3.1 Physiochemical characteristics of extract of anthocyanins concentrated by different processes

Physiochemical Characteristic	Crude Extract	Adsorption + MP* + FO†	Thermal Evaporation
Anthocyanin concentration (mg/L)	203.15	2,321.86	1,345.67
Sugars (µg/ml)	190.35	0.23	761.88
Total soluble solids (°Brix)	2	34	20
Degradation constant (day^{-1})	1.07	0.21	1.96
Nonenzymatic browning index (−)	0.54	0.18	0.75
Color density (−)	0.52	7.88	2.31
Polymeric color (%)	0.36	7.16	4.13

*MP—membrane pertraction. †FO—forward osmosis.

It can be observed from the table that the highest concentration (2,321.86 mg/L) and brix (34° B) of anthocyanins were obtained in case of integrated process. Concentration of sugars (which are the major cause for product degradation) was observed to decrease from an initial 190.35 to 0.23 µg/ml after purification. Evaluation of the colorant properties such as nonenzymatic browning index and degradation constant confirmed the high stability of anthocyanins in case of the anthocyanins obtained by integrated process. Nonenzymatic browning index was found to be 0.18 and 0.75 for the anthocyanins obtained by integrated process and thermal evaporation, respectively. Colour density was found to increase from 0.52 to 7.88 and 2.31 in case of the integrated process and thermal evaporation, respectively. Polymeric color was found to be highest (7.16) in case of integrated process. Hence, it can be concluded that integrated process involving adsorption followed by membrane processes is a potential alternative for the purification and concentration of anthocyanins.

3.4 CONCLUSIONS

An integrated process for purification and concentration of anthocyanins from jamun was developed. Adsorption could be successfully employed using Amberlite XAD7HP for the purification of anthocyanins. Dealcoholization after the purification was carried out employing membrane pertraction. Forward osmosis was employed successfully for the concentration of dealcoholized anthocyanins (2,321.86 mg/L, 34° Brix). Evaluation of the final product with respect to total sugars, degradation constant, nonenzymatic browning index, and color density indicated an increase in the stability of anthocyanins obtained by the integrated process.

KEYWORDS

- **Adsorption**
- **Anthocyanins**
- **Forward osmosis**
- **Integrated process**
- **Membrane pertraction**

REFERENCES

1. Francis, F. J.; *Crit. Rev. Food Sci. Nutr.* **1989**, *28*, 273–314.
2. De Moura, S. C. S. R.; Tavares, P. E. R.; Germer, S. P. M.; Nisida, A. L. A. C.; Alves, A. B.; and Kanaan, A. S.; *Food Biopro. Tech.* **2012**, *5*(6), 2488–2496.
3. Pala, C. U.; and Toklucu, A. K.; *Food Biopro. Tech.* **2013**, *6*(3), 719–725.
4. Giusti, M. M.; and Wrolstad, R. E.; *Biochem. Eng. J.* **2003**, *14*(3), 217–225.
5. . Achrekar, S.; Kakliji, G. S.; Pote, M. S.; and Kelkar, S. M.; *In vivo.* **1991**, *5*(2), 143–147.
6. Aqil, F.; Gupta, A.; Munagala, R.; Jeyabalan, J.; Kausar, H.; Sharma, R.; Singh, I. P.; and Gupta, R. C.; *Nutr. Cancer.* **2012**, *64*(3), 428–438.
7. Migliato, K. F.; *Caderno de Farmacia.* **2005**, *21*(1), 55–56.
8. Li, L.; Adams, L. S.; Chen, S.; Killian, C.; Ahmed, A.; and Seeram, N. P.; *J. Agri. Food Chem.* **2009**, *57*, 826–831.
9. Chandrasekhar, J.; Madhusudhan, M. C.; and Raghavarao, K. S. M. S.; *Food Bioprod. Proc.* **2012**, *90*, 615–623.
10. Chandrasekhar, J.; Naik, A.; and Raghavarao, K. S. M. S.; *Sep. Pur. Technol.* **2014**, *125*, 170–178.
11. Giusti, M. M.; and Wrolstad, R. E.; Characterization and measurement of anthocyanins by UV-Visible spectroscopy. In: R. E. Wrolstad (Ed.), Current Protocols in Food Analytical Chemistry. New York: John Wiley & Sons, Inc.; **2001**, F1.2.1–F1.2.13.

12. Helrich, K.; Percentages by weight corresponding to various percentages by volume at 15.56°C (60°F) in mixtures of ethyl alcohol and water, Food Composition; Additives; Natural Colorants, vol. II, Official Methods of Analysis, 15th ed.; **1990**, 1256.
13. Dubois, M.; Gilles, K. A.; Hamilton, J. K.; Rebers, P. A.; and Smith, F.; *Anal. Chem.* **1956**, *28*, 350–356.
14. Ibarz, A.; Garza, S.; and Pagan, J.; *Int. J. Food Sci. Tech.* **2008**, *43*, 908.
15. Brouillard, R.; Chemical structure of anthocyanins. In: P. Markakis (Ed.), Anthocyanins as Food Color. New York: Academic Press; **1982**, 1–40.
16. Patil, G.; Madhusudhan, M. C.; Babu, B. R.; and Raghavarao, K. S. M. S.; *Chem. Eng. Proc. Inten.* **2009**, *48*, 364–369.

CHAPTER 4

EFFECT OF PRESSURE DROP AND TANGENTIAL VELOCITY ON VANE ANGLE IN UNIFLOW CYCLONE

K. VIGHNESWARA RAO[1*], T. BALA NARSAIAH[2], and B. PITCHUMANI[3]

[1]Padmasri Dr. B.V. Raju Institute of Technology, Narsapur, A.P, India; *E-mail: vignesh.che@gmail.com

[2]Centre for Chemical Sciences & Technology, I.S.T, J.N.T.U.H, A.P, India;

[3]Department of Chemical Engineering, Indian institute of technology, New Delhi, India

CONTENTS

4.1	Introduction	38
4.2	Mathemaical Models	38
4.3	Experimental Work	40
4.4	Results and Discussion	41
4.5	Conclusions	45
Nomenclature		45
Keywords		46
References		46

4.1 INTRODUCTION

Cyclones have been widely used for separating air borne particles from gases in a variety of engineering applications like mineral processing, petroleum refining, food processing, pulp and paper making, environmental cleaning, etc. A Cyclone is relatively simple to fabricate and it requires low maintenance and cost effective. Reverse flow cyclones have wide applications in industries as dust collectors. The main drawback of this device is the large pressure drops and requires a high energy. Uniflow cyclone, developed at the University of Western Ontario and studied in detail by Sumner et al. [1], Vaughan [2], and Gauthier et al. [3], which offer low pressure drops by restricting flow reversal. In uniflow cyclone, the centrifugal energy is imparted to particles by guided vanes installed at the entrance of the cyclone. It comprises of vortex finder, Annulus, Central core, and Vanes. The vortex finder is kept concentric to the cyclone. The dust is collected from the annulus and dust free air leaves the cyclone from the central core. Akiyama et al. [4] experimentally correlated the design and operational variables such as inlet velocity, density, cyclone diameter, radius, pitch etc. to pressure drop and collection efficiency. Ramachandran et al. [5] pioneered an attempt to arrive at a mathematical model that describes pressure drop and collection efficiency as a function of design and operating variables. Ashwani Malhotra [6] attempted the Studies on fine particle collection and pressure drop in uniflow cyclone. Maynard [7] studied an axial flow cyclone under laminar flow conditions and developed mathematical model for low Reynolds number. As per this model the efficiency at the vanes is dependent on the dimensions and the number of vane turns. Zhang [8] developed a model for a vane axial cyclone to predict the particle separation efficiencies both under laminar and perfect mixing flow conditions. The tangential air velocity, the diameters of the inner and outer tubes, length of the separation chamber, and vane angle have been considered in this model. Ramchaval [9] carried out experimental investigation to understand the effect of vane angle on tangential velocity and pressure drop in uniflow cyclones. However, the range of vane angles studied is very limited. In the present study, vanes of different angles have been fabricated and developed a uniflow cyclone unit to predict the optimum vane angle for low pressure drop, high tangential velocity and developed a methodology for measuring the tangential velocity. The experiments were conducted for different inlet velocities ranging from 4 to 13 m/s and at different vane angles between 20° to 60°.

4.2 MATHEMAICAL MODELS

The following model equations have been used for calculating the pressure drop and tangential velocity. Ramachandran et al. [5] developed expressions to calculate the pressure drop, tangential velocity across the cyclone with the assumption of no radial component of velocity.

$$V_è = 2\pi R_c \frac{V_i}{P} \tag{4.1}$$

$$\Delta P = \frac{\rho V_i^2}{2}\left[\left(\frac{V_{avg}}{V_i}\right)^2\left(1+\frac{L_{hel}}{D_h}\right)f - 1\right] \tag{4.2}$$

Ashwani Malhotra [6] derived an equation for pressure drop based on the pressure drop equation for reverse flow cyclone, leit and litch [1971]. This is given below

$$\Delta P = \frac{V_i^2 \cdot \rho_g}{2g\rho_P}\Delta H \tag{4.3}$$

Where

$$\Delta H = 1 + 2\phi^2\left[2\frac{D}{D_v}-1\right] + 2\left[\frac{D^2}{D_v^2}\right] \tag{4.4}$$

$$\phi = \frac{\sqrt{\frac{D_v}{2D}+2Y}-\sqrt{\frac{D_v}{2D}}}{Y}, Y = \frac{2A_s}{A_i} \tag{4.5}$$

Maynard [7] studied an axial flow cyclone under laminar flow conditions and the expression for tangential velocity is given below

$$V_\theta = \frac{2\pi R_c V_i}{\left(P^2 + 4\pi^2 R_c^2\right)^{1/2}} \tag{4.6}$$

The experimental tangential velocity is calculated by measuring the rpm of vanes connected to the bearing which is placed in the radial flow of cyclone. The rpm is measured by noncontact tachometer and the tangential velocity is given by the expression below.

$$V_\theta = \frac{2\times\pi\times R_c \times n}{60} \tag{4.7}$$

4.3 EXPERIMENTAL WORK

4.3.1 PROCEDURE

The unit of a uniflow cyclone has been designed and fabricated which is shown in Figure 4.1. Vanes of different angles viz. 20°, 30°, 45°, 60° have been fabricated with four grooves on the cylindrical core where the blades are fixed in the groove by stick paste. One end of the cyclone is fixed with vanes and the other end is connected to a vacuum pump. Tangential velocity is measured with the developed Instrument. Inlet velocity is measured with anemometer. A valve is connected between the cyclone exit pipe and vacuum pump with PVC pipes. This valve can be used to control the air flow rate. Different inlet velocities were obtained by adjusting the valve position. A digital pressure meter is connected in between inlet and outlet across cyclone to measure the pressure drop.

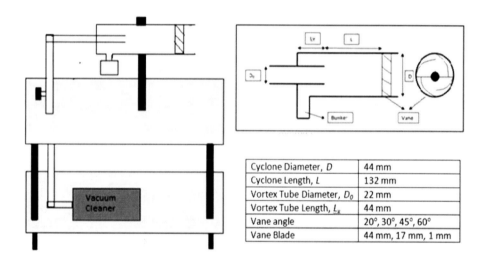

Cyclone Diameter, D	44 mm
Cyclone Length, L	132 mm
Vortex Tube Diameter, D_0	22 mm
Vortex Tube Length, L_x	44 mm
Vane angle	20°, 30°, 45°, 60°
Vane Blade	44 mm, 17 mm, 1 mm

FIGURE 4.1 Experimental Setup of Uniflow Cyclone.

4.3.2 TANGENTIAL VELOCITY MEASUREMENT

Experimental setup shown in Figure 4.2 consists of an arrangement for measuring the tangential velocity.

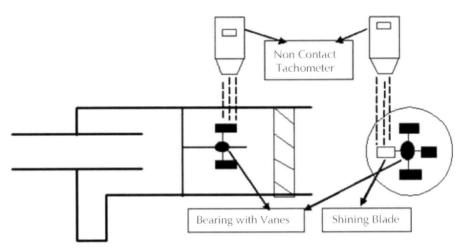

FIGURE 4.2 Tangential Velocity Measurement.

It has been made with flat vanes fixed to a small bearing which is rotated on a rod at center of cyclone. The vanes are rotated when there is swirl flow due to fixed angular vanes at the inlet of the cyclone. One of the flat rotating vanes is covered with reflected shining surface. The noncontact tachometer will focus the light on the shining blade of the rotated bearing and counted the number of revolutions per minute. The tangential velocity is then calculated by substitute the rpm in Eq. (4.7).

4.4 RESULTS AND DISCUSSION

Experiments were conducted at four angles using different vanes which are shown in Figures 4.3–4.7. The velocity is varied from 4 to 13 m/s and angles between 20° and 60°. Figures 4.3 and 4.4 shows the experiments results under different operating conditions. Figure 4.5 shows the effect of overall efficiency with inlet velocity. Figures 4.6 and 4.7 shows the comparison of experimental results with published model equations. Overall, the experimental data herein agreed with models very well for all experimental conditions. It has been observed from overall results the optimum vane angle for the uniflow cyclone is 45° which gives reasonably high tangential velocity, low pressure drop, and maximum efficiency for an inlet velocity of 12–13 m/s for all samples.

4.4.1 EFFECT OF TANGENTIAL VELOCITY

The variation of tangential velocity with inlet velocity for different vane angle is shown in Figure 4.3. The tangential velocity first increases then decreases with increase in vane angle, that is, from 20° to 30° the tangential velocity increases given the highest value at 30° but here the pressure is also high. Then from 30° to 60° tangential velocity decreases to a low value and the pressure drop also reduces. This is because of when the increase in inlet velocity increases the swirl flow induces a centrifugal force. As the centrifugal force increases tangential velocity increases, which increases the efficiency.

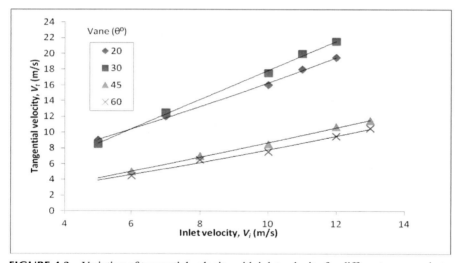

FIGURE 4.3 Variation of tangential velocity with inlet velocity for different vane angles.

4.4.2 EFFECT OF PRESSURE DROP

The effects of pressure drop with inlet velocity for different vane angle is illustrated in Figure 4.4. The pressure drop increases with increase in inlet velocity; this is because of increasing swirl flow inside the cyclone. The pressure drop increases with decrease in vane angle. It has been observed that when decrease in vane angle induces a more number of peaks inside the cyclone. Therefore the pressure is increases by the centrifugal force created from the no of turns of swirl flow due to no of peaks. The pressure drops given a low value at 60° due to least number of peaks and reduces the centrifugal force hence pressure drop reduced.

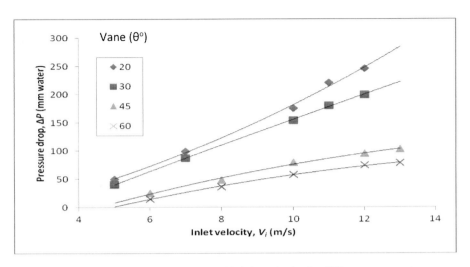

FIGURE 4.4 Variation of pressure drop with inlet velocity for different vane angles.

4.4.3 EFFECT OF OVERALL EFFICIENCY

The effects of overall efficiency with inlet velocity for different powder samples, that is, flyash, cement, and granite is illustrated in Figure 4.5.

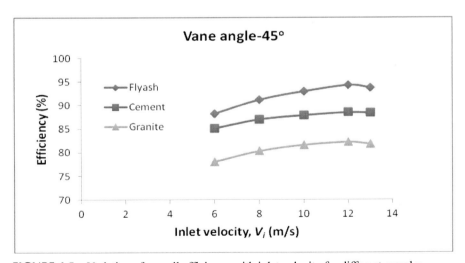

FIGURE 4.5 Variation of overall efficiency with inlet velocity for different samples.

The overall efficiency increases with increase in inlet velocity. This is so because as the inlet velocity increases the tangential velocity and centrifugal force increases therefore particle separation efficiency increases. Figure 4.5 shows the plot between overall efficiency and inlet velocity. In this figure the overall efficiency increases with inlet velocity that reaches a maximum value at 12 m/s and then starts to decrease. This happens because of the air velocity goes on increasing and it carry some of the particles along with air without being separated.

4.4.4 COMPARISON OF EXPERIMENTAL RESULTS WITH MODEL EQUATIONS

The experimental values of tangential velocity and pressure drops are compared with the model equations available in literature. The tangential velocity values are compared with Ramachandran et al. [5] and Maynard [7] and the pressure drop values compared with Ramachandran et al. [5] and Ashwani Malhotra [6]. The results are illustrated in Figures 4.6 and 4.7. The Figure 4.6 shows the experimental tangential velocity compared with the model equations for different inlet velocities. The values are well fitted for Maynard [7] with ±1 percent error. Figure 4.7 shows the experimental pressure drop compared with the model equation for different inlet velocities. The values are well fitted for Ramachandran et al. [5] with ±1 percent error.

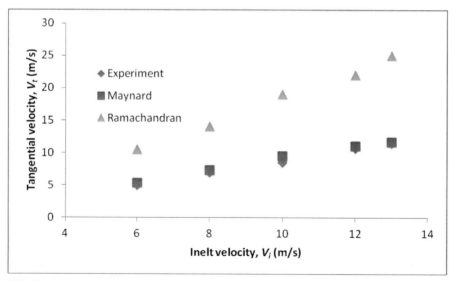

FIGURE 4.6 Comparison of tangential velocity with model equation at 45° vane angle.

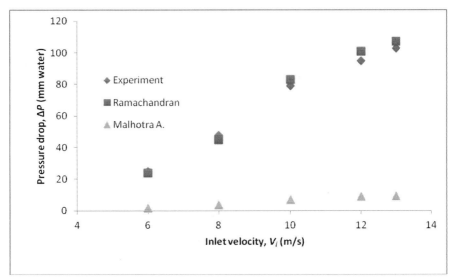

FIGURE 4.7 Comparison of pressure drop with model equations for 45° vane angle.

4.5 CONCLUSIONS

A simple unit of Uniflow cyclone was designed and fabricated. A suitable methodology was developed to measure tangential velocity. Several experiments were carried out to understand the effect of vane angle on tangential velocity and pressure drop by considering four different vane angles from 20° to 60°. It is observed that reasonably high tangential velocity, low pressure drop thus high efficiency are obtained at 45° vane angle.

The experimental values are validated with the model equations available in literature. The pressure drop values are well matched with Ramachandran et al. [5] model and tangential velocity values are well matched with Maynard [7] model by reasonably good accuracy. The overall efficiency increases with inlet velocity and achieved a maximum efficiency at an inlet velocity of 12–13 m/s for all the three samples.

NOMENCLATURE

V_i	Inlet axial velocity	[m/s]
V_{avg}	Average velocity of the gas in the helix	[m/s]
V_θ	Tangential velocity	[m/s]
D_h	Hydraulic diameter of the cyclone	[m]
f	Friction factor	[–]

L_{hel}	Average length of path of gas in the helix	[m]
ρ_g	Density of gas	[kg/m³]
ρ_p	Density of particle	[kg/m³]
P	Pitch	[m]
R_c	Cyclone radius	[m]
A_s	Cross section area of vane core	[m²]
A_i	Cross section area of cyclone body	[m²]
D	Diameter of cyclone	[m]
D_v	Diameter of vane core	[m]
n	Number of revolutions	[rpm]

KEYWORDS

- **Overall efficiency**
- **Pressure drop**
- **Tangential velocity**
- **Vane angle**
- **Vortex finder**

REFERENCES

1. Sumner, R. J., Briens, C. L., Bergougnou, M. A., (1987), Study of a novel Uniflow cyclone design for the ultra-rapid fluidized pyrolysis reactor, Can. J. Chem. Eng., 65, 470-475.
2. Vaughan, N. P., "The Design, Construction and Testing of an Axial Flow Cyclone Preseparator", J. Aerosol Sci., 18, 6, 789-791, 1987.
3. T. A. Gauthier, C.L. Briens, M. A. Bergougnou, P. Galtier, (1990), Uniflow cyclone efficiency study, Powder Technology, 62, 217-225.
4. Akiyama, T. and et al "Dust Collection Efficiency of a Straight Through Cyclone Effects of Duct Length, Guide Vanes and Nozzle Angle For Secondary Rotational Air Flow", Powder Technology, 58, 181-185, 1989.
5. Ramchandran, G. and et al "Collection Efficiency and Pressure Drop For a Rotary – Flow Cyclone", Filtration and Separation, 631-636 September/October, 1994.
6. Malhotra, A.; "Studies in Fine Particle Collection in Uniflow Cyclone", Ph.D. Dissertation, Department of Chemical Engineering, Indian Institute of Technology, New Delhi; **1995**.
7. Maynard. A. D, "A Simple Model of Axial Flow Cyclone Performance Under Laminar Flow Conditions," J. Aerosol Sci. Vol. 31, No. 2, pp 151 –167, 2000.
8. Zhang, Y.; *ASHRAE Trans*. 2003, *109*(2), 815–821.
9. Ramchaval S., "Experimental Investigation of Tangential Velocity and Pressure Drop in Uniflow Cyclone," M. S. thesis, Department of Chemical Engineering, Indian Institute of Technology, New Delhi; 1995.

CHAPTER 5

MULTIVARIATE STATISTICAL MONITORING OF BIOLOGICAL BATCH PROCESSES USING CORRESPONDING ANALYSIS

SUMANA CHENNA*, ANKIT MISHRA, and ABHIMANYU GUPTA

Process Dynamics and Controls Group, Chemical Engineering Sciences, Indian Institute of Chemical Technology, Hyderabad

*Email: sumana@iict.res.in, sumana.chenna@gmail.com

CONTENTS

5.1	Introduction	48
5.2	Materials and Methods	48
5.3	Penicillin-Fed Batch Fermentation Process	50
5.4	Results and Analysis	52
5.5	Conclusions	53
	Keywords	53
	References	54

5.1 INTRODUCTION

Biological batch processes have been gaining increasing interest from both industries as well as academia as they are found to be promising in providing alternative technologies which are greener and safer for the manufacture of several industrial grade products including fine chemicals, pharmaceuticals, polymers etc. However, maintaining consistent product qualities with minimum batch to batch variation, while ensuring the sterile, safe, and optimal conditions of a process is a daunting task. Therefore, it is essential to develop efficient on-line process monitoring methodologies. The presence of significant nonlinearities, complex growth dynamics of microorganisms of biological batch processes makes the task of online monitoring more complicated.

Generally, the quality of any product can be monitored by studying certain variables offline. But, offline monitoring generally involves the study of few variables linked with the quality of final product, which may not always reflect the true state of the process and mostly associated with time delays. Early detection and diagnosis of abnormal conditions is essential for avoiding the unsafe operation and off-spec products. Though there are several approaches available in literature on batch process monitoring, the scope of the present work in limited to process history based methods which uses the abundant multivariate data collected in the form of measurements for process monitoring. Principal Component Analysis (PCA) is a most popular multivariate method, used widely in signal processing, factor analysis, chemometrics, and chemical process data analysis. PCA has also been extensively applied for online monitoring of several continuous chemical and biochemical processes [1]. Nomikos and MacGregor [2] were introduced the concept of batch process monitoring using principal component analysis (PCA) and further its application has been improved and extended by others [2,3]. Recently, correspondence analysis (CA) with improved features over conventional PCA has been introduced for online monitoring and fault diagnosis [4, 5]. More recently, the application of CA has been extended to biological batch processes by Patel and Gudi [6] which involves the monitoring of process by evaluating multivariate statistical process control charts such as Hotelling T^2 and SPE statistics. In the present paper, an improved version of MCA methodology which is computationally less intensive has been proposed and applied to penicillin fed-batch fermentation process.

5.2 MATERIALS AND METHODS

5.2.1 BATCH PROCESS MONITORING—DATA UNFOLDING

The historic data for batch processes is represented as a three dimensional data matrix $X(I \times J \times K)$ where I represents the number of batch runs, J represents the number of measured variables in a single batch and K represents the sample points or time indices of a particular batch. So as to apply MCA the data needs to be unfolded such that an analysis could be carried out in two-dimensional representative form. The unfolding of the data is an important step of batch process monitoring which can be carried out in three basic ways. (i) *time wise unfolding* provide the analysis of variables concerned to different batches with pertinent to a single sample point (ii) *batch wise unfolding* provides the analysis of variable profiles for each specific batch (*iii*) *variable wise unfolding* results in profiles of each variable across different batches. In the present work, the data is unfolded time wise, that is, into a data matrix $X \in R^{I \times JK}$ and subjected to correspondence analysis so as to analyze variability among samples.

5.2.2 MCA METHODOLOGY

Correspondence analysis (CA) is a linear data dimensionality reduction technique which uses a measure of joint row column association as it's metric to form a lower dimensional representation of the data in contrast to PCA using variance as it's metric. CA assumes statistical independence so as to calculate expectation values and constructs a metric based on the distance of original data set from the expectation values and also employs a weight proportional to these expectation values. This metric when subjected to singular value decomposition decomposes into an uncorrelated subspace which classifies the data well.

5.2.2.1 SINGULAR VALUE DECOMPOSITION

The χ^2 index is a representative measure of the deviation from the condition of statistical independence and it can be constructed as follows:

$$C = \frac{X_{ij} - E_{ij}}{E_{ij}^{1/2}} \quad (5.1)$$

The vectors obtained by SVD of C matrix can be transformed into these as,

$$A = D_r^{1/2} L \quad ; \quad B = D_c^{1/2} M \quad (5.2)$$

where, L contains the eigenvectors of CC^T and M contains the eigenvectors of $C^T C$ such that $C = L D_\mu M^T$ and $D_\mu = \text{diag}(\mu)$; where μ contains the singular value of C.

5.2.2.2 DIMENSIONALITY REDUCTION

Since μ quantifies the amount of inertia expressed along a particular principal axis. The selection of appropriate number of reduced dimensions l which can contribute up to a prespecified fraction of total inertia L can be done based on the following criteria:

$$\frac{\sum_{i=1}^{l}\mu_i}{\sum_{j=1}^{m}\mu_j} \geq L \tag{5.3}$$

5.2.2.3 FAULT DETECTION SING MCA

Fault detection is performed by evaluating multivariate monitoring indices T^2 and SPE as follows.

$$T^2 = f^T D_\mu^{-2} f \quad ; \quad SPE = \left[Bf - \left(\frac{1}{r}x - c\right)\right]^T \left[Bf - \left(\frac{1}{r}x - c\right)\right] \tag{5.4}$$

The row coordinates and the column coordinates can be computed as,

$$F = D_r^{-1} AD_\mu \quad ; \quad G = D_c^{-1} BD_\mu \tag{5.5}$$

In this work a new monitoring index for MCA has been proposed which can be computed as,

$$CI = x\phi x^T \quad ; \quad \phi = \frac{\left(GD_\mu^{-1}\right)D_\mu^{-2}\left(GD_\mu^{-1}\right)^T}{T_\alpha^2} + \alpha \frac{\left(I - B\left(GD_\mu^{-1}\right)^T\right)^T \left(I - B\left(GD_\mu^{-1}\right)^T\right)}{Q_\alpha} \tag{5.6}$$

The confidence limits for all the above monitoring statistics can be calculated for given level of significance analytically from the normal operations such that they represent normal behavior, or they can be also be estimated using Kernel Density Estimation method.

5.3 PENICILLIN-FED BATCH FERMENTATION PROCESS

Production of secondary metabolites such as Penicillin can be done generally in a fed-batch mode. The dynamic simulator was downloaded from http://simulator.iit.edu/web/pensim/bground.html. Monitoring of such biological batch processes is very crucial in order to ensure the sustained product quality in different batches. The schematic of the process is given in Figure 5.1. More details of the simulator and the process can be found elsewhere [7]. The data required for training and testing are generated using the simulator. The measurements of 11 variables are considered for multivariate statistical model development. They are aeration rate, agitation power, substrate feed rate, substrate feed temperature, dissolved oxygen concentration, culture volume, CO_2 concentration, p^H, bioreactor temperature, generated heat and cooling water flow rate.

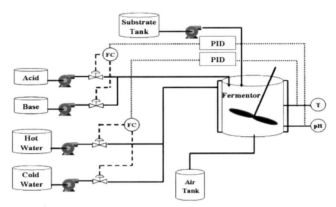

FIGURE 5.1 Schematic of penicillin fed-batch fermentation process.

5.3.1 DATA GENERATION

A set of 40 normal operating batch data consisting of measurements of 11 variables at every 200 sampling intervals has been considered for developing the model concerned to normal operating conditions (NOC) based on MCA. For testing purpose, data corresponding to faulty conditions are considered. The faulty batch data was simulated by introducing the disturbance at 50th hr and allowed to persist till the end of the batch duration, that is, 400 hr. Two types of faults are considered for the analysis whose details are given in Table 5.1.

TABLE 5.1 Fault definitions

S. No.	Variable	Disturbance Type	Disturbance Magnitude
Fault 1	Substrate feed flow rate	Ramp	−0.005
Fault 2	Agitation power	Step	−25%

5.3.2 MEASURES FOR PROCESS MONITORING

The ability of the given monitoring methodology can be quantified in terms of false alarm rates (Type-I error) and missed detection (Type-II error) rates. There is always a trade-off between the missed detection rates and false alarm rates. False alarm rates signify the ability of the monitoring method during normal operation whereas the missed detection rate signifies the ability during faulty batch operation.

5.4 RESULTS AND ANALYSIS

An offline multivariate statistical MCA model has been developed and considered as reference normal operating model for online process monitoring using the data generated from Penicillin fed-batch fermentation process. The detection results of proposed MCA method are compared with conventional PCA method.

The data generated from the simulator representing the normal batch operation is of dimension $40 \times 11 \times 200$ and it is considered for offline model development. After unfolding time wise, the resulted data matrix is of dimension 40×2200. Data scaling for PCA is done such that all the variables are mean centered and scaled to unit variance, whereas for CA, the variables are scaled between 0 and 100 of their operating range. At every sampling instant, local PCA and CA models are developed considering the respective data from all batches, thus resulted in 200 reference NOC models. The resulted reduced dimensions in case of CA are seven explaining about 86.1 percent inertia, and the reduced dimensions in case of PCA are six explaining about 75 percent of total variance. Ninety-five percent confidence limits were considered for all the monitoring statistics and their confidence limits are evaluated using kernel density estimation method. The false alarm rates exhibited by all the statistics both in case of MCA and MPCA methods are found to be zero. The missed detection rates of different statistics are presented in Table 5.2. From the comparison results, it is clearly noted that the fault detection ability of the proposed MCA method is much superior to that of conventional MPCA method. In case of Fault1, considering the performance of T^2 statistic, the proposed MCA method has shown an improved detection ability of 85 percent over the 62 percent ability of conventional MPCA method. Similar trend is observed in the performance of SPE statistic. Moreover, if we compare the results of individual statistics of MCA method, CI statistic has shown better detection performance than that of T^2 and *SPE* or Q statistics, which ensures the ability of the new CI proposed in this paper. These results are shown in Figure 5.2.

TABLE 5.2 Comparison of missed detection rates

S. No.	MPCA Method			Proposed MCA Method		
	T^2	SPE(Q)	CI	T^2	SPE(Q)	CI
Fault 1	0.38	0.17	0.1	0.15	0.13	0.04
Fault 2	0.98	0.59	0.79	0.89	0.49	0.48

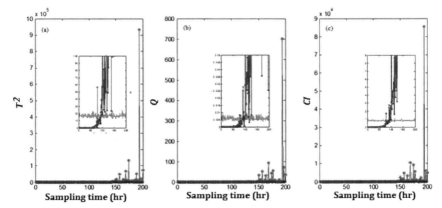

FIGURE 5.2 Detection results of MCA method for Fault 1.

5.5 CONCLUSIONS

In this paper, a novel online method based on multi way correspondence analysis has been proposed for biological batch process monitoring. Further an improved monitoring index for CA has been proposed. The application results on Penicillin fed-batch fermentation process revealed that the proposed MCA method outperforms over conventional MPCA method by exhibiting improved detection ability. In addition, it has also been proved in the paper that the proposed combined monitoring index for CA has shown improved performance over conventional T^2 and SPE statistics. Further the application of the proposed monitoring method can be easily extended to any other batch process.

KEYWORDS

- **Correspondence analysis**
- **Fault diagnosis**
- **Fed-batch fermentation**
- **Multivariate Process monitoring**

REFERENCES

1. Raich, A.; and Cinar, A.; *AIChE J.* **1996**, *42*(4), 995–1009.
2. Nomikos, P.; and MacGregor, J. F.; *AIChE J.* **1994**, *40*, 1361–1375.
3. Chen, J.; and Liu, K. C.; *Chem. Eng. Sci.* **2002**, *57*(1), 63–75.
4. Greenacre, M. J.; Theory and Application of Correspondence Analysis. London: Academic Press Inc.; **1984**.
5. Detroja, K. P.; Gudi, R. D.; and Patwardhan, S. C.; *Control Eng. Pract.* **2007**, *15*, 1468–1483.
6. Patel, S. R.; and Gudi, R. D.; Improved monitoring and discrimination of batch processes using correspondence analysis. In: American Control Conference. St. Louis, MO; June 10–12, 2009, 3434–3439.
7. Birol, G.; Undey, C.; and Cinar, A.; *Com. Chem. Eng.* **2002**, *26*, 1553–1565.

CHAPTER 6

EXTENDED KALMAN FILTER-BASED COMPOSITION ESTIMATION USING SIMPLE HYBRID MODEL APPROXIMATION FOR A REACTIVE BATCH DISTILLATION PROCESS

P. SWAPNA REDDY and K. YAMUNA RANI*

Process Dynamics and Control Group, Chemical Engineering Division, Indian Institute of Chemical Technology Hyderabad - 500 607

*E-mail: kyrani@iict.res.in

CONTENTS

6.1	Introduction	56
6.2	Process Description	56
6.3	Hybrid Model	58
6.4	Results and Discussion	60
6.5	Conclusions	65
	Nomenclature	65
	Keywords	66
	References	66

6.1 INTRODUCTION

State estimation techniques have been well developed for dynamic processes described by ordinary differential equations. Among them the Kalman filter (KF) is an optimal estimator for linear dynamical systems in the presence of state and measurement uncertainties [1, 2]. Many techniques have been reported for inferential estimation of compositions in continuous distillation columns [3–6]. State estimator design for batch distillation has to deal with the time varying nature of the batch column. Quintero-Marmolet al. [7] applied an extended Luenbergerobserver to predict compositions in multicomponent batch distillation from temperature measurements. In order to compensate for limited state variable measurements, the extended Kalman filter (EKF) has been extensively employed to estimate the unmeasured states and unknown parameters from measured states. Composition profiles and operating conditions may change over a wide range of values during the entire operation and the state estimators must be designed to deal with the time-varying nature of the batch columns [8, 9].

Reactive batch distillation (RBD) is an intrinsically dynamic process, in which the composition profiles and the operating conditions can change over a wide range of values during the entire operation. In this process, the compositions on all stages, the product and the slop cut compositions define the state variables. In the present study, an EKF-based state and parameter estimation is used in a continuous time framework with models incorporating a hybrid model which includes reduced-order model and data-driven model.

6.2 PROCESS DESCRIPTION

In the present study, production of butyl acetate from butanol and acetic acid has been considered as a simulation case study as this system involves typical complexities. Venimadhavanet al. [10] showed that reactive distillation in batch columns can be applied successfully to the esterification of butanol to butyl acetate, a powerful solvent that is mainly used in coatings. The four-component reactive system considered in the present study exhibitsone maximum-boiling homogeneous binary azeotrope at a temperature of 118.28°C (Acetic acid-Butanol), one minimum boiling homogeneous binary azeotrope at a temperature of 116.95°C (Butanol-Butyl acetate), two heterogeneous minimum boiling binary azeotropes (Water-Butanol, 92.83°C and Water-Butyl acetate, 91.18°C) and one minimum boiling heterogeneous ternary azeotrope at a temperature of 90.92°C (Water-Butanol-Butyl acetate). The ternary azeotrope, which has the minimum bubble point, is distilled overhead, butyl acetate and reaction water being thus separated from the reacting mixture.

The simulation is based on the first principles model of the RBD column consisting of unsteady state mass and component balance equations under the following assumptions: the column has N trays, excluding the condenser, decanter, and reboiler; the volumetric liquid holdup on all trays, the condenser, and the decanter is constant; the vapor holdup is neglected; the vapor and liquid phases on each tray are in equilibrium; the vapor flow is constant on all trays; chemical reactions occur only in the liquid phases. The reaction is considered to occur on 20 trays above the reboiler. The model equations including material and component balance equations are reported by Patel et al. [11], whereas the reaction kinetics is reported by Venimadhavan et al. [10]. A simplified method for vapor-liquid-liquid equilibria (VLLE) for this system is reported by Swapna and Rani [12].

In this study, the esterification of butanol and acetic acid is simulated with a binary feed of molar composition x_{BuOH}=0.51 and x_{HOAc}=0.49 in a RBD column. The distillation column and operational details Venimadhavan et al. [10] are reported in Table 6.1.

TABLE 6.1 Distillation column details and operational data

Number of trays	31
Number of trays on which reaction occurs	20
Initial mixture composition of acetic acid and butanol	0.49, 0.51
Initial hold up of the reboiler	2.475 kmol
Condenser hold up	0.1 kmol
Internal plate holdup	0.0125 kmol
Vapor boil-up rate	3 kmol. h^{-1}
Column operating pressure, P	1 atm

For the condenser-reflux drum system, the total mass balance and component-wise mass balance are

$$V_1 = L_0 + D + \frac{dA_0}{dt} - A_0 \sum_{i-1}^{c} R_{0,i} \qquad (6.1)$$

$$\frac{dx_{0,i}}{dt} = \frac{V_1}{A_0}\left(y_{1,i} - x_{0,i}\right) + R_{0,i} \quad (i=1 \text{ to } c) \qquad (6.2)$$

Where V_1 = vapor boil-up rate, L_0 = liquid rate, and A_0 = holdup in the condenser $R_{0,i}$ is the rate of reaction for each component in the condenser, $x_{0,i}$ is the liquid composition in the condenser, and $y_{1,i}$ is the vapor composition leaving the first tray.

For an arbitrary plate j ($1 \leq j \leq N$) in the column, the total mass balance, the component-wise mass balance, and the component balance yield respectively

$$L_j = V_{j+1} + L_{j-1} - V_j - \frac{dA_j}{dt} \qquad (j = 1 \text{ to } N) \qquad (6.3)$$

$$\frac{dx_{j,i}}{dt} = \frac{V_{j+1}(y_{j+1,i} - x_{j,i}) + L_{j-1}(x_{j-1,i} - x_{j,i}) \ V_j(y_{j,i} - x_{j,i}) + A_j R_{j,i}}{A_j} \qquad (j = 1 \text{ to } N; i = 1 \text{ to } c) \qquad (6.4)$$

where V_j = vapor boil-up rate, L_j = liquid rate, and A_j = holdup on the jth tray, $R_{j,i}$ is the rate of reaction for each component on the jth tray, $x_{j,i}$ is the liquid composition and $y_{j,i}$ is the vapor composition leaving the jth tray.

Similarly for reboiler, the total mass balance, the component-wise mass balance, and the component balance yield respectively

$$\frac{dA_{N+1}}{dt} = L_N - V_{N+1} \qquad (6.5)$$

$$\frac{dx_{N+1,i}}{dt} = \frac{L_N(x_{N,i} - x_{N+1,i}) - V_{N+1}(y_{N+1,i} - x_{N+1,i}) + A_{N+1}(R_{N+1,i})}{A_{N+1}} \qquad (i = 1 \text{ to } c) \qquad (6.6)$$

For each tray, the VLLE is described using UNIQUAC model. Therefore, for 31 trays with four components on each tray, the model consists of 124 differential equations, that is, material and component balance equations and 31 algebraic equations, that is, thermodynamic equations for VLLE.

6.3 HYBRID MODEL

The hybrid model considered in this study includes a reduced-order model part and a data-driven model part. Reduced-order model consists of mass balance and component balance equations for three components in the condenser and reboiler along with the kinetic expression and VLLE model. Data-driven model consists of time dependent composition profiles as piecewise constant polynomials, which are generated using the data obtained for the base case using the first principles model of the RBD process. In this study the first principles model of RBD process, that is, Eqs. (6.1–6.6) are considered as the plant and the hybrid model, that is, reduced-order model part, data-driven model part and measurement model part, is considered for state estimation.

6.3.1 REDUCED-ORDER MODEL PART

The reduced-order model considered in this study is:
For the condenser-reflux drum system, the component balance

$$\frac{dx_{0,i}}{dt} = \frac{V}{A_0}\left(\theta_i - x_{0,i}\right) + R_{0,i} \qquad (i=1 \text{ to } 3) \tag{6.7}$$

Similarly for reboiler the component balance

$$\frac{dx_{N+1,i}}{dt} = \frac{L\left(\theta_{3+i} - x_{N+1,i}\right) - V\left(y_{N+1,i} - x_{N+1,i}\right) + A_{N+1}\left(R_{N+1,i}\right)}{A_{N+1}} \qquad (i=1 \text{ to } 3) \tag{6.8}$$

Where V = vapor boil-up rate, L = liquid rate, A_0 = holdup in the condenser, A_{N+1} = holdup in the reboiler. The parameters θ in the above equations denote the vapor compositions of three components on the top tray, and the liquid compositions of three components on the tray just above the reboiler. The reduced-order model has the same assumptions as the first principles model.

6.3.2 DATA-DRIVEN MODEL PART

Data-driven model incorporated into the hybrid model includes time dependent parameter profiles as piecewise polynomial functions as follows:

$$\theta_i(t) = \sum_{k=1}^{np} \phi_k\left(t_s\right) \sum_{j=0}^{mk} a_{ijk} t_s^j \qquad (i=1,...,6) \tag{6.9}$$

$$\phi_k\left(t_s\right) = 1 \quad \text{if } t_{k-1} \leq t_s \leq t_k$$
$$= 0 \quad \text{otherwise}$$

where t_s denotes the scaled time, t_{k-1} and t_k denote the lower and upper scaled time limits of kth interval, np denotes the number of intervals for the ith parameter, mk denotes the degree of polynomial for the kth interval, and a_{ijk} represents the polynomial coefficient for the ith parameter, kth interval, and jth term in the polynomial.

6.3.3 MEASUREMENT MODEL PART

The measurement model incorporated into the hybrid model consists of two temperature measurements, that is, thermodynamic vapor liquid equilibria equations calculated using UNIQUAC method.

Thus, in the hybrid model we have six states, that is, component balance equations in the condenser for three components (Acetic acid, Butanol, and Butyl acetate) and component balance equations in the reboiler for three components, six parameters, and two temperature measurements as mentioned above.

6.4 RESULTS AND DISCUSSION

The complete first principles model equations for the production of Butyl acetate in a RBD column are used to carry out simulation studies. In the present work, reflux ratio is taken as the ratio of liquid flow rate through the column to the total vapor flow rate at the top (RR = L_0/V_1), where 0 denotes condenser and 1 as top tray. Thus reflux ratio lies between 0 and 1. Hybrid model incorporating reduced-order model and data-driven models described above is used in the EKF-based state and parameter estimation. Model equations consisting of component and material balance equations are considered as the states and time dependent parameter profiles developed from first principles model as the parameters in the EKF.

In this study we considered a base case study for reflux ratio of 0.880547 and vapor boil-up rate 3 kmol. h^{-1} for developing time dependent parameter profiles for condenser and reboiler using Eq. (6.3). The composition profiles in the condenser are divided into six intervals and into three intervals in the reboiler. A new variable is defined as a scaled time, t_s based on the the switch time, t_{switch}, that is, the time until which the first product (water) withdrawal is carried out corresponding to a given reflux ratio and vapor boilup rate and is given by the following equation:

$$\frac{dt_s}{dt} = \frac{t_{switch, ref}}{t_{switch}} \qquad (6.10)$$

where $t_{switch,ref}$ represents the switch time for the base case of reflux ratio and vapor boil-up rate, whereas t_{switch} denotes the switch time for any V or rr values. The switch time mainly depends on reflux ratio and vapor boil-up rate and can be approximated for any reflux ratio and vapor boil-up rate by the following equation:

$$t_{switch} = \frac{2.475 \times 0.49 \times 0.995}{V \times (1-rr)} \qquad (6.11)$$

where V = vapor boil-up rate, rr = reflux ratio. The basis for this equation is the approximate quantity of pure butyl acetate product produced based on 99.5 percent conversion with respect to the limiting reactant (represented by the numerator) and the denominator representing the distillate flow rate, and the ratio of these represents the time for which the top product composition remains constant at its azeotropic composition.

The detailed formulation of EKF is reported by Rani and Swetha [13]. Initially the covariance matrices for state and parameter estimation (**P, Q,** and **R**) are assigned as diagonal matrices and in this study different tuning parameters for covariance matrices have been studied. A conservative choice of covariance matrices (**P, Q**) of the order of 10^{-8} is chosen, whereas for **R** matrix, a value between 1 and 10 resulted in the best performance. In this study the EKF state estimator performance is evaluated by cumulative integral square deviation (ISE) between the process and estimated temperature measurements for range of reflux ratio and vapor boil-up rate and are reported in Table 6.2.

TABLE 6.2 Performance evaluation of EKF by ISE

S. No.	Reflux Ratio	Vapor Boil Up Rate	Integral Square Error for Condenser	Integral Square Error for Reboiler
1.	0.880547	3	0.03814842	0.000315646
2.	0.880547	4	0.06720225	0.000433268
3.	0.9	3	0.04175612	0.007658897
4.	0.9	4	0.0847097	0.001653304
5.	0.92	3	0.04820368	0.023576266
6.	0.92	4	0.10365635	0.008780123

The composition profiles obtained using EKF-based state and parameter estimation with different reflux ratio and vapor boil-up rate are shown in the Figures 6.1–6.5. The states and parameters based on EKF are in accordance with the actual process. In order to evaluate the suitability of the proposed hybrid-model-based state estimation in the closed-loop control implementation, a time-varying reflux ratio and vapor boil-up rate resembling closed-loop operation are considered. The composition profiles are shown in Figures 6.6 and 6.7 which show that with time-varying reflux ratio, and vapor boil-up rate, the estimated states are in accordance with the actual process. This ensures that the hybrid model can be used in the closed-loop control operation.

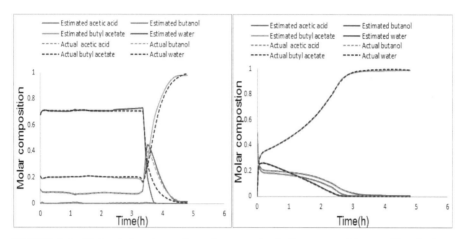

FIGURE 6.1 Estimated and actual condenser and reboiler composition profiles for base case values: reflux ratio 0.880547 and vapor boil-up rate 3 kmol. h^{-1}.

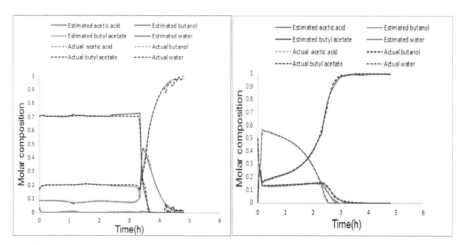

FIGURE 6.2 Estimated and actual top tray and bottom tray time dependent parameter profiles for reflux ratio 0.880547 and vapor boil-up rate 3 kmol. h^{-1}.

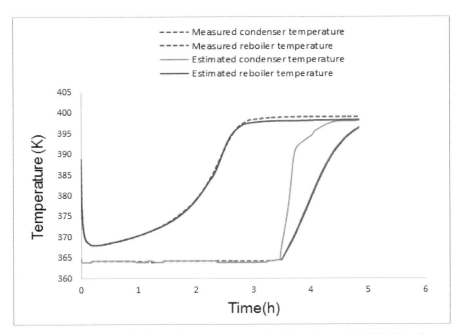

FIGURE 6.3 Estimated and actual temperature profiles for reflux ratio 0.880547 and vapor boil-up rate 3 kmol. h^{-1}.

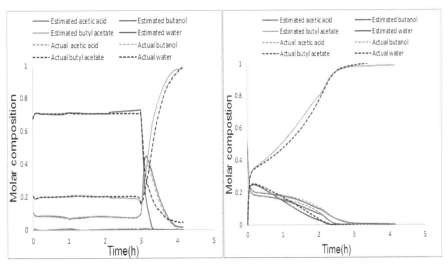

FIGURE 6.4 Estimated and actual condenser and reboiler composition profiles for reflux ratio 0.9 and vapor boil-up rate 4 kmol. h^{-1}.

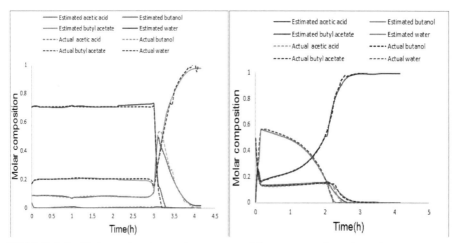

FIGURE 6.5 Estimated and actual top tray and bottom tray time dependent parameter profiles for reflux ratio 0.9 and vapor boil-up rate 4 kmol. h^{-1}.

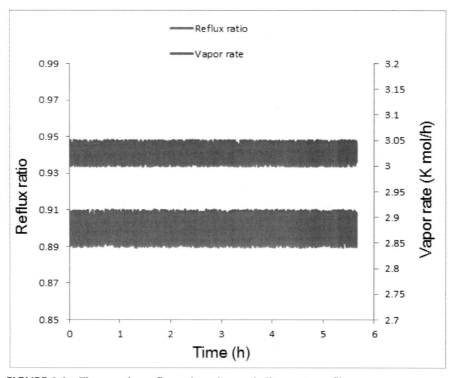

FIGURE 6.6 Time varying reflux ratio and vapor boil-up rate profiles.

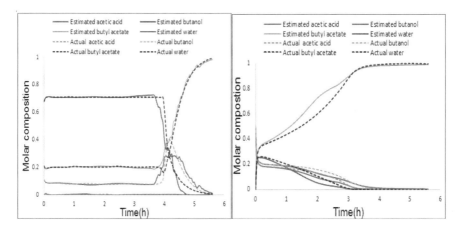

FIGURE 6.7 Estimated and actual condenser and reboiler composition profiles for time-varying reflux ratio and vapor boil-up rate.

6.5 CONCLUSIONS

In the present study, EKF-based state and parameter estimation has been used for RBD process for butyl acetate production with the help of a hybrid model. The state and parameter composition profiles generated using EKF are in accordance with the actual process, that is, first principles model. The performance of the EKF is evaluated by the Integral square deviation values and the obtained values describe the efficacy of the EKF by considering the hybrid model. The key contribution of this study is the flexibility of the data-driven model in estimating the states and parameters for any reflux ratio and vapor boil-up rate. The evaluation of the proposed hybrid model for time-varying reflux ratio and vapor boil-up rate ensures that this approach can be used in inferential composition control based on available temperature measurements in the real time context.

NOMENCLATURE

A_j	molar holdup on jth plate (mol)
D	distillate flow rate (mol/s)
BuAc	butyl acetate
BuOH	butanol
HAc	acetic acid
H_2O	water

L_j	liquid flow rate from jth plate (mol/s)
N	number of stages excluding reboiler and condenser = 31
Q_0	heat duty on condenser-reflux drums (J/s)
Q_{n+1}	heat duty on reboiler (J/s)
R_{ij}	rate of reaction of ith component on jth plate(s^{-1})
t	time (s)
V	vapor flow rate (mol/s)
x_{ji}	molefraction of ith component in liquid phase on jth plate
y_{ji}	molefraction of ith component in vapor phase on jth plate

KEYWORDS

- Data-driven parameter model
- Extended Kalman filter
- Reactive batch distillation
- Reduced order model
- State and parameter estimation

REFERENCES

1. Gelb, A.; Applied Optimal Estimation. Cambridge: MIT Press; **1988**.
2. Sorenson, H.; Kalman Filtering Theory and Applications. New York: IEEE Press; **1985**.
3. Joseph, B.; and Brosilow, C. B.; *AIChE J.* **1978**, *24*, 485–492.
4. Mejdell, T.; and Skogestad, S.; *Ind. Eng. Chem. Res.* **1991**, *30*, 2543–2555.
5. Yang, D. R.; and Lee, K. S.; *Comput. Chem. Eng.* **1997**, *21*, S565–S570.
6. Kano, M.; Miyazaki, K.; Hasebe, S.; and Hashimoto, I.; *J Process Control.* **2000**, *10*, 157–166.
7. Quintero-Marmol, E.; Luyben, W. L.; and Georgakis, C.; *Ind. Eng. Chem. Res.* **1991**, *30*, 1870–1880.
8. Mujtaba, I. M.; and Macchietto, S.; *Ind. Eng. Chem. Res.* **1997**, *36*, 2287–2295.
9. Oisiovici, R. M.; and Cruz, S. L.; *Chem. Eng. Sci.* **2000**, *55*, 4667–4680.
10. Venimadhavan, G.; Malone, M. F.; and Doherty, M. F.; *Ind. Eng. Chem. Res.* **1999**, *38*, 714–722.
11. Patel, R.; Singh, K.; Pareek, V.; and Tade, M. O.; *Chem. Prod. Process Model.* **2007**, *2*, 1934–2659.
12. Swapna, P.; and Rani, K. Y.; *Ind. Eng. Chem. Res.* **2012**, *51*, 10719–10730.
13. Rani, K. Y.; and Swetha, M.; *Chem. Eng. Commun.* **2006**, *193*, 1294–1320.

CHAPTER 7

ESTIMATION OF SOLUBILITY IN DILUTE BINARY SOLUTIONS USING MOLAR REFRACTION, DIPOLE MOMENT, AND CHARGE TRANSFER FUNCTIONS

N. V. K. DUTT, K. HEMALATHA, K. SIVAJI, and K. YAMUNA RANI*

Process Dynamics and Control Group, Chemical Engineering Sciences, Indian Institute of Chemical Technology, Hyderabad 500 607, India

*E-mail: kyrani@iict.res.in

CONTENTS

7.1	Introduction	68
7.2	Methodology	68
7.3	Results and Discussion	70
7.4	Conclusion	73
	Keywords	74
	References	74

7.1 INTRODUCTION

Solid liquid equilibrium (SLE) studies are essential to the industrial areas such as the safe operation of pipe lines and the design of crystallizers. The solubility of a solid in a liquid solvent is determined from a knowledge of the heat of fusion (ΔH_m), melting point (T_m) and the activity coefficient, γ^∞ at any temperature, T (in K). Of the several methods of estimating γ^∞ (such as Wilson and NRTL models), the UNIFAC model based on group contributions is the most commonly used. However, the approaches used in the calculation of solubilities such as molecular dynamic simulations, equations of state and UNIFAC, although provide reasonable estimates for individual binary mixtures, are too complicated for the development of a generalized method, using larger SLE data bases. Our present work is a step in this direction.

7.2 METHODOLOGY

A survey of literature on the various methods of estimating solubilities of solids in liquids revealed the important role played by the properties dependent on the attractive interactions, group contributions constituting the molecular structures, and the electron charge distribution within the atoms and molecules. In the present work, we selected molar refraction, R_m, which depicts the dispersive (attractive and repulsive) forces, dipole moment, μ_P and charge transfer functions (CTF) calculated from the elemental contributions of electronegativity (c) and chemical hardness (η) as input properties needed in the calculation of solubilities. The charge transfer function for a solute (A) – solvent (B) binary system is defined by:

$$\text{CTF} = \frac{(\chi_A - \chi_B)}{2(\eta_A + \eta_B)} \tag{7.1}$$

The numerator of the Eq. (7.1) denotes the driving force needed for the electron exchange between the solute (A) and solvent (B); the denominator denotes the maximum stability condition of the binary described as the addition of the chemical hardness terms.

The term P_p (termed as polar or orientation polarization) is defined as:

$$P_p = \frac{6096.5 \mu_P^2}{T} \tag{7.2}$$

where μ_p is the dipole moment expressed in Debyes, and T is the temperature in K.

Estimation of Solubility in Dilute Binary Solutions

The electronegativity (c) and chemical hardness (η) of an element is defined by:

$$\chi = 0.5\,(IP + EA) \tag{7.3}$$

$$\eta = 0.5\,(IP - EA) \tag{7.4}$$

where IP and EA are the ionization potential and electron affinity. The values of IP and EA are reported in CRC Handbook [1].

The electronegativity of solute (A) or solvent (B) molecule is calculated from the elemental contributions using the equalization method by means of the equation:

$$\chi = \frac{N_t}{\Sigma(C_i/N_i)} \tag{7.5}$$

where N_t is the total number of atoms in the molecule and C_i and N_i are the contribution and the number of atoms of the type i.

The solubilities (X) of solids in binary solutions (in the dilute region) are obtained using a simple method of using pure compound molar refractions (R_M), dipole moments (μ_p) and charge transfer functions (CTF) for a number of binary systems at 282 data points and at various temperatures.

The proposed method is based on the following equations for the calculation of the solubility

$$\ln \gamma^\infty = 64.137 \log R_M + 4.6482\, R_M^2 + 9.415938 \log P_p - 1.65579 P_p^2 \\ - 236.75 \log (CTF) + 20.40981 (CTF)^2 \tag{7.6}$$

$$\ln X = \frac{\ln \gamma^\infty - \left\{1.323518\left[\frac{\Delta H_m}{RT}\left(\frac{T}{T_m}-1\right)+0.79691\right]\right\}}{\left\{0.06459\left[\frac{\Delta H_m}{RT}\left(\frac{T}{T_m}-1\right)-0.851767\right]\right\}} \tag{7.7}$$

where X is the solubility, ΔH_m is the heat of fusion, T is the temperature in K, T_m is the melting point, and γ^∞ is the activity coefficient at infinite dilution. Equation (7.7) is an explicit approximation for the following implicit equation for X in terms of ΔH_m, T_m, and γ^∞.

$$\ln \gamma^{\infty} = \frac{\left[\frac{\Delta H_m}{RT}\left(\frac{T}{T_m}-1\right)-\ln X\right]}{(1-X)^2} \tag{7.8}$$

7.3 RESULTS AND DISCUSSION

The inputs used in the present method, namely R_M, P_p (containing the value of dipole moment, μ_p) and the CTFs using the elemental values of electronegativity (χ) are extensively used in property correlations by chemical engineers and chemists. The data consisting of temperature, solubility, R_M, P_p, CTF, and $\frac{\Delta H_m}{RT}\left(\frac{T}{T_m}-1\right)$ and $\ln(\gamma^{\infty})$ at 282 data points is reported in Table 7.1, and comprises of 20 solutes and 31 solvents.

After evaluating different combinations of the variables considered, Eq. (7.6) has been obtained as a form with minimum number of coefficients and exhibiting reasonable accuracy. Calculation of solubility from the knowledge of γ^{∞} using Eq. (7.8) requires an iterative procedure for each data point. Therefore, within the range of $\frac{\Delta H_m}{RT}\left(\frac{T}{T_m}-1\right)$, γ^{∞}, and X values, an approximation has been developed for Eq. (7.8), and is reported as Eq. (7.7), which has resulted in coefficient of determination (R^2) as 0.999. The approximation is not based on the actual data. But on the basis of the equation structure, and therefore, percent absolute deviation for this relation is not reported.

TABLE 7.1 Calculations for various solutes with different solvents at various temperature ranges

Solutes & Corresponding solvents	Temp. Range °K	X*10⁴ Mol.frn. range	$\frac{\Delta H_m}{RT}\left(\frac{T}{T_m}-1\right)$ range	$R_{m,sol}$-$R_{m,solv}$ range	$P_{p,sol}$-$P_{p,solv}$ range	CTF range	$\ln \gamma^{\infty}$ of solute range
Pthalic Anhydride Solv.: Cyclohexane, Hexane, Heptane, i-Octane, Decane, Do-Decane, Hexadecane, Tri chloro-methane	298.15	6.9 - 630	-2.48	-2.58-5.44	564.60-547.47	0.15--0.21	4.8-0.32

Estimation of Solubility in Dilute Binary Solutions

TABLE 7.1 *(Continued)*

Solutes & Corresponding solvents	Temp. Range °K	$X*10^4$ Mol.frn. range	$\dfrac{\Delta H_m}{RT}\left(\dfrac{T}{T_m}-1\right)$ range	$R_{m,sol}$-$R_{m,solv}$ range	$P_{p,sol}$-$P_{p,solv}$ range	CTF range	$\ln \gamma^\infty$ of solute range
Carbazole Solv.: Cyclohexane, Hexane, Heptane, i-Octane, Decane, Do-Decane, Hexadecane, Tri chloro-methane	298.15	1.8 - 37	-5.04	-23.18 -31.48	83.43- 66.29	0.09- -0.27	3.58- 0.56
Picric acid Solv.: Cyclohexane, Hexane, Heptane, i-Octane, Decane, Do-Decane, Hexadecane, Tri chloro-methane	298.15	0.44 - 44	-1.7	-29.74- 24.92	46.01- 21.27	0.2- -0.15	8.45- 3.76
2-Hydroxy benzoic acid Solv.: Cyclohexane, Hexane, Heptane, i-Octane, Decane, Do-Decane, Hexadecane, Trichloro-methane, Tetra chloro-methane, Benzene.	298.15	3.8-169	-2.45	-44.9-9.76	143.59- 118.85	0.08- -2.25	5.31- 1.69
3-Hydroxy benzoic acid Solv.: Trichloro-methane, Tetrachloro methane, Benzene, Cyclohexane	298.15	1.9-0.001	-3.33	10.95- 2.93	116.8- 92.06	-0.28- 0.08	5.24- 12.79
4-Hydroxy benzoic acid Solv.: Trichloro-methane, Tetrachloro methane, Benzene, Cyclohexane	298.15	0.82-0.04	-3.65	10.95- 2.93	131.02- 155.76	-0.28- 0.08	5.76- 10.17
TetraDecanoic acid Solv.: Nitromethane	323.15 -293.15	124.1- 18.6	-0.21- -1.93	55.91	-214.67- -236.64	-0.02	4.28– 4.4
Hexadecanoic acid Solv.:Ethylacetate, Heptane, Nitromethane, Diethylether, Acetone	333.15– 243.15	0.19– 96.7	-0.15– -7.33	43.09– 65.14	12.48– -238.19	-0.01– 0.01	0.07– 4.59

TABLE 7.1 *(Continued)*

Solutes & Corresponding solvents	Temp. Range °K	X*10⁴ Mol.frn. range	$\frac{\Delta H_m}{RT}\left(\frac{T}{T_m}-1\right)$ range	$R_{m,sol}$-$R_{m,solv}$ range	$P_{p,sol}$-$P_{p,solv}$ range	CTF range	$\ln \gamma^\infty$ of solute range
9-Octadecenoic acid Solv.: Ethylacetate, Heptane, Methanol, Diethylether, Acetone, Toluene	273.15–223.15	0.11–28	-4.56– -9.37	48.76–-67.14	67.78–-155.63	-0.03–0.02	2.95–-0.02
Oleic acid Solv.: EthylAcetate, Methanol, Diethylether, Acetone, Toluene	253.15–213.15	190–0.34	-2.18– -6.86	52.26–67.14	50.93–-175.27	-0.03–0.02	3.36–0.47
Stearic acid Solv.: Acetonitrile, Dichloromethane, 2,2dimethylbutane, Ethanol, Ethyl acetate, Heptane, Methanol, Toluene, 2-Methylbutane, Methylcyclo hexane, i-Octane, i-Hexane, Nitroethane, Nitro methane, Diethylether, Acetone, Pentane	333.15–253.25	257.7–0.1	-0.6 – -7.59	78.64–4.37	74.6 –-246.03	-0.04–-0.02	6.08–0.16
O-Chloro Benzoic acid Solv. : Benzene	336.75–298.2	631–91	-1.71– -2.89	10.28	108.67-122.72	0.06	1.21–1.84
2-MethylBenzoic acid Solv.: Hexane	331.85–298.1	855–128	-0.84– -1.58	8.15	53.09 –59.1	0.06	1.94–2.85
4-Methyl Benzoic acid Solv.: Benzene	340.15–295.65	473–71	-1.97– -3.16	12.87	75.32–86.66	0.01	1.19–1.82
4-Methoxy Benzoic acid Solv.:Benzene	330.85–297.65	67–14	-2.92– -4.09	12.24	103.5–115.05	0.01	2.12–2.48
6-Octadecenoic acid Solv.: Ethylacetate, Heptane, Methanol, Toluene, Ethylether, Acetone	273.15–223.15	46–0.071	-4.3– -10.24	75.07–48.76	10.27–-206.02	-0.03–0.02	4.29–-1.62
6-Octadecenoic acid Solv: Ethylacetate, Heptane, Methanol, Toluene, Diethylether, Acetone	263.15-223.15	23–0.32	5.31– -10.24	75.07–48.76	10.27–-215.26	-0.03–0.02	1.50–-3.13

TABLE 7.1 *(Continued)*

Solutes & Corresponding solvents	Temp. Range °K	$X*10^4$ Mol.frn. range	$\dfrac{\Delta H_m}{RT}\left(\dfrac{T}{T_m}-1\right)$ range	$R_{m,sol}$ - $R_{m,solv}$ range	$P_{p,sol}$ - $P_{p,solv}$ range	CTF range	$\ln \gamma^\infty$ of solute range
Eicosanoic acid Solv:DiIodomethane, Freon-113	338.65– 298.55	262–5.0	-0.68– -3.98	63.49- 70.06	-0.46– 5.64	-0.16– -0.68	5.09– 1.58
Docosanoic acid Solv:Ethylacetate, Methanol, Toluene, Diethylether, Acetone	283.15– 263.15	10–0.01	-6.49– -8.96	74.3-89.18	42.41– -146.62	-0.01 – 0.02	4.86– -0.15
Heptadecanoic acid Solv.: Acetone, Isopropylether	303.15– 278.55	975.7– 10.6	-5.97– -1.55	50.48- 66.09	-1.35– -154.93	0.01	0.53– 2.74

The proposed method consisting of a combination of Eqs. (7.6) and (7.7) yields a relative deviation $(\Delta \bar{X}/\bar{X})$ of 0.6675 where $\Delta \bar{X}$ is the average absolute deviation in X and $\bar{X}$ is the average value of the solubility and R^2 value of 0.50. These values are compared to that reported by Ksiazczak and Anderke [2] (0.745 for relative deviation and 0.28 for R^2) who used a more complicated approach over a narrower range of 42 data points. It is observed that the present method yielded lower relative deviation and a much higher R^2.

7.4 CONCLUSION

The solubilities of solids in dilute solutions are modeled by a simple method proposed using properties such as pure component molar refraction, dipole moment and charge transfer functions. This method yielded average absolute deviations comparable to more complicated methods reported in the literature. The obtained results point out the predictive significance of the present method as a simple alternative for complex methods using pure component properties.

KEYWORDS

- **Charge transfer function**
- **Dipole moment**
- **Molar refraction**
- **Solid liquid equilibrium**
- **Solubility**

REFERENCES

1. David, R. L.; CRC Handbook of Chemistry and Physics, 78th edn. Boca Raton, FL: CRC Press;1997.
2. Ksiazczak, A.; and Andredo, A.; *Fluid Phase Equilib.* 1987, *35*, 127–151.

CHAPTER 8

KINETIC STUDY OF BECHAMP PROCESS FOR NITROBENZENE REDUCTION TO ANILINE

UMESH SINGH, AMEER PATAN, and NITIN PADHIYAR*

Department of Chemical Engineering, Indian Institute of Technology Gandhinagar, India; *E-mail: nitin@iitgn.ac.in

CONTENTS

8.1	Introduction	76
8.2	Reaction Mechanism	77
8.3	Chemical and Experimental Set-Up	80
8.4	Mathematic Model	80
8.5	Result and Discussion	81
8.6	Conclusion	88
Keywords		88
References		88

8.1 INTRODUCTION

Aniline has numerous commercial applications. The largest application of aniline until September 2007 is for the preparation of methylene diphenyldiisocyanate (85 percent), an intermediate material in the production of rigidpolyurethane [1]. Other application of aniline is in rubber industry (9 percent), agrochemicals (i.e. herbicides, 2 percent), dye making and pigments manufacturing (2 percent) [2]. The principal use of aniline is in the dye industry as an intermediate component for producingazo dyes. NH_2 group in aromatic amines is one of the chromophore groups in the dye formulation; and NH_2 is also the member of auxochrome group which is necessary to impart the solubility and to increase the adherences property [3]. Preferable methods for nitroaromatic reductions are catalyst hydrogenation and Bechamp process.

The Bechamp reduction process was first described by Antoine Bechamp for reducing nitrobenzene to aniline in 1854 [3]. In Bechamp process, hydrogen ions for the reduction are produced by the reaction of acid and water along with zero-valent iron. These hydrogen ions react with aromatic nitro compound on Fe surface to convert into aromatic amines. Also, the side reaction in the Bechamp process is the oxidation of Fe to $Fe(OH)_2$ and other hydroxides. Further, the overall Bechamp reaction is an exothermic reaction and hence, it is quite crucial to control the temperature during the process for better yield of the aromatic amine compounds. Bechamp process and the quality of amine product from this process are sensitive to parameters such as physical state of iron, amount of water used, amount and type of acid used, agitation speed, reaction temperature, and the use of various catalysts or additives [4]. Thus, the quality of the aromatic amine can be controlled by the above mentioned parameters. The Bechamp reduction can be summarized as follows.

8.1.1 BECHAMP REACTION

H^+ Generation:

$$6HCOOH + 6H_2O \rightarrow 6HCOO^- + 12H^+ + 6OH^- \tag{8.1}$$

Overall Reaction:

$$4ArNO_2 + 4H_2O + 9Fe^0 \rightarrow 4ArNH_2 + 3Fe_3O_4 \tag{8.2}$$

Bechamp process is the oldest commercial process for preparation of amines, but in more recent years it has been largely replaced by catalytic hydrogenation. Nevertheless, the Bechamp reduction is still used in the dye industry for the production of small volumes of aromatic amines and for the production of iron oxide for pigments [1]. Few problems found in catalytic hydrogenation can be avoided in Bechamp process. For example, using Bechamp process we can produce pure aniline from nitrobenzene without any trace of the catalyst. Hence, pharmaceutical grade of aniline is usually produced by Bechamp process. Though, due to the slow reaction rate in Bechamp process and costly downstream separation methods, their applications are limited to batch process [5]. Further, Bechamp process is more economical alternative process. For example small volumes amine production, catalytic regeneration and catalyst losses lead to a costly process compare to Bechamp process. For large volumes Bechamp process loses its edge over catalytic reduction. Other applications of this process involve treatment of groundwater, degrading non-biodegradable nitrobenzene to quite less hazardous, and biodegradable aniline at low concentrations. Hence, Bechamp process is used to convert aromatic nitro to aromatic amine in ground water and making them easily biodegradable [6].

The kinetic study in the literature [6] have the following limitations, (1) the kinetic study was carried out at only one temperature 15°C, (2) the application was for very dilute nitrobenzene concentration in ground water, and (3) the effect of rpm was studied in the range of 4–50. We have attempted to address these three issues and present the effect of temperature, initial composition and rpm at high value (200–600) in this work. Reduction of nitrobenzene to aniline by Bechamp process is carried out in a 500 ml batch reactor in this work. Gas Chromatograph (GC) is used for the sample analysis. A GC method has been developed with toluene as the solvent for determining the compositions of various reaction components.

8.2 REACTION MECHANISM

Reduction mechanism for aromatic nitro compounds can be explained by considering it as a three step process, namely (1) generation of H^+ ions by water-acid reaction, (2) reaction of H^+ with Nitrobenzene, and (3) oxidation of Fe^0 by H^+. These reactions [6] are explained in the following section,

H^+ Generation:

First step in the Bechamp process is the production of hydrogen ion, which is produced by reaction between acid and water. Different acids have been used for this purpose in the literature [7], but most popular used acids are formic acid and hydrochloric acid. Since, halogen group in hydrochloric acid is quite reactive and hence may react with the aromatic ring, formic acid is more popular than HCl.

8.2.1 ACID—WATER REACTION

$$6HCOOH + 6H_2O \rightarrow 6HCOO^- + 6OH^- + 12H^+ \quad (8.3)$$

$$6H^+ + 6HCOO^- \rightarrow 6HCOOH \quad (8.4)$$

Overall Reaction:

$$6H_2O \rightarrow 12H^+ + 6OH^- \quad (8.5)$$

The overall reaction constitutes the decomposition of water molecule suggesting little amount of acid is sufficient to carry out the reaction. It is reported in the literature that amine is not detected in the strong acidic conditions. Further, it is also suggests that 5–6.9 is the optimum pH in the Bechamp process [6].

8.2.2 REDUCTION OF NITROBENZENE

Reduction of Nitrobenzene can be summarized as a three step process, adsorption of PNT from bulk phase to iron surface, reaction of the same on iron surface, and desorption of the reaction products from iron surface to the bulk phase.

8.2.3 ADSORPTION OF NITROBENZENE ON THE FE^0 SURFACE

$$\text{Nitrobenzene is adsorbed on zero-valent iron} \rightarrow ArNO_2 + Fe^0 \quad (8.6)$$

8.2.4 REACTION BETWEEN NITROBENZENE AND HYDROGEN ION ON FE SURFACE

This is a complex reaction, a result of series and parallel reactions, with generation of one or more intermediates and byproducts.

$$ArNO_2 + Fe^0 + H^+ \rightarrow ArNH_2 + Fe^{2+} \quad (8.7)$$

The reduction of nitrobenzene to aniline is truly a nonelementary reaction. Several authors had proposed reaction mechanism for this reduction reaction involving the following steps. Firstly, Nitro benzene reduces to nitrosobenzene followed by nitrosobenzene reduction to phenylhydroxylamine, and finally the reduction of phenylhydroxylamine to Aniline. Reduction of nitro benzene can be summarized in three steps as follows,

$$ArNO_2 + Fe^0 + 2H^+ \rightarrow ArNO + Fe^{2+} + H_2O \qquad (8.8)$$

$$ArNO + Fe^0 + 2H^+ \rightarrow ArNHOH + Fe^{2+} \qquad (8.9)$$

$$ArNHOH + Fe^0 + 2H^+ \rightarrow ArNH_2 + Fe^{2+} + H_2O \qquad (8.10)$$

First reaction corresponds to two electron reduction of Nitrobenzene to nitrosobenzene; the second one corresponds to four electron reduction of nitrosobenzene to phenylhydroxylamine and third corresponds to two electron reduction of phenylhydroxylamine to aniline.

8.2.5 OXIDATION OF FE^0

Above mentioned all the three reactions result in the oxidized state of Fe, that is, Fe^{2+}. This in turn reacts with OH^- ion that was produced from water dissociation and produces ferrous hydroxide, $Fe(OH)_2$. This $Fe(OH)_2$ can still provide free surface for adsorption of *p*-nitrobenzene, that is similar to zerovalent iron surface. Further, on the surface of ferrous hydroxide, *p*-nitrobenzene reduces to the amine counterpart with further oxidation of $Fe(OH)_2$ to $Fe(OH)_3$. This can be summarized as follows [6, 7]

$$3Fe^{2+} + 6OH^- \rightarrow 3Fe(OH)_2 \qquad (8.11)$$

$$ArNO_2 + 6Fe(OH)_2 + 4H_2O \rightarrow 6Fe(OH)_3 + ArNH_2 \qquad (8.12)$$

$$Fe(OH)_2 + 2Fe(OH)_3 \rightarrow Fe_3O_4 + 4H_2O \qquad (8.13)$$

Though, the above mechanism results into only two Fe compounds, namely $Fe(OH)_2$ and Fe_3O_4, a mixture of different iron-oxide reactions are reported in the literature [6, 7]. Composition of iron-oxide depends on the reaction conditions, though not significant experimental results have been demonstrated to support this claim.

An overall reaction including nitro reduction and Fe oxides generation can be shown below:

$$4H_2O + 4ArNO_2 + 9Fe^0 \rightarrow 4ArNH_2 + 3Fe_3O_4 \qquad (8.14)$$

8.3 CHEMICAL AND EXPERIMENTAL SET-UP

Nitrobenzene (yellowish oily liquid), aniline (colorless liquid), formic acid and toluene having high purity of more than 99.5 percent were purchased from Merck. Iron powder with purity 99.5 percent and having 6–9 µ size ranges was purchased from sigma-Aldrich. Double RO water was used in all the experiments which is free from dissolved solids and impurities. Toluene of HPLC and/or GC grade was used as a solvent for Gas chromatography analysis. The experimental setup consists of a five- neck conical flask reactor, electrical heater, agitator, and a reflux condenser.

8.4 MATHEMATIC MODEL

Reduction of Nitrobenzene (NB) by Bechampprocess under several operating conditions gives aniline along with certain amount of intermediates/byproducts. Our GC results of liquid phase compositions show only three peaks of NB, nitrosobenzene, and Aniline, with RTD of 4.78 min, 3.048 and 3.67 min, respectively. Hence, to ignore the complexity in modeling we avoid the intermediates/byproducts. The liquid form compositions of aromatic compounds were found using GC making assumption that there is no accumulation on the surface of iron.

It is believed that the conversion from NB to Aniline is a result of the following three steps in series: (1) adsorption of NB on the iron surface followed by, (2) reaction on iron surface, and (3) desorption of the product [8].

However, in the reduction of aromatic nitro compounds the adsorption step is a limiting step. Hence we make a hypothesis of adsorption as the limiting step for NB to Aniline reduction. The hypothesis is verified with the help of experimental results, failure leads to change in the hypothesis.

With adsorption as the rate limiting step,

$$-r_A = k_m \left(C_{Ab} - C_{As} \right) \tag{8.15}$$

where, $-r_A$ is the rate of adsorption of A(NB), k_m is mass transfer constant, C_{Ab} is the concentration of A in Bulk liquid, and C_{As} is the concentration of A on the iron surface. According to the hypothesis of adsorption being the slowest step, C_{As} can be assumed to be very small compare to C_{Ab}. Thus, ignoring C_{As} in Eq. (8.15),

$$-r_A = k_m \left(C_{Ab} - 0 \right) \tag{8.16}$$

Hence,

$$\frac{dC_A}{dt} = k_m \times C_{Ab} \tag{8.17}$$

Or upon integration

$$\ln\left(\frac{C_{A0}}{C_A}\right) = k_m \times t \tag{8.18}$$

Where, C_{A0} is the initial concentration in the bulk phase. Thus, if adsorption is the limiting step, Eq. (8.17) or (8.18) should be satisfied. Please note that a first order of reaction with reaction being the limiting step may also satisfy the above Eq. (8.17). Hence, to confirm the mass transfer being the limiting step, we conducted experiments at different rpm in the range of 200–600. Further, Frossling *correlation* [9] of rpm and the rate constant shown in Eq. (8.19) can be fitted in the case of mass transfer as the limiting step.

$$Sh = 2 + 0.6\,Re^{1/2}\,Sc^{1/3} \tag{8.19}$$

Where, Sh is the Sherwood number (kd_p/D_{AB}), Sc is Schmidt number (v/D_{AB}), k is the mass transfer constant, d_p is the iron particle diameter, D_{AB} is the diffusivity coefficient, V is angular velocity, and v = kinematic viscosity. As can be observed from the above Frossling equation, rate constant is proportional to the square root of rpm at constant temperature and particle diameter.

$$k \propto (rpm)^{1/2} \tag{8.20}$$

8.5 RESULT AND DISCUSSION

The experimental study was carried out considering the effect of RPM and temperature.

8.5.1 EFFECT OF RPM

Experiments were performed with nitrobenzene having initial compositions of 0.5 g in 300 ml of water. Reaction Temperature and pH were fixed at 100°C, 5–5.5, respectively and rpm ranging from 200 to 600. These experiments were performed to study the effect of rpm on nitrobenzene reduction at high temperature, 100°C and high concentrations.

We had observed only two peaks in GC chromatogram, indicating no intermediate/by product in the bulk phase in measurable quantity. Thus, the overall reaction can be represented by,

$$C_6H_5NO_2 + Fe \xrightarrow{HCOOH} C_6H_5NH_2$$

Where, $C_6H_5NO_2$: nitrobenzene
$C_6H_5NH_2$: aniline

For the operating conditions mentioned above for nitrobenzene reduction, the GC analysis results in the nitrobenzene and aniline compositions graph shown in Figure 8.1

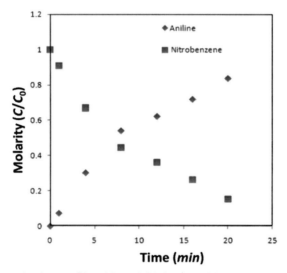

FIGURE 8.1 Reduction profile with an initial composition of 0.5 g of NB in 300 ml of water with 2 g of iron at 200 rpm.

Figure 8.2 shows a plot of $\ln(C_{A0}/C_A)$ vs. time the slope of which gives the mass transfer coefficient, k. It has been observed from the temporal plot of Figure 8.2 that it follows rate of adsorption discussed in Section 4.4 with the best fit found at mass transfer coefficient $k = 0.0894$ min^{-1} and the R^2 value of 0.9867.

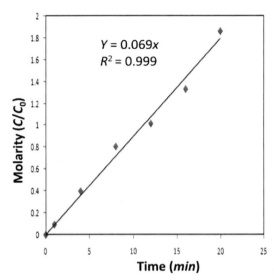

FIGURE 8.2 Temporal plot of composition of nitrobenzene for the first order kinetics.

The reaction kinetics shows that the reaction time for 85 percent conversion of nitrobenzene is 20 min. Rate constant was computed by the method of integration.

The data fit with R^2 value of 0.9867 suggests that mass transfer controlled reaction is a valid hypothesis. Please note that a hypothesis of first order of reaction with reaction being the limiting step may also be justified with the above fit. Though, if that is true, there should not be an effect of rpm on the overall rate of reaction. Thus, to assure that the adsorption rate is the limiting step, we carry out a study of the effect of rpm on the overall rate.

Figure 8.3 summarizes the effect of rpm on the overall rate of reduction of nitrobenzene at three different rpms, 200, 400, and 600. As can be observed, increasing rpm from 200 to 600, an increase in the rate of conversion was found indicating the adsorption being the rate limiting step.

Thus, increase in the agitation speed enhanced the supply rate of the reactant on the solid surface for the reaction to occur.

Further, the graph of mass transfer coefficient k vs $\sqrt{\text{rpm}}$ should be straight line according to Eq. (8.20). We conducted experiments at three different rpm 200, 400, and 600 and get the rate constant 0.0894, 0.118, and 0.1479 min^{-1}. The least square fit of the above model for these data is shown in Figure 8.4. We found an equation for graph: $k = 0.005\sqrt{\text{rpm}} + 0.006$ with an R^2 fit of 0.995.

Thus, a satisfactory fit of the Frossling correlation was obtained to represent the experimental data. Thus, it can be stated that the reduction of nitrobenzene to aniline is a mass transfer controlled reaction.

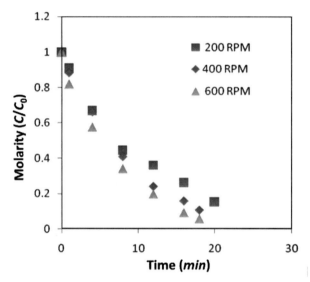

FIGURE 8.3 Effect of rpm on the reduction rate of NB for rpm range 200–600 at 100°C.

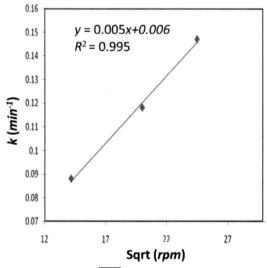

FIGURE 8.4 Kinetic constant k vs $\sqrt{rpm}$ for rpm range of 200–600.

8.5.2 EFFECT OF TEMPERATURE

To study the effect of temperature on reaction kinetics, we carried out experiments from 30 to 100°C with initial nitrobenzene compositions of 0.5 g in 300 ml of water. The rpm was fixed at 400 and pH was maintained at 3 by adding formic acid. The experimental results of NB reduction at 100, 90, 80, 60, and 30°C are shown in Figures 8.5–8.9. Here, we had observed one additional peak for nitrosobenzene with the retention time of 3.048 min including nitrobenzene and aniline. This is an intermediate in the Bechamp reaction also suggested by Agrawal and Tratnyek [6]. Please note that we had observed negligible reaction conversion at a pH value of 5 at 30°C. Hence, in this subsection, we have lowered the pH value to 3 for obtaining comparable reaction conversion.

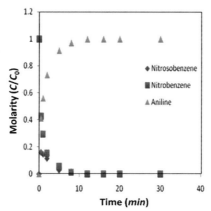

FIGURE 8.5 Effect of temperature at 100°C with 0.5 ml acid and 0.5 g NB in 300ml water at 400 RPM.

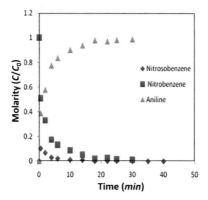

FIGURE 8.6 Effect of temperature at 90°C with 0.5 ml acid and 0.5 g NB in 300 ml water at 400 RPM.

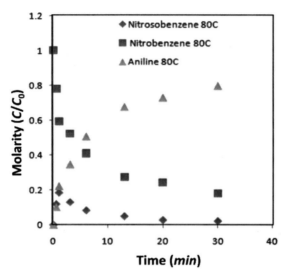

FIGURE 8.7 Effect of temperature at 80°C with 0.5ml acid and 0.5 g NB in 300 ml water at 400 RPM.

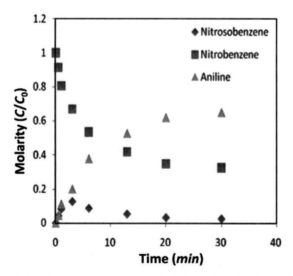

FIGURE 8.8 Effect of temperature at 60°C with 0.5ml acid and 0.5 g NB in 300 ml water at 400 RPM.

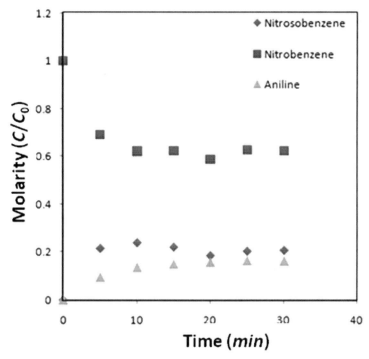

FIGURE 8.9 Effect of temperature at 30°C with 0.5 ml acid and 0.5 g NB in 300 ml water at 400 RPM.

Except nitrosobenzene no other intermediate/byproduct was observed in this temperature range. This fact is consistent with the results of Agrawal andTratnyek [6]. At 30°C we see that intermediate nitrosobenzene is highly stable and higher in composition in comparison to one that is observed at 100°C. The mass transfer coefficient values at various temperatures considered in this subsection are summarized in Table 8.1.

TABLE 8.1 Summary of rate constants for Temperature range of 30-100 °C

S. No	Temperature (°C)	k (min^{-1})
1	30	0.00432
2	60	0.0471
3	80	0.0612
4	90	0.2309
5	100	0.5752

8.6 CONCLUSION

Experimental study of Bechamp reduction is carried out in acidic medium for the reduction of nitrobenzene to aniline in this work. The experiments were conducted for the temperature range of 30–100°C and agitator speed of 200–600 RPM. Nitrobenzene reduces to aniline with different by products depending upon the reaction conditions. In nitrobenzene reduction, we had found higher selectivity of aniline or higher production of aniline compared to negligible production of other side products.

The overall conversion of nitrobenzene to aniline is best described by three steps in series, adsorption of nitrobenzene on iron surface, reaction of nitrobenzene to aniline on Fe surface, and desorption of aniline from the Fe surface to the bulk liquid phase. Among that Adsorption of nitrobenzene from the bulk liquid to the iron surface was found to be the controlling step of the overall reduction rate in the operating conditions considered in this work. Experimental results show that reduction of nitrobenzene to aniline was mass transfer controlled. The rate constants are found for three different RPM values in the range of 200–600 and five different temperature values in the range of 30–100°C. Higher RPM and higher temperature were favorable for the reduction on NB. Though, in a separate study a reverse trend was observed at 800 RPM (results not shown here), where the centrifugal force dominates. The rpm effect on the rate constant was satisfactorily verified using Frossling correlation in this work.

KEYWORDS

- Bechamp process
- Frossling correlation
- Nitrobenzene reduction

REFERENCES

1. Windholz; and Annsville; National Service Center for Environmental Publications. *Aniline fact sheet*; **1992**.
2. Contreras, S.; Rodryguez, M.; Chamarro, E.; and Esplugas, E. V; UV/Fe*(III)-enhanced* ozonation of nitrobenzene in aqueous solution. *J. Photochem. Photobiol.* **2001**, 142, 79-83.
3. Becham A. J.; Bechamp Reduction. *Am. Chem. Phys.* **1854**, 42, 186.
4. Hindle, K. T.; Jackson, S. D.; Stirling, D.; and Webb, G.; The Hydrogenation of para-Toluidine over Rh/Silica: The Effect of Metal Particle Size and Support Texture. *J. Catal.* **2006**, 241, 417–425.

5. Wang, Z.; Bechamp reduction. *Compr. Org. Name Reactions Reag.*; **2010**, 284–287.
6. Agrawal, A.; and Tratnyek, P. G.; Reduction of Nitro Aromatic Compounds by Zero-valent Iron Metal. *J. Environ. Sci. Technol.* **1996**, 30, 153–160.
7. Othmer, K.; Amine by Reduction. *Enyclopindia.* **2004**, 2, 476–490.
8. Khan, F. A.; and Sudheer, C.; Oxygen as moderator in the zinc-mediated reduction of aromatic nitro to azoxy compounds. *Tetrahedron Lett.* **2009**, 50, 3394–3396.
9. Fogler, H. S.; *Elements of Chemical Reaction Engineering.* **2011**, 253–310.
10. McCabe, W. L.; Smith, J. C.; and Harriott, P.; *Unit Operations of Chemical Engineering, 7th Ed. New Delhi: McGraw-Hill.* **2005**, 249–251.
11. Choe, S.; Liljestrand, H. M.; and Khim, J.; Nitrate reduction by zero-valent iron under different pH regimes. *J. Appl. Geochem.* **2004**, 19, 335–342.

CHAPTER 9

FLOW BEHAVIOR OF MICROBUBBLE-LIQUID MIXTURE IN PIPE

RAJEEV PARMAR and SUBRATA KUMAR MAJUMDER

Department of Chemical Engineering, Indian Institute of Technology Guwahati, Guwahati-781039, Assam, India.

E-mail: r.parmar@iitg.ernet.in; skmaju@iitg.ernet.in

CONTENTS

9.1	Introduction	92
9.2	Experiment Setup and Procedure	92
9.3	Theoretical Analysis	93
9.4	Results and Discussion	94
9.5	Conclusions	98
Nomenclature		98
Keywords		98
References		98

9.1 INTRODUCTION

Recently microbubble has gained a lot of attention. Microbubbles are the bubbles having diameter in the order of 50 µm. They possess a lot of spectacular properties which makes them superior over the conventional bubbles. According to Young–Laplace equation the excess pressure inside a bubble is inversely proportional to its size [1]. The internal pressure depends on the intermolecular and molecule-wall interactions. If the diameter of the bubble increases, the intermolecular interaction and the molecule-surface interaction decrease whereas it increases as the size of the bubble decreases. The microbubble, due to their small size exhibit high pressure and thus provides a high gas dissolution rate. They are also beneficial as they provide large gas liquid interfacial area. They experience a low degree of buoyance force, due to their small size, which gives them a long residence time in the liquid as compared to micro- and macrobubbles. They find a large number of applications in engineering as well as medical field. Microbubbles can increase the efficiency of gas–liquid contacting devices for various applications in chemical, petrochemical, and biochemical processes [2–5]. Despite of such extreme uses, the flow behavior of microbubble-liquid mixture has not been analyzed properly. During the simultaneous flow of microbubble-liquid mixture through a device, various hydrodynamic characteristics such flow behavior, pressure drop, friction factor, interfacial stress have to be analyzed. In the present chapter the rheological behavior and friction factor of microbubble-liquid mixture is studied in pipe.

9.2 EXPERIMENT SETUP AND PROCEDURE

The experimental setup mainly consists of a tank, a pump, microbubble generator, pressure transducer with data logger system, one rotameters, valvesas shown in Figure 9.1. Pressure drop is measured in straight horizontal pipe having diameter of 6×10^{-3} m. The length of pipe is taken about 2.5 m. The length taken is sufficient enough, so that accurate measurements of the pressure drop could be acquired with the pressure transducers as well as fully developed flow obtained throughout the pipe. The test section is also designed well to minimize any losses due to contraction and expansion. The working fluids werewater with different surfactant (Sodium dodecyl sulfate) concentrations and air. The density of the fluid is calculated by specific gravity bottle and the surface tension is measured by tensiometer (KRUSS GmbH Co., model- K9-MK1, Germany). All the experiments were carried out at room temperature. The detail of the physical property of the system is given in Table 9.1.

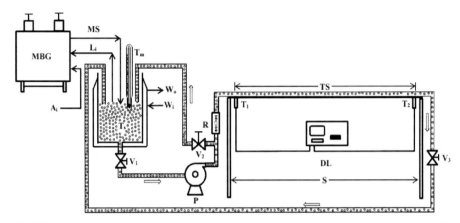

FIGURE 9.1 Schematic representation of experimental setup. Legend: A_i: Air inlet; CP: Pump; DL: Data logger; L_i: Liquid inlet; MBG: Microbubble generator; MS: Microbubble suspension; P_{1-4}: Pipes; R: Rotameter; S: Self angle support; T_{1-2}: Pressure transducers; TS: Test section; T_s: Microbubble suspension tank with cooling jacket; T_m: Thermometer; V_{1-11}: Flow control valve; W_i: Water inlet; W_o: Water outlet.

TABLE 9.1 Physical properties of system at 25±1°C

Liquid	Density (kg/m³)	Surface Tension (mN/m)
Water without microbubbles	999.68	71.20
Microbubble-liquid mixture without surfactant	998.00	71.20
Microbubble-liquid mixture at 3 ppm surfactant	998.09	69.00
Microbubble-liquid mixture at 7 ppm surfactant	969.15	65.17
Air	1.18	–

9.3 THEORETICAL ANALYSIS

The rheological behavior of a real fluid can be analyzed by using Ostwald–de Waelepower law [6] relationship. It can be expressed mathematically as:

$$\tau_w = \mu_{eff} \gamma_a = K \gamma_w^n \qquad (9.1)$$

where τ_w, γ_w, and γ_a represents wall shear stress, wall shear rate, and apparent shear rate respectively. μ_{eff} is the effective viscosity of fluid. K and n are flow consistency index and flow behavior index respectively.

The wall shear rate and apparent shear rate for a circular pipe can be calculated by using Eqs. (9.2) and (9.3) respectively [7]

$$\tau_w = \frac{D\Delta P}{4L} \qquad (9.2)$$

$$\gamma_a = \frac{8U_f}{D} \qquad (9.3)$$

where U_f is average velocity of fluid mixture in pipe. The friction factor (f) for microbubble-liquid mixture can be calculated as:

$$f = \frac{2\tau_w}{\rho_f U_f^2} \qquad (9.4)$$

9.4 RESULTS AND DISCUSSION

In this section the experimental results obtained are discussed in detailed. The variation of wall shear stress with apparent shear rate of a fluid shows its rheological nature. The typical primary experimental data with estimated properties is shown in Table 9.2. The effect of apparent shear rate on wall shear stress on log-log plot in the present system is shown in Figure 9.2. It is observed that the shear stress vs shear rate for the microbubble-liquid mixture shows a power law dependency. It is also observed that the flow behavior index of microbubble-liquid mixture is less than unity (~0.945) indicating that microbubbles-liquid mixture behaves as non-Newtonian fluid. The rheology of suspension can mainly be affected by the gas holdup but other factors such as size of bubble, presence of other particles, presence of electrical charge and physiochemical properties of liquid. The addition of surfactant further increases the non-Newtonian behavior of the fluid [8]. The variation of effective viscosity of microbubble-liquid mixture with shear rate is shown in Figure 9.3. The surface tension of mixture decreases on addition of surfactant which increases gas holdup of system. This increase in gas holdup affects the pressure difference of system, due which the shear stresses changes, causing the effective viscosity of the system to decrease. The effect of Reynolds number on friction factor is shown in Figure 9.4. In the present work it observed that the friction factor of microbubble-liquid mixture is lower than single phase fluid without microbubble. Microbubble significantly decreases the friction factor. The addition of surfactant in microbubble-liquid mixture further decreases the friction factor. There are many studies available in literatures [9, 10] which showed that friction factor decreases in presence of microbubble however the extent of reduction of friction may depends on Reynolds number, bubble size, and gas holdup. In the present study in entire range of Reynolds number the friction factor found to be decreased with increase in Reynolds number as well as surfactant concentration in the liquid. The dependency of friction factor with Reynolds number for microbubble-liquid mixture can be presented by developing correlation based on present experimental data as:

$$f_{mb} = \frac{15 - 0.021 C_s^2 - 0.031 C_s}{\text{Re}_{mb}} \qquad (9.5)$$

where C_s is the concentration of surfactant in liquid and Re_{mb} is the Reynolds number of fluid of microbubble suspension. The correlation coefficient of Eq. (9.5) is found to be 0.97. To test the validity of correlation proposed for friction factor (f), the experimental and predicted value of friction factor is plotted as shown in Figure 9.5. It is observed that the developed correlation is good in agreement within the range of variables.

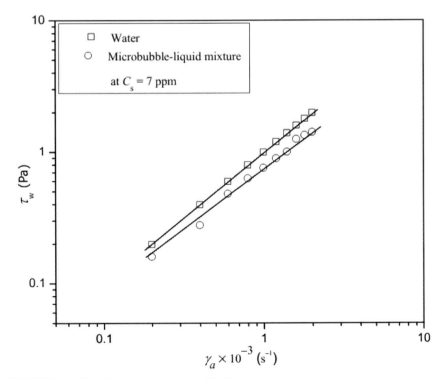

FIGURE 9.2 Variation of shear stress with shear rate.

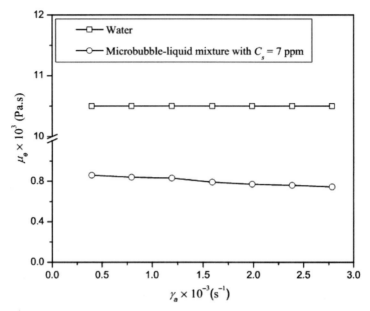

FIGURE 9.3 Variation of effective viscosity with apparent shear rate.

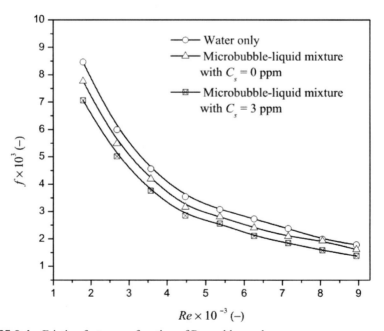

FIGURE 9.4 Friction factor as a function of Reynolds number.

Flow Behavior of Microbubble-Liquid Mixture in Pipe

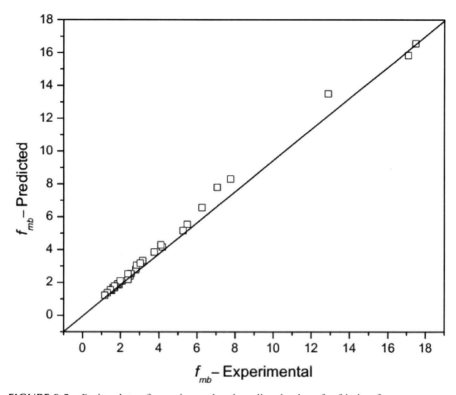

FIGURE 9.5 Parity plots of experimental and predicted values for friction factor.

TABLE 9.2 Primary experimental data 25±1°C

Average Fluid Velocity (U_f)	Pressure Drop (ΔP)	Surfactant Concentration (C_s)	Flow Behavior Index (n)	Flow Consistency Index (k)
0.149	312.498	3	0.965	0.0010
0.298	516.666	3	0.965	0.0010
0.447	866.666	3	0.965	0.0010
0.596	1100.000	3	0.965	0.0010
0.149	291.666	7	0.945	0.0011
0.298	463.333	7	0.945	0.0011
0.447	800.000	7	0.945	0.0011
0.596	1050.000	7	0.945	0.0011

9.5 CONCLUSIONS

The present chapter mainly emphases on the rheological behavior of microbubble suspension in horizontal pipe. The following conclusions can be drawn from the experimental data. The effective viscosity and wall shear stress of microbubble-liquid mixture found to be decreasing with apparent shear rate. Friction factor found to be decrease inversely with the Reynolds number. A correlation has also been developed to analyze the effect of surfactant and Reynolds number of friction factor of microbubble-liquid mixture. It is also observed from experimental results that the presence of microbubbles reduces the frictional resistance and the flow of microbubble-liquid mixture shows the non-Newtonian behavior during flow.

NOMENCLATURE

C_s	Concentration of surfactant (ppm)
D	Pipe diameter (m)
K	Consistency of fluid (Pa.s^n)
L	Length (m)
n	Flow behavior index (–)
ΔP	Pressure drop (N/m^2)
Re	Non-Newtonian liquid Reynolds number (–)
U_f	Fluid velocity (m/s)
ρ_f	Density of fluid (kg/m^3)
τ_w	Wall Shear stress (Pa)
γ_w	Wall shear rate (s^{-1})
γ_a	Apparent shear rate (s^{-1})

KEYWORDS

- Microbubble
- Pressure drop
- Rheology
- Friction factor

REFERENCES

1. Matsumoto, M.; and Tanaka, K.; *Fluid Dyn. Res.* **2008**, *40*, 546–553.
2. Bredwell, M. D.; and Worden, R. M.; *Biotechnol. Prog.* **1998**, *14*, 31–38.

3. Tsuge, H.; *Bull. Soc. Sea Water Sci. (Jpn.).* **2010**, *64*, 4–10.
4. Xiaohui, D.; Jingyu, X.; Yingxiang, W.; and Donghong, L.; *Gongye Shuichuli.* **2011**, *31*, 89–91.
5. Dikjmans, P. A.; Juffermans, L. J. M.; Musters, R. J. P.; Wamel, A. V.; Cate, F. J. T.; Gilst, W. V.; Visser, C. A.; Jong, N. D.; and Kamp, O.; Microbubbles and ultrasound: from diagnosis to therapy. *Eur. J. Echocardiogr.* **2004**, *5*, 245–256.
6. Enzendorfer, C.; Harris, R. A.; Valko, P.; Econmides, M. J.; Fokker, P. A.; and Davies, D. D.; Pipe viscometry of foams. *J. Rheol.* **1995**, *39*, 345–356.
7. Chhabra, R. P.; and Richardson, J. F.; Non-Newtonian Flow in the Process Industries: Fundamentals and Engineering Applications, 1st edn. Oxford: Butterworth-Heinemann; **1999**.
8. Jeffrey, D. J.; and Acrivos, A.; *AIChE J.* **1976**, *22*, 417–432.
9. Kodama, Y.; Kakugawa, A.; Takahashi, T.; and Kawashima, H.; *Int. J. Heat Fluid Flow.* **2000**, *21*, 582–588.
10. Madavan, N. K.; Deutsch, S.; and Merkle, C. L.; *Phys. Fluids.* **1984**, *27*, 356–363.

CHAPTER 10

METHANE PRODUCTION BY BUTANOL DECOMPOSITION: THERMODYNAMIC ANALYSIS

BRAJESH KUMAR[1], SHASHI KUMAR[2], and SURENDRA KUMAR[3*]

[1,2,3]Department of Chemical Engineering, Indian Institute of Technology Roorkee, Roorkee, Uttarakhand 247667, India; *E-mail: skumar.iitroorkee@gmail.com

CONTENTS

10.1 Introduction 102
10.2 Thermodynamic Analysis 103
10.3 Results and Discussion 104
10.4 Conclusion 106
Keywords 106
References 107

10.1 INTRODUCTION

The depletion of fossil fuels and growing demand for energy generation dictates the production of simple fuels by using bio-renewable feedstocks. Butanol is produced by fermentation process of feedstocks such as sugar beets, sugarcane, corn, wheat, ligno-cellulosic biomass, and more recently by micro-algae. Hence, recently butanol has been proposed as an alternative feedstock to produce fuel. Methane is produced from various renewable sources such as methanol, ethanol, glycerol, etc. It is the simplest alkane and the main component of natural gas. The relative abundance of methane makes it an attractive fuel. Methane is used in industrial chemical processes as fuel (Natural gas, liquid methane rocket fuel, and chemical feedstocks) [1–4].

Butanol as a higher member of the series of straight chain alcohols contains 4-carbon atoms. Butanol has many advantages including higher content of hydrogen (13.51 wt%) as compared to ethanol (13.04 wt%) and methanol (12.50 wt%), better safety aspects because it is combustible but not flammable like ethanol, lower vapor pressure, greater tolerance to water contamination [5]. Besides, butanol is used as potential biofuel with 85 percent strength in an I.C. engine, as a solvent in various organic synthesis, coating, chemical and textile processes, and as a component of hydraulic and brake fluids. The isomers of butanol and their applications [6] are listed in Table 10.1.

TABLE 10.1 Applications of butanol isomers

Butanol isomers	Applications
n-butanol	Additives of gasoline
	Chemical intermediate- for butyl ethers or butyl esters, etc.
	Plasticizers
sec-butanol	Solvents-for paints, resins, dyes related items, etc.
	Solvent
	Industrial cleaners- paint removers, etc.
	Chemical intermediate- for butanone, etc.
	Cosmetics like perfume, or in artificial flavors, etc.
iso-butanol	Additives of gasoline
	Ink ingredient
	Industrial cleaners—paint removers, etc.
	Solvent and paint additives

TABLE 10.1 *(Continued)*

Butanol isomers	Applications
tert-butanol	Solvent
	Composite for ethanol
	Industrial cleaner-paint removers, etc.
	Additives of gasoline for octane booster and oxygenated intermediate for methyl *tert*-butyl ether, ethyl *tert*-butyl ether, *tert*-butyl hydroperoxide

10.2 THERMODYNAMIC ANALYSIS

The thermodynamic equilibrium calculations are usually performed via two approaches: stoichiometric approach and non-stoichiometric approach. In stoichiometric approach, the system is described by a set of stoichiometric reactions and the equilibrium analysis is based on equilibrium constants for these reactions. While, in non-stoichiometric approach, the equilibrium compositions of products are derived by the direct minimization of Gibbs free energy [7]. In present research work, thermodynamic analysis of butanol decomposition has been performed to produce mainly methane by using non-stoichiometric approach. In this approach, the chemical species resulting in atomic combinations of constituent elements of feed species which may coexist as potential products at equilibrium are defined.

10.2.1 REACTIONS NETWORK

The possible reactions of butanol decomposition are given below [8]:

$$C_4H_{10}O \rightleftharpoons C_2H_4 + CH_4 + CO + H_2$$

$$C_2H_4 \rightleftharpoons C_2H_2 + H_2$$

$$CO + 3H_2 \rightleftharpoons CH_4 + H_2O$$

$$CO + H_2O \rightleftharpoons CO_2 + H_2$$

$$CO + H_2 \rightleftharpoons C + H_2O$$

$$CO + 2H_2 \rightleftharpoons C + 2H_2O$$

$$CH_4 \rightleftharpoons 2H_2 + C$$

10.2.2 MINIMIZATION OF GIBBS FREE ENERGY

In present research work, the method of direct Gibbs energy minimization is used to determine the equilibrium concentrations of reaction species. In case of butanol decomposition, eight products namely H_2, CO, CO_2, H_2O, CH_4, C_2H_2, C_2H_4, and carbon are considered. The equilibrium compositions of the products are estimated by minimization of Gibbs free energy using R-Gibbs reactor in Aspen Plus (Version 2006.5) software.

10.3 RESULTS AND DISCUSSION

10.3.1 EFFECT OF TEMPERATURE AND PRESSURE

The thermodynamic equilibrium investigations have been carried out for pressure ranging from 1 to 10 atm and temperature 300–1,200°C. The butanol flow rate of 1 kmol/s has been taken in the feed. The simulation results show the complete conversion of butanol is achieved at all temperature and pressure conditions.

Figure 10.1 (a)-(f) illustrate the effect of pressure and temperature on equilibrium flow rates of the products. It is clear from Figure 10.1 (a) that at all pressure conditions the production of CH_4 is maximum (2.02 kmol/sec) at temperature between 300 and 700°C. Above 700 °C, temperature does not favour the production of CH_4. On the contrary, higher pressure favours the production giving minimum at 1 atm.

Similar trends has been found in case of C_2H_4 production [Figure 10.1 (b)]. The maximum C_2H_4 flow rate (0.5 kmol/sec) has been obtained upto temperature of 500 °C at all pressures. The production trends of C_2H_4 and H_2 are found to be identical [10.1(c) and 10.1 (d)]. At the temperature $\leq$ 600 °C, no production of these gases has been found at all pressure conditions. The higher temperature (> 600 °C) and low pressure promote the production of C_2H_2 and H_2.

The Figure 10.1 (e) depicts that the flow rate of CO_2 becomes negligibly small at temperature $\geq$ 700 °C at all pressure conditions. Higher pressure and lower temperature favour the CO_2 production.

Reverse trends have been found in case of CO production [Figure 10.1 (f)]. Maximum flow rate of CO (1 kmol/sec) is achieved at temperature ≥ 700 °C. Lower temperature and high pressure suppresses CO production.

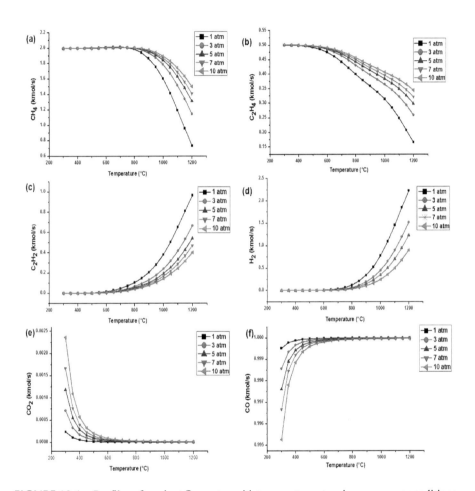

FIGURE 10.1 Profiles of product flow rates with temperature at various pressure conditions

Table 10.2 summarizes the results of CH_4 formation at various temperatures and pressure conditions with percent increase in its value with respect to 1 atm. This table clearly indicates that percent increase in CH_4 production with respect to 1 atm is significant at 3 atm or higher pressure and temperatures above 700°C.

TABLE 10.2 CH_4 formation by butanol decomposition and its percent increase (PI) with respect to 1 atm pressure

Pressure	1 atm	3 atm		5 atm		7 atm		10 atm	
Temperature (°C)	Flow Rate (kmol/s)	Flow Rate (kmol/s)	PI (%)	Flow Rate (kmol/s)	PI (%)	Flow Rate (kmol/s)	PI (%)	Flow Rate (kmol/s)	PI (%)
700	2.02	2.01	−0.50	2.01	−0.50	2.01	−0.50	2.01	−0.50
750	2.01	2.01	0.00	2.01	0.00	2.01	0.00	2.01	0.00
800	1.99	2.00	0.50	2.00	0.50	2.00	0.50	2.00	0.50
850	1.94	1.98	2.06	1.99	2.58	1.99	2.58	1.99	2.58
900	1.87	1.94	3.74	1.96	4.81	1.97	5.35	1.98	5.88
950	1.76	1.88	6.82	1.91	8.52	1.93	9.66	1.95	10.80
1,000	1.60	1.79	11.88	1.84	15.00	1.87	16.88	1.90	18.75
1,050	1.41	1.67	18.44	1.75	24.11	1.79	26.95	1.83	29.79
1,100	1.19	1.52	27.73	1.63	36.97	1.69	42.01	1.74	46.22
1,150	0.96	1.34	39.58	1.48	54.17	1.56	62.50	1.63	69.79
1,200	0.74	1.15	55.41	1.31	77.03	1.41	90.54	1.50	102.70

10.4 CONCLUSION

The thermodynamic analysis of butanol decomposition was carried out by using Gibbs free energy minimization method to produce methane. The effect of temperature and pressure on CH_4 production was analyzed. The thermodynamic analysis results revealed that most favorable conditions for butanol decomposition is 300°C and 1 atm pressure at which 1 kmol/s of butanol had the capacity to produce 2.02 kmol/s of CH_4 and 1 kmol/s of CO. Methane (CH_4) with carbon monoxide (CO) can be used as a fuel for Molten-Carbonate Fuel Cell (MCFC) and Solid-Oxide Fuel Cell (SOFC) applications.

KEYWORDS

- Butanol
- Gibbs free energy minimization
- Methane production
- Thermodynamic analysis

REFERENCES

1. Qureshi, N.; *Food Biotechnol.* **2005**, *5*, 25–51.
2. Qureshi, N.; and Ezeji, T. C.; *Biofuel Bioprod. Bioref.* **2008**, *2*, 319–330.
3. Qureshi, N.; Li, X. L.; Hughes, S.; Saha, B. C.; and Cotta, M. A.; *Biotechnol. Prog.* **2006**, *22*, 673–680.
4. Jesse, T.; Ezeji, T. C.; Qureshi, N.; and Blaschek, H. P.; *J. Ind. Microbiol. Biotechnol.* **2002**, *29*, 117–123.
5. Nahar, G. A.; and Madhani, S. S.; *Int. J. Hydrogen Energy.* **2010**, *35*, 98–109.
6. Silva, A. L.; Malfatti, C. F.; and Müller, I. L.; *Int. J. Hydrogen Energy.* **2009**, *34*, 4321–4330.
7. Katiyar, N.; Kumar, S.; and Kumar, S.; *Int. J. Hydrogen Energy.* **2013**, *38*, 1363–1375.
8. Wang, W.; and Cao, Y.; *Int. J. Hydrogen Energy.* **2011**, *36*, 2887–2895.

CHAPTER 11

NOVEL METHOD FOR INCREASING THE CLEAVAGE EFFICIENCY OF RECOMBINANT BOVINE ENTEROKINASE ENZYME

A. RAJU[1,2], K. NAGAIAH[2*], and S. S. LAXMAN RAO[3]

[1]Jawaharlal Nehru Technological University Hyderabad, Kukatpally-500085, Hyderabad;

[2]Gland Pharma Limited, D.P. Pally(Narsapur Road),Near Gandimaisamma 'X' Road, Hyderabad- 500043; E-mail: arajubio@gmail.com

[3]Sreenidhi Institute of Science and Technology, Hyderabad

CONTENTS

11.1	Introduction	110
11.2	Materials and Methods	110
11.3	Results and Discussion	112
11.4	Conclusion	116
Keywords		116
References		116

11.1 INTRODUCTION

Enterokinase is the protease of choice for N-terminal fusions, since it specifically recognizes a five amino-acid polypeptide (D-D-D-D-K-X1) and cleaves at the carboxyl site of lysine. Sporadic cleavage at otherresidues was observed to occur at low levels, dependingon the conformation of the protein substrate [1]. Biochemical analyses haveshown that the cleavage efficiency depends on the aminoacid residue X1 downstream of the D4K recognition site [2]. Recombinant hybrids containing apolypeptide fusion partner, termed affinity tag, to facilitatethe purification of the target polypeptides are widelyused. Many different proteins, domains, or peptides canbe fused with the target protein. The advantages of usingfusion proteins to facilitate purification and detection ofrecombinant proteins are well-recognized [3].All tags, whether large or small, have the potential to interfere with the biological activity of a protein, impede its crystallization, or otherwise influence its behavior.Consequently, it is usually desirable to remove the tag [4]. Therefore various proteases like factor Xa, recombinant bovine, enterokinase, etc., were used for tag separation.Protease activity is also regulated by metal ion cofactors that reversibly bind to proteases and affect their final activity, often in an allosteric manner [5]. Most of the first row transition metals (the only exceptions being scandium, titanium, and perhaps chromium) as well asmolybdenum; tungsten and magnesium are known to function as cofactors in enzymatic catalysis. Typically, these metal-ionic cofactors are bound to the enzyme viacoordination of amino acid side-chains. Enzymes with more looselybound metal ion cofactors, commonly called metalactivatedenzymes, require the presence of the appropriatemetal ion in the buffer in order to observe maximalcatalytic activity [6]. The metal cofactors, Mg^{2+} and Mn^{2+}, appear to stabilize two distinct conformational states of the enzyme which differ in response to varying substrate and effector concentrations [7]. The aim of the present investigation is to study the effect of cofactor Magnesium ion (Mg^{2+}) ion on the activity of recombinant bovine enterokinase.

11.2 MATERIALS AND METHODS

Fusion protein CBD-PTH (1–34), Trizma Base (Sigma), Sodium Hydroxide (Merck India), recombinant Bovine Enterokinase (USV Limited-Mumbai), Acetonitrile (Merck India), Trifluoro Acetic Acid (Merck India), Magnesium sulfate (SRL Laboratories, Mumbai), Terrific broth (Sigma), D(+) Glucose, Isopropyl thiogalactoside (IPTG) (SRL Laboratories, Mumbai), Protein estimation reagent (Bio-Rad, USA), and Capto MMC cation exchange resin-GE Healthcare, USA.

11.2.1 RECOMBINANT ESCHERICHIA COLICULTURING AND PRODUCTION OF FUSION PROTEIN CBD-PTH(1–34)

One milliliter of recombinant *E. coli* stored at −70°C with aexpression gene of fusion protein Parathyroid Hormone,PTH (1–34) with a fusion tag Chitin binding domain (CBD) and a recognition sequence for protease enzyme recombinant bovine enterokinase {(Asp)$_4$-Lys} in theplasmid pET24a was inoculated into 500ml of presterilized terrific broth media with 2 percentdextrose in 1L conical flask. The inoculated flasks were incubated at 37°C in orbital shaker at 250 rpm. After 3–4 hr of incubation when the optical density (OD$_{600}$) reaches 0.8–1.0, the culture is induced with 1mM IPTG and the incubation was continued for another 6–8 hr until the OD$_{600}$reaches 2.5–3.0.The culture was centrifuged at 10,800g for 15min and the supernatant was discarded. The cells were lysed using high pressure homogenizer at 1,200 bar pressure. The lysate was centrifuged and the supernatant containing fusion protein was subjected to Cation-Exchange chromatography using the resin Capto MMC for the initial isolation of fusion protein. The fusion protein so obtained is of 60–70 percentpure.

11.2.2 SEPARATION OF FUSION TAG CBD FROM PTH(1–34) USING RECOMBINANT BOVINE ENTEROKINASE

The total protein obtained in the elution was determined using protein estimation kit. The pH of the solution was adjusted to 8.0 using 10 percent Sodium hydroxide solution. Various concentrations of recombinant bovine enterokinase enzyme were added per milligram of total protein after lysis and clarification containing fusion protein in soluble form and the incubation was carried out at different temperatures and durations. Magnesium sulfate was added at different concentrations to enhance the cleavage efficiency and the percentage of cleavage or PTH (1–34) liberated was checked using RP-HPLC.Details were furnished in Table 11.1.

TABLE 11.1

Recombinant Bovine Enterokinase (IU/mg)	Incubation Time(hr)	Incubation Temperature (°C)	% PTH Released by Area	MgSo$_4$ (mM)
2	6	22	12.1	0
2	10	22	26.2	0
1	5	22	17.38	0
1	5	24	26.69	1
1	5	24	26.55	5
0.2	30	24	11.16	0
0.2	16	24	28.21	5
0.05	16	24	31.19	5

11.2.3 ANALYSIS OF CLEAVAGE/PERCENTAGE OF RPTH(1–34) LIBERATED BY RP-HPLC

This method uses a gradient high-performance reversed-phase liquid chromatography (HPLC) separation procedure coupled with UV absorbance detection for determining the purity of rPTH(1–34). Purity determination of rPTH(1–34) is based on a peak area vs.total area calculation. Mobile phase A contains 10 parts of acetonitrile and 90 parts of 0.2M sulfate buffer (pH 2.3) and mobile phase B contains 50 parts of acetonitrile and 50 parts of 0.2M sulfate buffer (pH 2.3). The run time is of 55 min with any of the following gradient profiles.

PROFILE 1

Time (min)	A (%)	B (%)	Time (min)	A (%)	B (%)
0.01	100	0	45	0	100
5	65	35	45.01	100	0
35	60	40	55	100	0

PROFILE 2

Time (min)	A (%)	B (%)	Time (min)	A (%)	B (%)
0	100	0	45	10	90
5	65	35	45.01	100	0
35	50	50	55	100	0

11.3 RESULTS AND DISCUSSION

11.3.1 OPTIMIZATION OF THE CLEAVAGE EFFICIENCY OF RECOMBINANT BOVINE ENTEROKINASE

Purification of the proteins expressed in recombinant *E. coli* is a challenging task for the reason that the lysate contain large number of impurities and the percentage of the protein of interest is very small. To ease the purification process various fusion tags were used to express along with the protein of interest which allows the purification by affinity chromatographybased on the specific interaction between immobilized ligand and target proteins [8]. These affinity-tag systems share the following features: (a) one-step adsorption purification;(b) a minimal effect on tertiary structure and biological activity; (c) easy and specific removal to produce the native protein; (d) simple and accurate assay of the recombinant protein during purification; and (e)applicability to a number of different proteins [3]. At the same time, the removal of the fusion tag to get the pure protein of desired activity employs enzymes to cleave the affinity tag. But, the enzymes remarkably add to the cost of process. To increase the cleavage efficiency of the bovine enterokinase,two samples were taken Initially and recombinant bovine enterokinase was added at a concentration of 2IU/mgof the total protein and incubated while stirring at 22°C for 6and 10 hr, and the percentage of rPTH (1–34) liberated after cleavage was found to be 12.1 and 26.2 (Figure 11.1). But to reduce the cost of the process, increase in the efficiency of the enzyme is necessary. Hence different approaches were adopted.The concentration of the recombinant bovine enterokinase/mg of the total protein was reduced to half, that is, 1IU/mg and incubated at 22°C for 5hr and the percentage of rPTH(1–34) liberated after cleavage was found to be17.38. To increase the cleavage further,magnesium sulfate ($MgSO_4$) was added to the concentration of 1mM which supplies Magnesium ion (Mg^{2+}) to the sample containing recombinant bovine enterokinase of 1IU/mg of total protein that is expected to act as cofactor,incubated while stirring at 24°C for 5 hr. The percentage of rPTH(1–34) liberated after cleavage was found to be 26.69 which is nearly equivalent to the yield at 2IU/mg for 10 hr. Keeping the recombinant enterokinaseconcentration at 1IU/mg of the total protein,the concentration of magnesium sulfate was increased to 5mM and the cleavage experiment was conducted by incubating at 24°C while stirring for 5 hr,the percentage of rPTH (1–34) liberated after cleavage was found to be 26.55 which is not much different from that of 1mm magnesium sulfate. Continuing the optimization, the concentration of the enzyme was reduced by 10 times,that is, 0.2IU/mg of total protein; the concentration of the (Mg^{2+}) was increased to 5mM from 1mm and was incubated while stirring at 24°C for 16hr. This was done to give enough time for the cleavage to happen. The percentage of the cleavage was found to be 28.21 (Figure 11.2) which was higher than any of the above. When the same experiment was repeated without (Mg^{2+}) ion,the cleavage percentage did not increase more than 11.16 (Figure 11.3) even after 30hr. This clearly indicates the role of (Mg^{2+}) as cofactor for recombinant bovine enterokinase. Now the IU of the enterokinase per milligram of total protein containing fusion protein CBD-PTH (1–34) was further reduced to 0.05 and was incubated at 24°C for 16hr with 5mm Magnesium Sulphate and the percentage of rPTH (1–34) liberated after cleavage was found to be 31.19 (Figure 11.4) which is highest. Therefore the addition of (Mg^{2+}) increased the efficiency and recovery of the protein of interest rPTH (1–34). Now the concentration of the recombinant bovine enterokinase required for the cleavage of the fusion protein had decreased from 2IU/mg to 0.05IU reducing the input cost of the enzyme by 40 times. Results were summarized in Table 11.1.

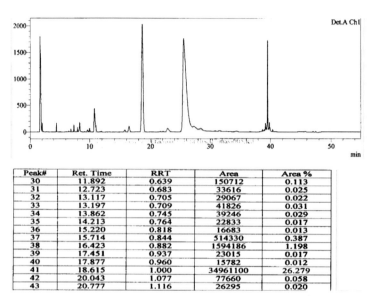

Peak#	Ret. Time	RRT	Area	Area %
30	11.892	0.639	150712	0.113
31	12.723	0.683	33616	0.025
32	13.117	0.705	29067	0.022
33	13.197	0.709	41826	0.031
34	13.862	0.745	39246	0.029
35	14.213	0.764	22833	0.017
36	15.220	0.818	16683	0.013
37	15.714	0.844	514330	0.387
38	16.423	0.882	1594186	1.198
39	17.451	0.937	23015	0.017
40	17.877	0.960	15782	0.012
41	18.615	1.000	34961100	26.279
42	20.043	1.077	77660	0.058
43	20.777	1.116	26295	0.020

FIGURE 11.1 Cleavage of fusion protein CBD-PTH (1–34) using recombinant bovine enterokinase, 2 IU/mg of total protein, 10 hr of incubation at 24°C.

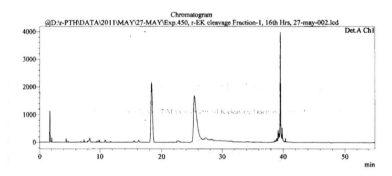

Peak#	Ret. Time	RRT	Area	Area %
30	13.015	0.709	44950	0.033
31	13.685	0.745	24907	0.018
32	14.075	0.766	12931	0.009
33	14.223	0.774	41143	0.030
34	15.024	0.818	13689	0.010
35	15.482	0.843	593305	0.430
36	15.925	0.867	8449	0.006
37	16.236	0.884	1010527	0.733
38	17.240	0.939	16433	0.012
39	17.629	0.960	14586	0.011
40	18.369	1.000	38907903	28.214
41	19.847	1.080	81541	0.059
42	20.658	1.125	28747	0.021
43	21.405	1.165	190662	0.138
44	21.888	1.192	138109	0.100

FIGURE 11.2 Cleavage of fusion protein CBD-PTH (1–34) using recombinant bovine enterokinase, 0.2 IU/mg of total protein, 16 hr of incubation at 24°C.

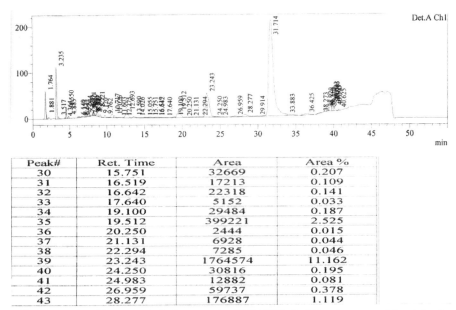

Peak#	Ret. Time	Area	Area %
30	15.751	32669	0.207
31	16.519	17213	0.109
32	16.642	22318	0.141
33	17.640	5152	0.033
34	19.100	29484	0.187
35	19.512	399221	2.525
36	20.250	2444	0.015
37	21.131	6928	0.044
38	22.294	7285	0.046
39	23.243	1764574	11.162
40	24.250	30816	0.195
41	24.983	12882	0.081
42	26.959	59737	0.378
43	28.277	176887	1.119

FIGURE 11.3 Cleavage of fusion protein CBD-PTH (1–34) using recombinant bovine enterokinase, 0.2 IU/mg of total protein, 30 hr of incubation at 24°C.

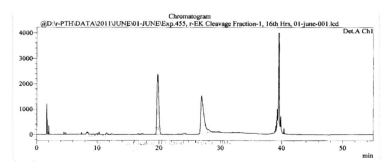

Peak#	Ret. Time	RRT	Area	Area %
44	18.626	0.940	4390	0.003
45	19.059	0.962	14126	0.009
46	19.815	1.000	47098287	31.198
47	21.192	1.069	13859	0.009
48	21.433	1.082	54491	0.036

FIGURE 11.4 Cleavage of fusion protein CBD-PTH (1–34) using recombinant bovine enterokinase, 0.05 IU/mg of total protein, 16 hr of incubation at 24°C.

11.4 CONCLUSION

A novel method was developed to increase the cleavage efficiency of recombinant bovine enterokinase using magnesium ion as cofactor. Recombinant bovine enterokinase is very often used in the process of protein purification to cleave the fusion tag which is also a protein from protein of interest. But the cost of the enzyme is so high that the cost of the enzyme alone accounts for 30–40 percentof the total purification cost. Studies were done to reduce the cost by increasing the efficiency of the enzyme to do the same work at low concentration.Magnesium sulfate which supplies (Mg^{2+}) ion that can act as cofactor was added to the concentration of 5mM and the results show that the yield/efficiency had doubled and the enzyme concentration had decreased by 40 times.

KEYWORDS

- Cofactor
- Enterokinase
- Enzyme activity
- Fusion protein
- Parathyroid Hormone (1–34)

REFERENCES

1. Choi, S. I.; Song, H. W.; Moon, J. W.; and Seong, B. L.; *Biotechnol. Bioeng.* **2001**, *75*, 718–724.
2. Hosfield, T.; and Lu, Q.; *Anal. Biochem.* **1999**, *269*, 10–16.
3. Terpe, K.; *Appl. Microbiol. Biotechnol.* **2003**, *60*, 523–533.
4. Waugh, D. S.; *Trends Biotechnol.* **2005**, *23*(6), 316–320.
5. Turk, B.; *Nat. Rev. Drug Discov.* **2006**, *5*, 785–799.
6. Broderick, J. B.; Coenzymes and cofactors. In: Encyclopedia of Life Sciences. Nature Publishing Group; **2001**.
7. Brown, D. A.; and Cook, R. A.; *Biochemistry.* **1981**, *20*(9), 2503–2512.
8. Lee, W. C.; and Lee, K. H.; *Anal. Biochem.* **2004**, *324*(1), 1–10.

CHAPTER 12

APPLICATION OF THE THOMAS MODEL FOR CESIUM ION EXCHANGE ON AMP-PAN

CH. MAHENDRA[1*], P. M. SATYA SAI[2], C. ANAND BABU[3], K. REVATHY[1], and K. K. RAJAN[1]

[1]Fast Reactor Technology Group, Indira Gandhi Centre for Atomic Research, Kalpakkam, Tamil Nadu, *E-mail: mahendra.ch6@gmail.com

[2]Waste Immobilization Plant, Bhaba Atomic Research Centre Facilities, Kalpakkam, Tamil Nadu

[3]PSG College of Technology, Coimbatore, Tamil Nadu

CONTENTS

12.1	Introductions	118
12.2	Materials and Methods	118
12.3	Results	120
12.4	Conclusions	123
	Keywords	124
	References	124

12.1 INTRODUCTIONS

Cesium-137 is one of the important fission products from the nuclear fission of uranium. The removal of radioactive cesium from radioactive liquid waste is important from the point of view of their biotoxicity and therefore has been of interest for radioactive waste management.

Ammonium molybdophosphate (AMP) has been extensively investigated and found to be highly effective for the removal of cesium from acidic liquids [1]. Ammonium molybdophosphate becomes soluble in solutions with pH >4, and therefore, is only suitable for application to acidic solutions. Ammonium molybdophosphate (AMP) immobilized on polyacrylonitrile (PAN) is an engineered form of the cesium ion selective material developed at Czech Technical University [2–5].

In the present study, AMP-PAN is used for uptake of cesium from acidic solution in column experiments. Many models were proposed for the prediction of breakthrough curves, such as Thomas model, Clark model, Yoon-Nelson model, etc., This paper examines the application of the Thomas model on experimental break through curves in order to evaluate its use in prediction of shape and position of breakthrough curves for the chosen experimental conditions.

12.2 MATERIALS AND METHODS

12.2.1 AMP-PAN

AMP-PAN is obtained from M/s. THERMAX Ltd., Pune, India. The properties of AMP-PAN are listed in Table 12.1.

TABLE 12.1 Properties of AMP-PAN

Properties of AMP-PAN	
Trade name	AMP-PAN
Nature	Cationic resin
Particle Size	0.4 mm
Shape	Spherical
Pore type	Macro porous
Bulk density	1.5 g/cm^3
AMP loaded on resin	65%
Moisture content	55%

12.2.2 COLUMN EXPERIMENT

The column made of glass tube of 2 cm inner diameter and 50 cm height is used for the study. The schematic of experimental setup is shown in Figure 12.1.

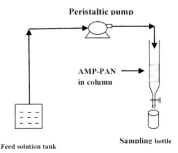

FIGURE 12.1 Experimental set-ups for the measurement of breakthrough curve.

A fixed bed of 14.5 ml (20g) of AMP-PAN was packed in the column with a layer of glass wool at the bottom. The column performance of Cs ion exchange on AMP-PAN was studied at different Cs concentration (500–1,000 mg/l), bed height (14.5–21.5cm) and flow rate (5–16 ml/min). The bed depths taken were 14.5cm (20 g), 18.2cm (25g), and 21.5 cm (30g). The Cs solution was pumped to the column in a down-flow direction by a peristaltic pump. Samples were collected at regular intervals and its concentrations were analyzed by Atomic Absorption Spectrophotometer (AAS).

12.2.3 THOMAS MODEL

For the successful design of a column ion exchange process, it is important to predict the breakthrough curve for different parameters. Various kinetic models have been developed to predict the dynamic behavior of the column. Thomas model [6–8] is based on the assumption that the process follows Langmuir kinetics with no axial dispersion; the rate driving force follows second order reversible reaction kinetics. The expression of Thomas model for a column operation is given as follows:

$$\frac{C}{C_0} = \frac{1}{1 + \exp\left[K_{Th}/Q(q_0 X - C_0 V_{eff})\right]}$$

where K_{Th} is the Thomas rate constant (ml/min mg); q_0 is the maximum solid-phase concentration of solute (mg/g). X is the amount resin in the column (g); V_{eff} is the effluent volume (l); Q is the flow rate (ml/min); C is effluent concentration (mg/l); and C_0 is influent concentration (mg/l).

The linearized form of the Thomas model is as follows:

$$\ln\left(\frac{C_0}{C}-1\right)=\frac{K_{Th}q_0 X}{Q}-\frac{K_{Th}C_0 V_{eff}}{Q}$$

A plot of $\ln(C_0/C-1)$ versus V_{eff} gives a straight line with a slope of $(K_{Th}C_0/Q)$ and an intercept of $(K_{Th}q_0 X/Q)$.

12.3 RESULTS

12.3.1 EFFECT OF FEED INLET CONCENTRATION

The breakthrough curves obtained by the initial cesium concentration from 500 to 100 mg/l at 5 ml/min flow rate and 14.5 cm bed depth are given in Figure 12.2.

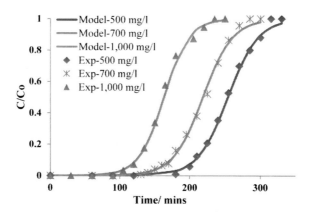

FIGURE 12.2 Comparison between experimental and model breakthrough curves for different inlet concentrations.

As expected, a decrease in cesium concentration gave a later breakthrough curve; the treated volume was greatest at the lowest transport due to a decreased diffusion coefficient or mass transfer coefficient. Breakthrough time (C/C_0=0.05) occurred after 200 min at 500 mg/l initial cesium concentration while the breakthrough occurred after 130 and 90 min for 700 and 1,000 mg/l respectively. At higher concentration the availability of the ions for the ion exchange sites is more, which leads to higher uptake of Cs at higher concentration even though the breakthrough time is shorter than the breakthrough time of lower concentrations. The breakthrough time decreased with increasing cesium concentration as the binding sites became more quickly saturated in the column.

12.3.2 EFFECT OF FEED FLOW RATE

The effect of feed flow rate on breakthrough was found by passing 700 mg/l of Cs in 4N HNO_3 with different flow rates (5, 10, and 16 ml/min) until no further cesium removal was observed. The breakthrough curve for a column was determined by plotting the ratio of the C/C_0 against time, as shown in Figure 12.3.

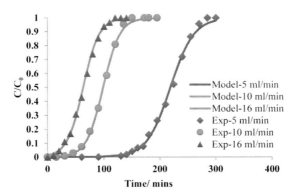

FIGURE 12.3 Comparison between experimental and model breakthrough curves for different velocities.

The column performed well at the lowest flow rate (5 ml/min). Earlier breakthrough and exhaustion times were achieved, when the flow rate was increased from 5 to 16 ml/min. The column breakthrough time ($C/C_0=0.05$) was reduced from 150 to 10min, with an increase in flow rate from 5 to 16 ml/min. This was due to a decrease in the residence time, which restricted the contact of cesium solution to the AMP-PAN. At higher flow rates the cesium ions did not have enough time to diffuse into the pores of the AMP-PAN and they exited the column before equilibrium occurred.

12.3.3 EFFECT OF RESIN BED HEIGHT

The uptake of cesium in a fixed-bed is dependent on the quantity of resin inside the column. In order to study the effect of bed height on cesium retention, three different bed heights, viz. 14.5, 18.2, and 21cm, were used. A feed solution of fixed concentration (700 mg/l) was passed through the fixed-bed column at a constant flow rate of 5 ml/min. As depicted by Figure 12.4 the breakthrough time varied with bed height.

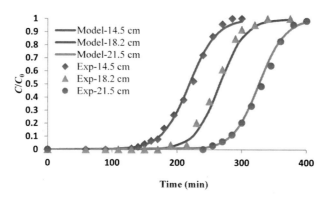

FIGURE 12.4 Comparison between experimental and model breakthrough curves for different bed heights.

It can be seen that the breakthrough time and exhaustion times increased with increasing bed height. The variation in bed height was proportional to the surface area available per each bed depth. At higher bed depth of 21.5 and 18.2cm, more mass of resin was residing in the column thereby providing larger service area for ion exchange of the solute to the resin. At low bed depth, the cesium ions do not have enough time to diffuse into the surface of the resin, and a reduction in breakthrough time occurs.

12.3.4 APPLICATION OF THE THOMAS MODEL

Thomas model has been used by many researchers to study packed bed ion exchange kinetics [9–11]. Application of the Thomas model to the data at C/C_0 ratios higher than 0.01 and lower than 0.99 enabled the determination of the kinetic coefficients and maximum uptake capacities in the system. The kinetic rate constant, K_{Th} and capacity of the bed, q_0 were determined from the plot of ln[(C_0/C) − 1] against V_{eff} and the results of K_{Th}, R^2, and q_0 are given in Table 12.2.

TABLE 12.2 Parameters of Thomas model for Cs exchange onto AMP-PAN

Flow rate Q (ml/min)	initial conc. C_0 (mg/L)	bed height H (cm)	Thomas constant (K_{Th}) (ml/min.mg)	q_0 mg/g	R^2
5	500	14.5	9.350 ×10^{-2}	31.82	99.4
5	700	14.5	7.385 ×10^{-2}	36.12	99.3
5	1000	14.5	5.535 ×10^{-2}	40.88	98.7
10	700	14.5	9.605 ×10^{-2}	34.51	99.8
16	700	14.5	9.910 ×10^{-2}	32.99	99.6
5	700	18.2	7.701 ×10^{-2}	37.06	98.1
5	700	21.5	8.107 ×10^{-2}	38.12	99.1

Figures 12.2–12.4 illustrate both experimental breakthrough curves and predicted breakthrough curves from the Thomas model at all operation stages of the packed bed column. It was observed that when the feed concentration increased from 500 to 1,000 mg/l, Thomas rate constant decreased from 0.0935 to 0.0553 ml/min.mg and capacity of the resin increased from 31.8 to 40.88 mg/g.

The Thomas model gave a good fit of the experimental data, at all the flow rates examined, which would indicate the external and internal diffusions were not the rate limiting step. The rate constant (K_{Th}) increased from 0.07385 to 0.0991 with increasing flow rate (5–16 ml/min) which indicates that the mass transport resistance decreases. The capacity of the resin decreased from 36.22 to 32.99 mg/g due to the less contact time as explained above. The Thomas constant remained almost constant when the bed height was increased from 14.5 to 21.5 cm and capacity of the resin increased from 36.12 to 38.12 mg/g. As indicated in Table 12.2, the Thomas model was found in a suitable fitness with the experimental data ($R^2 > 0.98$).

12.4 CONCLUSIONS

The effect of different operational parameters such as initial concentration, feed flow rate, bed height on break through curves of cesium ion exchange on AMP-PAN were examined in a packed column. It revealed that by increasing the feed concentration, flow rate and decreasing the bed height the breakthrough time decreased. Thomas rate constant, K_{Th} is dependent on flow rate, initial ion concentration and bed height. The maximum ion exchange capacity, q_0 increased with increase in initial ion concentration and bed height but decreased with increase in flow rate.

KEYWORDS

- **AMP-PAN**
- **Breakthrough curve**
- **Cesium**
- **Ion exchange**
- **Thomas rate constant**

REFERENCES

1. Van, R.; and Smith, J.; The AMP process for cesium separation, research report. National Chemical Research Laboratory, Pretoria, South Africa, **1963**.
2. Šebesta, F.; and Štefula, V.; *J. Radioanal. Nucl. Chem.* **1990**, *140* (1), 15–21.
3. Brewer, K. N.; Todd, T. A.; Wood, D. J.; Tullock, P. A.; Šebesta, F.; John, J.; and Motl, A.;*Czech. J. Phys.***1999**, *49*(S-1), 959–964.
4. Todd, T.A.; Mann, N. R.; Tranter, T. J.; Šebesta, F.; John, J.; and Motl, A.; *J. Radioanal. Nucl. Chem.* **2002**, *254*(1), 47–52.
5. Tranter, T. J.; Herbst, R. S.; Todd, T. A.; Olson, A. L.; and Eldredge, H. B.; *Adv. Environ. Res.* **2002**, *6*, 107–121.
6. Zheng, H.; Han, L.; Ma, H.; Zheng, Y.; Zhang, H.; Liu, D.; and Liang, S.; *J. Hazard. Mater.* **2008**, *158*, 577–584.
7. Hanen, N.; and Abdelmottaleb, O.; *J. Chem. Eng. Process Technol.* **2013**, *4*(3), 153.
8. Baek, K. W.; Song, S. H.; Kang, S. H.; Rhee, Y. W.; Lee, C. S.; Lee, B. J.; Hudson, S.; and Hwang, T. S.; *J. Ind. Eng. Chem.* **2007**, *13*(3), 452–456.
9. Thomas, H. C.; *J. Am. Chem. Soc.* **1944**, *66*(10), 1664–1666.
10. Sivakumar, P.; and Palanisamy, P. N.; *Int. J. ChemTech. Res.* **2009**, *1*(3), 502–510.
11. Han, R.; Ding, D.; Xu, Y.; Zou, W.; Wang, Y.; Li, Y.; and Zou, L.; *Bioresour. Technol.* **2008**, *99*, 2938–2946.

CHAPTER 13

MINIMIZING POSTHARVEST DAMAGE IN CITRUS

HARBANT SINGH[*], GHASSAN AL-SAMARRAI,
MUHAMMAD SYARHABIL, SUE YIN-CHU, and BOONBENG-LEE

School of Bioprocess Engineering, University Malaysia Perlis (UniMAP), Kompleks Pusat Pengajian Jejawi 3, 02600 Arau, Perlis, Malaysia;
*E-mail: harbant@unimap.edu.my

CONTENTS

13.1 Introduction 126
13.2 Materials and Methods 126
13.3 Results and Discussion 129
13.4 Conclusion 135
Keywords 135
References 135

13.1 INTRODUCTION

In agriculture, citrus fruits suffer from postharvest deterioration due to the fungi rot caused by *Penicilium* spp., *Aspergillus* spp., and *Fusarium*spp. and dehydration process [1]. This process subsequently causes weight losses, color changes, softening, surface pitting, browning, loss of acidity, and microbial spoilage in citrus. Among factors which contribute to the deterioration rate are intrinsic characteristics of the product and storage conditions, for example, temperature, relative humidity, and storage atmosphere composition [2].

Currently, development of the active packaging on the basis of mixing plant derivatives with Modified Atmosphere Packaging (MAP) is introduced with the use of eco-friendly materials to decrease the amount of fruit quality deterioration in storage after harvest [3]. Several researchers have already employed plant-based products as substitute for synthesis fungicide to control soft-rot in plant [2, 4–6]. The use of pure and eco-friendly derivatives from plant extract is a good substitute for synthetic fungicides on fruits, particularly for highly perishable ones with reduced shelf life [7].

Basically, bioactive packaging technique is a technique employing eco-friendly biodegradable materials in packaging that can increase the shelf life postharvest [8, 9] and has the potential for commercial application in postharvest. The present study explores the antifungal activity concentration of plant extracts that is incorporated in fruit wrapping with biodegradable materials to minimize postharvest damage in citrus.

13.2 MATERIALS AND METHODS

13.2.1 PLANT MATERIALS

13.2.1.1 PREPARATION OF PLANT EXTRACT

Sea mango (*Cerberaodollam* L.) and Neem (*Azadirachtaindica* L.) leaves were collected from trees growing by the river side park at Kangar, Malaysia. Hot chilli fruits (*Capsicum frutescene* L.), leaves, and stem of Lemon grass (*Cymbopogonnardus* L.) and Ginger (*Zingiberofficinale* L.) stem were purchased from wet market at Kangar. They were washed under running water to get rid of dirt. The samples were dried overnight in oven at 40°C. Dried materials were then pulverized using electric mixer and preserved in sterile sealed glass bottles. The technique used was modified from Ruch and Worf's method [10]. Later, the powder materials were treated with 500ml of 95 percent alcohol. The solutions were filtered twice; once through cheese-cloth gauze and later through Whitman's No.2 filter paper before being evaporated using rotary evaporator at 60°C. Dark spongy materials were dried in oven at 37°C for two days. Dried materials were stored in sterilized 10ml screw-capped glass bottles and were kept in the refrigerator at 4°C until further usage.

13.2.1.2 PREPARATIONS OF PLANT EXTRACT DILUTIONS

Powder extracts from plants were removed from refrigerator and brought to lab for the preparation of extract dilution. Dilutions were prepared by dissolving (0.5, 1, 2, 3, 4, 5g) from each plant powder in Dimethyl sulfoxide (DMSO 95.5 percent/1ml/1g) and topped up to 1 l of distilled water in volumetric flask to obtain the concentrations required (500, 1,000, 2,000, 3,000, 4,000, and 5,000ppm).

13.2.1.3 PATHOGENS

Penicilium digitatum L., *Aspergillus niger* L., and *Fusarium* sp. were identified using taxonomic and morphological reference. Highly aggressive, single spore isolates of these fungi were isolated from soft-rot damaged citrus fruits obtained from the market. These isolates were propagated in sterilized 10cm-Petri dishes containing PDA media at 25°C for seven days.

13.2.1.4 IN VITRO SCREENING OF MYCELIUM INHIBITION

PDA media was incorporated into 50 ml glass flasks and autoclaved for 20 min at 121°C. After cooling down, 5 ml of each plant extract (500, 1,000, 2,000, 3,000 ppm) was added to the flasks using pipette and were gently agitated for 2 min. Media cultures were amended into 10-cm Petri dishes and 250 mg of Chloramphenicol [4] was added to prevent bacterial growth. Later, 1 ml of fungal spore suspensions was added by pipette onto the center of the amended PDA extracts and the inoculated plates were incubated at 25°C for 10 days. Petri dish inoculated without extract concentration (water) was used as negative control. Colony diameter was determined by measuring the average radial growth. The inhibition zone was measure by the formula below [11]:

$$\% \text{Mycelia inhibition} = \frac{\text{Mycelial growth(control)} - \text{Mycelial growth(treatment)}}{\text{Mycelial growth(control)}} \times 100$$

13.2.1.5 LONGEVITY STUDY OF PLANT EXTRACTS

The best concentration obtained from the *in vitro* mycelium inhibition by the five plants was selected to study the longevity effectiveness of the antifungal activity weekly until the fourth week (W4). Fresh sample from each extracts was used as control for stored samples. Samples were stored at 25°C and 65–70 percent RH for four weeks. Fungal inhibition zone (%) was analyzed using *in vitro* screening mention above.

13.2.1.6 CYTOTOXICAL SCREENING (LC50) FROM CRUDE EXTRACTIONS (BRINE SHRIMP TEST)

The eggs of the brine shrimp were collected from a fish shop in Kangar, Malaysia. 5 mg of *Artemiasalina* (Leach) eggs were added into conical flask containing 50 ml of sea water. The flasks were kept under inflorescent bulb for 48 hr to allow the eggs to hatch into shrimp larvae. 6.4 mg of each dried crude plant extract was treated with 0.1 ml of DMSO and mixed with 5 ml of water into a 10 ml volumetric flask. The flasks were stirred manually for 5 min to get a fully dissolved homogeneous state of the extract (640µg/ml). A series of dilution (using equation $M_1V_1=M_2V_2$) was prepared to obtain concentrations of 5, 10, 20, 40, 130, 260, 390, and 520µg/ml. DMSO and fungicide (Guazatine) were used as negative and positive controls. The concentration treatments were then transfer into the conical flasks containing hatched larvae. After 24 hr, the numbers of dead larvae were counted and collected data analyze using Excel Word 2007.

13.2.2 FRUIT WRAPPING WITH BIODEGRADABLE MATERIALS AND PLANT EXTRACT

Healthy freshly citrus fruit were washed with tap water, air dried and then sterilized by immersion in 70 percent ethanol for 1 min before spraying [8]. The replication consists of 15 fruits with 5 fruits for each treatment. The fruits were randomly divided into nine equal groups and all the groups were wounded in depth of 5.0mm with a 1.25mm diameter needle at the equator. All treatments were inoculated by spraying in suspension of *P. digitatum*, *A.niger*, and *Fusarium* sp. 1×10^6 spores ml^{-1} and left for about 1 hr to stabilize the spores. After that, the fruits were sprayed with treatments and wrapped separately. The treatments were carried out with spraying 5000ppm of plant extract and 1000ppm of Guazatine on fruits. The untreated fruits were used as healthy control. After that, fruits were wrapped with polylactic acid, pectin, and newspaper under room conditions. Treated fruits were packed and stored in carton boxes and incubated at 25°C ±2 (RH 65–70 percent) under room conditions. Evaluations of the treatments were done after four weeks of data collection.

13.3 RESULTS AND DISCUSSION

13.3.1 EFFECT OF PLANT EXTRACTS ON DEVELOPMENT MYCELIUM GROWTH OF POSTHARVEST FUNGI IN VITRO

The study showed significant difference between treatment and control ($P \leq 0.05$) for the three fungi, *P.digitatum*, *A.niger*, and *Fusarium* sp. (Table 13.1). Sea mango showed complete inhibition of *P.digitatum* (almost 100 percent inhibition) when applying crude plant extracts concentration at 3,000ppm. This result agrees with previous studies [4, 12] regarding the effectiveness of plant extracts as antifungal against the fungi, *P. digitatum*, besides showing better result than when compared with Behnaz et al. [5] (reduced fruit contamination up to 17 percent) in fungal inhibition's percentage.

TABLE 13.1 Effect of plant extract (ppm) on inhibition colony diameter (cm) of *Penicilium digitatum*, *Aspergillus niger*, *Fusarium* sp. compare with control (Guazatine and DMSO) in PDA media inoculated at 27°C for seven days

Treatments	Comparisons (Concentrations ppm)				Mean	Remarks
	500	1,000	2,000	3,000		
	Colony Diameter (cm)					
Peniciliumdigitatum						
Cerberaodollam L.	2.76[b]	1.99[a]	1.46[a]	0.00[a*]	1.55[a]	Significant
Capsicum frutescence L.	2.90[b]	2.96[b]	2.10[c]	0.63[b]	2.14[c]	Significant
Azadirachtaindica L.	3.90[d]	2.90[b]	2.33[d]	1.33[c]	2.63[d]	Good
Cymbopogonnardus L.	3.90[d]	2.96[b]	2.33[d]	1.33[c]	2.63[d]	Good
Zingiberofficinale L.	4.16[f]	2.67[b]	2.23[d]	1.38[c]	2.61[d]	Good
Aspergillusniger						
Cerberaodollam L.	3.06[b]	2.07[a]	1.77[a]	0.73[a]	1.90[a]	Significant
Capsicum frutescence L.	2.96[b]	2.16[b]	1.95[b]	0.86[a]	1.98[a]	Significant

TABLE 13.1 *(Continued)*

Treatments		Comparisons(Concentrations ppm)				Mean	Remarks
		500	1,000	2,000	3,000		
		Colony Diameter (cm)					
Azadirachtaindica L.		4.00^c	3.03^c	2.73^d	2.53^c	3.07^c	Moderate
Cymbopogonnardus L.		4.04^c	3.06^c	2.59^d	2.38^c	3.01^c	Moderate
Zingiberofficinale L.		4.53^d	2.94^c	2.59^d	2.26^c	3.08^c	Moderate
***Fusarium*sp.**							
Cerberaodollam L.		2.06^a	2.02^a	1.92^a	0.71^a	1.76^a	Significant
Capsicum frutescence L.		2.76^b	2.10^a	1.98^a	0.86^b	1.92^b	Significant
Azadirachtaindica L.		2.99^b	2.63^b	2.44^b	2.35^d	2.60^c	Moderate
Cymbopogonnardus L.		3.11^{bc}	2.63^b	2.59^{bc}	2.40^{de}	2.85^d	Moderate
Zingiberofficinale L.		3.53^c	2.83^b	2.64^{bc}	2.40^{de}	2.87^d	Moderate
Control	Guazatine	2.20^a	1.90^a	1.86^b	2.00^d	1.99^b	Good
	DMSO	8.55^i	9.00^f	8.55^h	8.55^i	8.66^g	No effect

Alphabets different in the same column show significant difference using Duncan's Multiple Range test ($P \leq 0.05$) and average is calculated from three replicates.% inhibition = $CD_{control} - CD_{treatment}/CD_{control} \times 100$, where colony diameter (CD) (cm).

13.3.2 EFFICACY STORED PLANT EXTRACT ON FUNGAL INHIBITION ZONE UNDER FRIDGE CONDITION

Fungal inhibition zones (%) were calculated for four weeks after contact with 3,000ppm extraction solutions (shown in Figure 13.1). Each extracts tested maintained their efficacy (≤60 percent inhibition zone) until third bioassay. This is because storage at 4°C could decrease the rate of biological reactions compare to higher temperature. Result of current study agreed with previous studies by Fernando et al. [13], and Wang et al. [14], that storage cool condition had better quality for longer storage. Statistical analysis (ANOVA, $P \leq 0.05$) showed significant difference between Sea mangos with other plants.

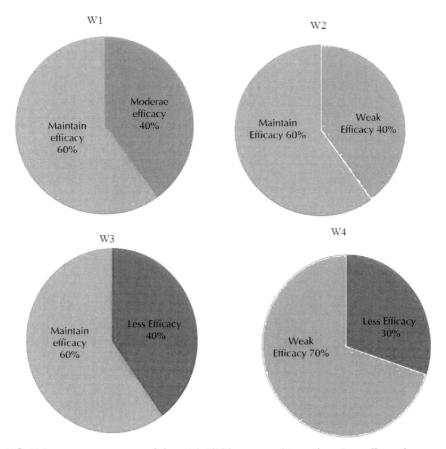

FIGURE 13.1 Percentage of fungal inhibition zone (%) against *Penicillium digitatum*, *Aspergilius niger*, and *Fusarium* sp. at 3000ppm for five crude plant extracts stored at 4°C and 85–90 percent RH for four weeks.

13.3.3 BIOTOXICITY OF PLANT EXTRACT ON BRINE SHRIMP TEST (LC_{50})

From Figure 13.2, all the extracts showed positive results indicating that the test samples are biologically active. Among the five plant extracts, Sea mango and chilli recorded toxic values of 5 µg/ml and 20 µg/ml and is nontoxic to human

[15] (≤2 µg/ml is safe to human).Results of this study agree with the studies of Meyer et al. [16] which also utilizing brine shrimp for natural product research.

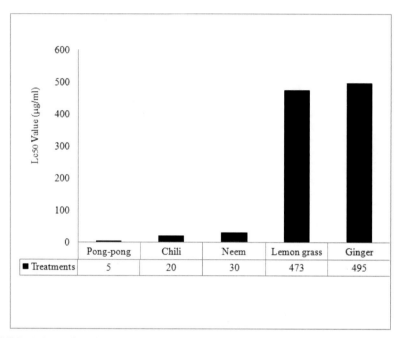

FIGURE 13.2 Brine Shrimp Test (BST) toxicity (µg/ml) of crude plant extracts under study. Guazatine (PC) is 326 µg/ml; DMSO (NC) is >1000 µg/ml; toxicity level reference—high = 0.2 µg/ml, medium = 0.2–2 µg/ml, low = 2–20 µg/ml, safe = ≥20 µg/ml [15].

13.3.4 EFFECT OF PLANT EXTRACTS ON FRUIT DECAY IN VIVO

Results presented in Figure 13.3 showed almost complete inhibition (0 percent or Clean) on fruit decay after 21 days of storage using 5,000ppm of plant extracts, which statistically outperformed other treatments. Sea mango, chilli, and Neem extract recorded 0 percent decay when compared with average 7.4 percent decay on fruits treated with synthetic Guazatine. This result appears to be among the first to use spraying method to determine the activity of plant extracts against *P. digitatum*, *A. niger*, and *Fusarium* sp.

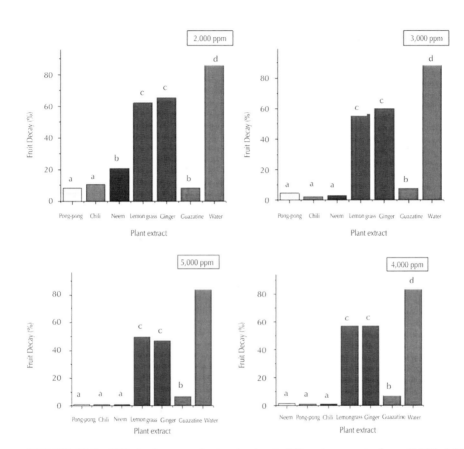

FIGURE 13.3 Effect of spray plant extracts at different concentrations (2,000–5,000 ppm) on the development of mycelium growth *Penicilliumdigitatum*, *Aspergiliusniger*, and *Fusarium*sp. on the orange surface during storage for 21 days at 25°C±2 and 65–70 percent RH.

13.3.5 DEVELOPMENT OF FUNGI ROT IN FRUIT WRAPPING

ANOVA statistical analysis (Figure 13.4) showed fruits sprayed with plant extract in the current study and wrapped materials (Polylactic acid, Pectin, and Newspaper) decreased fungal decay for stored fruits at ±30°C. They recorded a lower average value (≤11.97 percent damage) when compared with the untreated fruits (17.02 percent damage) in controlling infection of Green mold, Black rot, and Fusarium rot. Polylactic acid plus Sea mango, chili, and Neem was better (0 percent damage) when compared with Pectin and Newspaper wrapper. Treatments (extract + wrappers) tend to reduce oxygen and carbon dioxide levels during storage and did improve the MAP, thereby decrease citrus rot by fungi. This result is in agreement with studied carried out by Serrano et al. [9] who used plant derivatives to MAP during storage fruits after harvest.

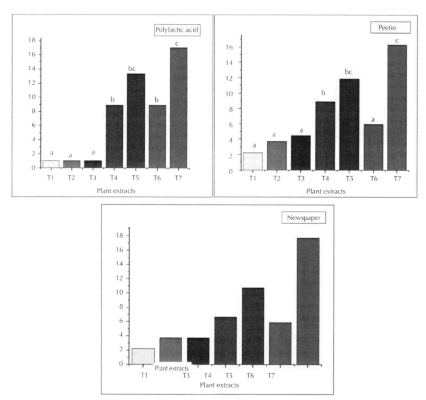

FIGURE 13.4 Treatment of fungal decay (%) using three type of the biodegradable materials (Polylactic acid, Pectin, andnewspaper) on the orange surface at 27°C and RH 65–70 percent for 21 days. T1: Sea mango; T2: Chili; T3: Neem; T4: Lemon grass; T5: Ginger; T6: Guazatine; T7: Untreated fruit.

13.4 CONCLUSION

The plants extracts in the current research showed antifungal property on post-harvest fungi (*Penicilium digitatum* L., *Aspergillus niger* L., and *Fusarium* sp) and recorded safe Lc$_{50}$ values. Wrapping the citrus fruits with biodegradable polylactic acid (PLA), pectin and newspaper incorporated with plant extracts inhibited post-harvest fungal infection. This trend showed that plant extracts have a great potential in the post-harvest fungal spoilage technology. Future work on the isolation characterization of active fungal biocompound from Sea mango is in progress.

KEYWORDS

- Chili
- Citrus
- Neem
- Postharvest damage
- Sea mango

REFERENCES

1. Adaskaveg, J. E.; Forster, H.; and Sommer, N. F.; Principle of post-harvest pathology and management of decays of edible horticultural crops. In: A. Kader (Ed.), Post-Harvest Technology of Horticultural Crops. Oakland, CA: University of California Publication; **2002**, 163–195.
2. Lurie, C.; and Crisosto, H.; *Postharvest Biol. Technol.* **2005**, *37*, 195–208.
3. Makino, Y.; *J. Agric. Eng. Res.* **2001**, *78*, 261–271.
4. Obagwu, J.; and Korsten, L.; *Plant Pathol.* **2002**, *109*, 221–225.
5. Behnaz, S.; Sadrollah, R.; Majid, R.; and Mohammad, J.; *Adv. Environ. Biol.* **2009**, *3*(3), 249–254.
6. Tajidin, N. E.; Ahmad, S. H.; Rosenani, A. B.; Azimah, H.; and Munirah, M.; *Afr. J. Biotechnol.* **2012**, *11*(11), 2685–2693.
7. Viudamartos, M.; Ruiz-Navajas, Y.; Fernandez-Lopes, J.; and Perez-Alvarez, J.; *J. Food Saf.* **2007**, *27*, 91–101.
8. Serrano, M.; Martinez-Romero, D.; Castillo, S.; Guilen, F.; and Valero, D.; *Innov. Food Sci. Emerg. Technol.* **2005**, *6*, 115–123.
9. Serrano, M.D.; Martinez-Romero, S.; and Valero, D.; *Innov. Food Sci. Emerg. Technol.* **2008**, *6*, 115–123.
10. Ruch, B.W.; and Worf, R.; Processing of Neem for Plant Protection Simple and Sophisticated Standardized Extracts. Uberaba: University of Uberaba; **2001**, 499.

11. Ogawa, J. M.; Dehr, E.I.; Bird, G.W.; Ritchie, D.F.; Kiyoto, V.; and Uyemoto, J. K.; Compendium of Stone Fruit Disease. St. Paul, MN: APA Press; **1995**.
12. Mossini, S. A. G.; Carla, C.; and Kemmelmeier, C.; *Toxins*. **2009**, *1*, 3–13.
13. Fernando Ayala-Zavala, J.; Wang, S. Y.; and Wang, C. Y.; *LWT-Food Sci. Technol*. **2004**, *37*(7), 687–695.
14. Wang, G.; and Cao, R.L.; *J. Agric. Food Chem*. **1996**, *44*, 701–705.
15. Anonymous Toxicity of Pesticide Pesticide Safety Fact Sheet. Information and Communication Technologies in the College of Agricultural Sciences; **2006**.
16. Meyer, B. N.; Ferrigni, N. R.; Putnam, J. E.; Jacobsen, L. B.; Nichols, D. E.; and McLaughlin, J. L.; *Planta Med*. **1982**, *45*, 31–34.

CHAPTER 14

DEVELOPMENT OF PROCESS USING IONIC LIQUID-BASED ULTRASOUND ASSISTED EXTRACTION OF URSOLIC ACID FROM LEAVES OF *VITEX NEGUNDO LINN*

ANUJA CHAVHAN, MERLIN MATHEW, TAPASVI KUCHEKAR, and SUYOGKUMAR TARALKAR*

Chemical Engineering Department, MIT Academy of Engineering, Alandi(D), Pune-412105, Maharashtra, India; *E-mail: suyogkumartaralkar@yahoo.co.in, svtaralkar@chem.maepune.ac.in

CONTENTS

14.1 Introduction 138
14.2 Materials and Methods 139
14.3 Results and Discussion 140
14.4 Conclusion 145
Keywords 145
References 146

14.1 INTRODUCTION

Vitex Negundo Linn (Verbenaceae family) is an important medicinal plant usually found in the south east Asian countries. The extract from leaves and roots has important applications in the field of medicine. Decoction of leaves is considered as tonic and vermifuge [1]. The leaf extract has seen application in Ayurveda and Unani [2] medicine. Its mosquito repellent [3], antiarthritic [4], analgesic [5], hepatoprotective [6], anti-inflammatory, and antiallergic [6, 7] activities are commendable. The plant has insecticidal [1, 8], antibacterial, and antifungal activities [9] also. A long series of compounds, for example, essential oils [10], vitamin-C, carotene [11], b-sitosterol, iridoid glycosides [12], flavonoids, flavone glycosides [13, 14], and triterpenoids-ursolic acid and betulinic acid [6] are also identified from the leaves and seeds of this plant.

Ursolic acid is one of the important pentacyclic triterpenoids useful in medicinal applications. It is known to have anti-inflammatory, antitumor, antimicrobial, antibacterial, and antifungal activity [6]. The structure of ursolic acid is as shown in Figure 14.1.

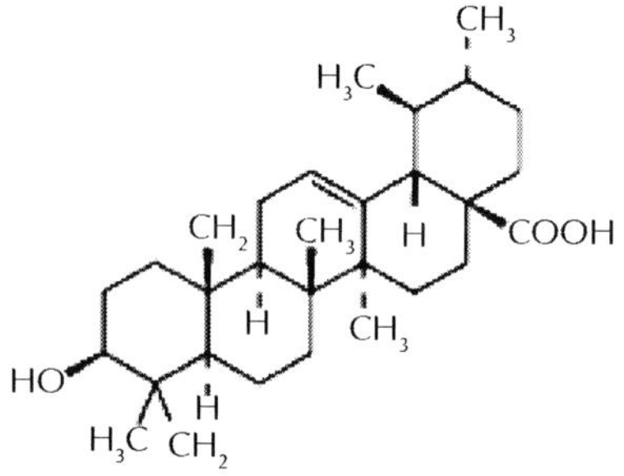

FIGURE 14.1 Structure of ursolic acid.

An efficient method for separating these compounds from natural sources is warranted in view of the beneficial effects of the active components of *Cortex fraxinus*. The conventional extraction methods, such as ethanol heat reflux extraction [15–17], methanol ultrasonic-assisted extraction [18–20], surfactant-ultrasonic-assisted extraction [21], and ethanol maceration extraction [15], are laborious, time-consuming, discommodious, and usually require large volumes of toxic, flammable organic solvents. It is necessary to find a rapid and effective method to alternate the traditional techniques in consideration of environmental protection and economy. In recent years, environment friendly techniques become more attractive with the development of the Green Chemistry. Ionic liquids have been proposed as greener alternatives to volatile organic solvents. Lei Yang et al. [22] proposed a method for optimization of the process of ionic liquid-based ultrasonic-assisted extraction of aesculin and aesculetin from *Cortex fraxini* by response surface methodology.

The aim of the present paper is development of a rapid and effective method, ionic liquid-based ultrasonic-assisted extraction, to extract ursolic acid from leaves of *Vitex Negundo Linn* (Nirgudi).

14.2 MATERIALS AND METHODS

14.2.1 MATERIALS

The leaves of *Vitex Negundo Linn* were collected locally. Leaves were washed with water and shade dried (at 30°C for 168hr) and powdered. The powdered mass was then sorted using screening technique and 0.106–0.42, 0.42–0.60, and 0.60–0.85 mm sizes were prepared. Standard ursolic acid (90 percent) was obtained from Fluka, USA, were used for calibration. Solvents used for extraction study and UV analysis were of AR grade from Merk. The ionic liquid 1-Butyl-3-methylimidazolium bromide (>97 percent) was obtained from Fluka.

14.2.2 BATCH EXTRACTION EXPERIMENT

Batch experimentation was directed toward estimation of extraction kinetics and to analyze influence of the operating parameters. 5gm leaves were taken in a four baffled 150 ml stirred borosilicate glass vessel (5 cm ID, and 9.5 cm height) along with 100 ml solvent. Four bladed turbine type agitator (2 cm diameter) was used to stir the mass at 700 rpm and the revolution was monitored through an online speedometer. All experiments were conducted at 30°C (±1°C) unless otherwise stated to minimize evaporation losses of solvent. Figure 14.2 shows the schematic of batch extraction setup.

Effect of speed of agitation (500, 700, 950 rpm), particle sizes (0.106–0.42, 0.42–0.60, 0.60–0.85 mm), percent solid loading (5, 7.5, and 10 percent), and temperature (30, 40, 50, and 60°C) were investigated to select the optimum conditions for each parameter. Samples were collected at various time intervals and extraction was carried out for 120 min to ensure no further improvement in yield of ursolic acid.

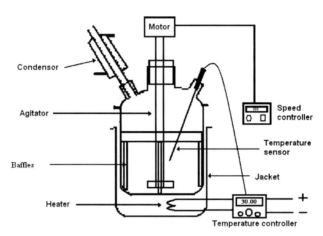

FIGURE 14.2 Schematic of batch extraction setup

14.2.3 IONIC LIQUID BASED EXTRACTION

The ionic liquid 1-Butyl-3-methylimidazolium bromide with 0.5 percent solution in methanol was used as solvent for extraction of ursolic acid from leaves of *Vitex Negundo Linn*. The same procedure was carried out as discussed in previous Section 14.2.2.

14.2.4 ULTRASONIC EXTRACTION

The ultrasonic extraction of ursolic acid from leaves of *Vitex Negundo Linn* was carried out in ultrasonic bath (capacity 3.5 L and frequency 50 Hz) with methanol as solvent and in presence of ionic liquid.

14.2.5 ANALYSIS

Standards and all unknown samples of ursolic acid concentrations were measured using bauble beam UV spectrophotometer (Thermo 2700) at 210 nm wavelength.

14.3 RESULTS AND DISCUSSION

14.3.1 EFFECT OF AGITATION SPEED

The extraction of ursolic acid was studied with methanol as a solvent at room temperature (30°C) at varying speeds of agitation 500, 700, and 950 rpm and

Figure 14.3 shows that percentage extraction increases with increase in speed. The rise is considerable from 500 to 950 rpm. Increased turbulence reduces interface mass transfer resistance and attains a limiting value at and beyond 950 rpm, indicating thereby pore diffusion controls, and there is no further increase in mass transfer. Therefore, speed of agitation was selected at 950 rpm for all subsequent studies. Rakotondramasy-Rabesiaka et al. [23] had investigated the effect of speed of agitation (100, 200, 400, 600 rpm) and reported that at lower rpm because of sedimentation of particles extraction rate of protopine is slow while between 200 and600 rpm no difference in extraction rate occurs.

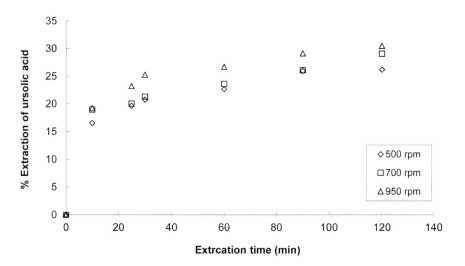

FIGURE 14.3 Influence of the speed of agitation on the extraction of ursolic acid at 5 percent solid loading, 30°C, particle size 0.106–0.42 mm and methanol as a solvent.

14.3.2 EFFECT OF PARTICLE SIZE

Batch extractions with methanol as solvent, 950 rpm speed, and 30°C temperature were performed using different particle sizes (0.106–0.42, 0.42–0.60, 0.60–0.85 mm). As particle size is reduced, the rate of extraction for ursolic acid increased (Figure 14.4). For smaller particles, diffusional resistance path decreases, and surface area increases, both of these help in increased recovery of ursolic acid. Similar observations were reported by Wongkittipong et al. [24] while investigating effect of particle size (0.1–0.3, 0.45–0.6, 0.6–0.8 mm) on the extraction of andrographolide.

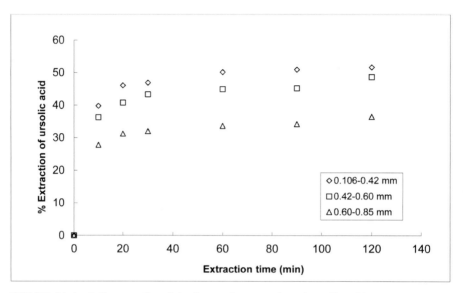

FIGURE 14.4 Influence of particle size on the extraction of ursolic acid at 5 percent solid loading, 950 rpm, 30°C and methanol as a solvent

14.3.3 EFFECT OF SOLID LOADING

Influence of percentage solid loading (g/ml) on extraction of ursolic acid was investigated in methanol at 950 rpm speed, and 0.106–0.42 mm particle size at 30°C (±1°C). Figure 14.5 shows that, ultimate value (at 120 min) of extracted ursolic acid (%) increases with decrease in solid loading. Initial 30 min percentage extraction is very similar for all types of loading, this is because at the beginning the solvent is free of any solute thus acid molecules experiences a large driving force difference to come out and get dissolved in solvent and slowly the concentration difference decreases. Therefore, release is quite rapid initially, and then slows down and tries to attain saturation equilibrium. The extracted equilibrium concentration of acid while compared with maximum amount of acid (0.19 percent dry basis) present in the dry solid taken, the yield is found to decrease with higher loading. As time increases beyond 50 min the percentage recovery becomes distinct. At higher solid loading, concentration gradient at the solid liquid interface decreases with time as the amount of solvent remains unchanged. Therefore, after 120 min the percentage of ursolic acid recovered reaches minimum for the 10 percent loading and gradually increases for 7.5 and 5 percent.

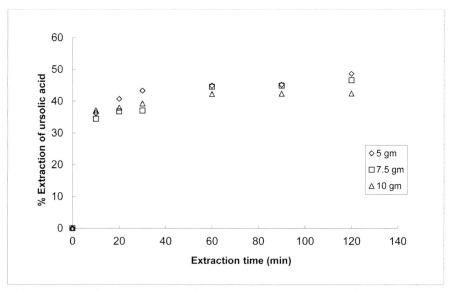

FIGURE 14.5 Influence of solid loading on extraction of ursolic acid with methanol as solvent at 950 rpm agitation speed and 0.106–0.42 mm particle size at 30°C (±1).

14.3.4 EFFECT OF TEMPERATURE

Effect of temperature on recovery of ursolic acid was investigated from 30 to 60°C with methanol as solvent at 950 rpm and 0.106–0.42 mm particle size. Figure 14.6 shows that with increase in temperature percentage recovery of ursolic acid increases with respect to time. This increase may easily be attributed to higher solubility of ursolic acid in methanol at higher temperature. The kinetic energy as well as diffusivity of solvent molecules increases with increasing in temperature, and thus solvent penetrate better inside the cellulose matrix leading thereby easy release of the acid molecules.

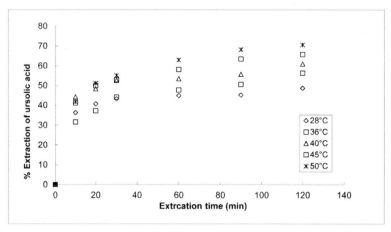

FIGURE 14.6 Influence of temperature on extraction of ursolic acid with methanol as solvent at 950 rpm agitation speed and 0.106–0.42 mm particle size

14.3.5 EFFECT OF IONIC LIQUID

The effect of ionic liquid on extraction of ursolic acid is studied with 0.5 percent concentration of ionic liquid in methanol as solvent. Figure 14.7 shows the effect of ionic liquid for extraction as compared with without ionic liquid. The ionic liquid with methanol as solvent shows increment of extraction than that of methanol only. This increase extraction may be due to increase in solubility of ursolic acid in presence of ionic liquid.

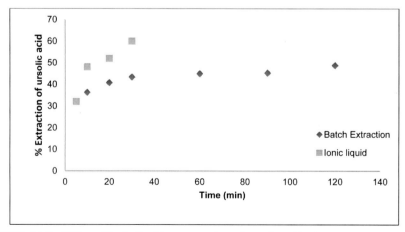

FIGURE 14.7 Effect of ionic liquid on extraction of ursolic acid with methanol as solvent at 950 rpm agitation speed and 0.106–0.42mm particle size.

14.3.6 EFFECT OF ULTRASOUND

The effect of ultrasound on extraction of ursolic acid has investigated. Figure 14.8 represent the effect of ultrasound on extraction compared with batch extraction. The ultrasonic extraction enhances the rate of extraction and reduces the extraction time significantly.

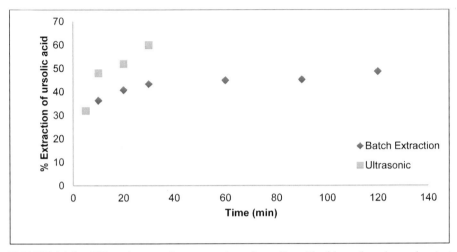

FIGURE 14.8 Effect ultrasound on extraction of ursolic acid with methanol as solvent at 950 rpm agitation speed and 0.106 – 0.42 mm particle size.

14.4 CONCLUSION

The present study shows that effect of operating parameters on extraction of ursolic acid from leaves of *Vitex Negundo Linn*. Presence of ionic liquid in solvent attributes increase in extraction is more interesting. The use of ultrasound enhances the extraction efficiency and reduces the extraction time.

KEYWORDS

- **Ionic liquid**
- **Ultrasonic-assisted extraction**
- **Ursolic acid**
- ***Vitex Negundo Linn***

REFERENCES

1. Chadha, Y. R.; The Wealth of India: A Dictionary of Indian Raw Material and Industrial Products. New Delhi: CSIR (Council of Scientific & Industrial Research); **1976**, 522–524.
2. Kapur, V.; Pillai, K. K.; Hussain, S. Z.; and Balani, D. K.;*Indian J. Pharm. Sci.* **1994**,*26*, 35–40.
3. Hebbalkar, D. S.; Hebbalkar, G. D.; Sharma, R. N.; Joshi, V. S.; and Bhat, V. S.; *Indian J. Med. Res.* **1992**, *95*, 200–203.
4. Tamhankar, C. P.; and Saraf, M. N.; *Indian J. Pharm. Sci.* **1994**, *56*(1), 158–159.
5. Gupta, M.; Mazumdar, U. K.; and Bhawal, S. R.; *Indian J. Exp. Biol.* **1999**, *37*, 143–146.
6. Chawla, A. S.; Sharma, A. K.; Handa, S. S.; and Dhar, K. L.; *J. Nat. Prod.* **1992**, *55*(2), 163–167.
7. Jana, U.; Chattopadhyay, R. N.; and Shaw, B. P.; *Indian J. Pharmacol.* **1999**, *31*(3), 232–233.
8. Aswal, B. S.; Goel, A. K.;Kulshrestha, D. K.; Mehrotra, B. N.; and Patnaik, G. K.; *Indian J. Exp. Biol.* **1996**, *34*(5), 444–467.
9. Ragasa, C. Y.; Morales, E.; and Rideout, J. A.; *Philipp J. Sci.* **1999**, *128*,21–29.
10. Taneja, S. C.; Gupta, R. K.; Dhar, K. L.; and Atal, C. K.;*Indian Perfumer.* **1979**, *23*,162–163.
11. Basu, N. M.; Ray, G. K.; and De, N. K.; *J. Indian Chem. Sci.* **1947**, *24*, 358–360.
12. Dutta, K. P.; Chowdary, S.; Achari, B.; and Pakrashi, C. S.;*Tetrahadran.* **1993**, *39*, 3067–3072.
13. Liu, J.; *J. Ethnopharmacol.* **1995**, *49*, 57–68.
14. Prema, M. S.; and Misra, G. S.; *Indian J. Chem.* **1978**, *16*, 615–616.
15. Zhang, Z.; Hu, Z.; and Yang, G.; *Chromatographia.* **1997**, *44*, 162–168.
16. Liu, R.; Sun, Q.; Sun, A.; and Cui, J.; *J. Chromatogr.* **2005**, *1072*, 195–199.
17. Zhou, L.; Kang, J.; Fan, L.; Ma, X. C.; Zhao, H. Y.; Han, J.; Wang, B. R.; Guo, D. A.; *J. Pharm. Biomed. Anal.* **2008**, *47*(1), 39–46.
18. Zhang, H.; Li, Q.; Shi, Z.; Hu, Z.; and Wang, R.; *Talanta.* **2000**, *52*, 607–621.
19. Bo, T.; Liu, H.; and Li, K. A.; *Chromatographia.* **2002**, *55*, 621–624.
20. Li, C.; Chen, A.; Chen, X.; Ma, X.; Chen, X; and Hu, Z.; *Biomed. Chromatogr.* **2005**, *9*(9), 696–702.
21. Shi, Z.; Zhu, X.; and Zhang, H.; *J. Pharm. Biomed. Anal.* **2007**, *44*(4), 867–873.
22. Yang, L.; Liu,Y.; Zu, Y.G.; Zhao, C. J.; Zhang, L.; Chen, X. Q.; and Zhang, Z. H.; *Chem. Eng. J.* **2011**, *175*(1), 539–547.
23. Rakotondramasy-Rabesiaka, L., Havet, J. L., Porte, C., Fauduet, H.; Sep. and Puri. Tech. **2008**, 59, 253-261.
24. Wongkittipong, R., Prat, L., Damronglerd, S., Gourdon, C.; Sep. and Puri. Tech. **2004**, 40, 147-152.

CHAPTER 15

EXTRACTION OF CHITIN AND CHITOSAN FROM FISHERY SCALES BY CHEMICAL METHOD

S. KUMARI[1*] and P. RATH[2]

[1]Department of Chemical Engineering, Research Scholar, National Institute of Technology, Rourkela India; *E-mail: suneetak7@gmail.com

[2]Department of Chemical Engineering, Professor, National Institute of Technology, Rourkela India

CONTENTS

15.1 Introduction 148
15.2 Materials and Method 149
15.3 Characterization of Chitin and Chitosan 151
15.4 Results and Discussion 151
15.5 Conclusion 154
Keywords 154
References 154

15.1 INTRODUCTION

Chitin is usually distributed in marine invertebrates, insects, fungi, and yeast. Chitin is a bio polymeric substance derived from crustaceous shell a homopolymer of β (1→4) linked N- Acetyl-D-glucosamine[1]. In spite of the presence of nitrogen it may be regarded as cellulose with hydroxyl at position C-2 replaced by an acetamido group [2]. Like, cellulose functions a structural polysaccharides (Figure 15.1).Commonly, the shell of selected crustacean consists of 30–40 percent protein, 30–50percent calcium carbonate and calcium phosphate and 20–30 percent chitin [3]. The prime source is shellfish water such as shrimps crabs and crawfish [4]. It isalso obtained naturally in a few species of fungi. Chitin is formed a linear chain of acetylglucosamine groups while chitosan is recovered by removing enough acetyl groups (CH_3-CO) from chitin therefore the chitin molecule and the resultant product is found to be soluble in most diluted acids. The actual variation between chitin and chitosan is the acetyl content of the polymer. Chitosan having a free amino group is the most useful of chitin [5]. Chitosan is a non toxic biodegradable polymer of high molecular weight. Chitosan is a one of the promising renewable polymeric materials for their broad application in the pharmaceutical and biomedical industries for enzyme immobilization [6]. Chitosan is used in the chemical wastewater treatment and food industrial for food formulation as binding, gelling, thickening, and stabilizing agent [7].

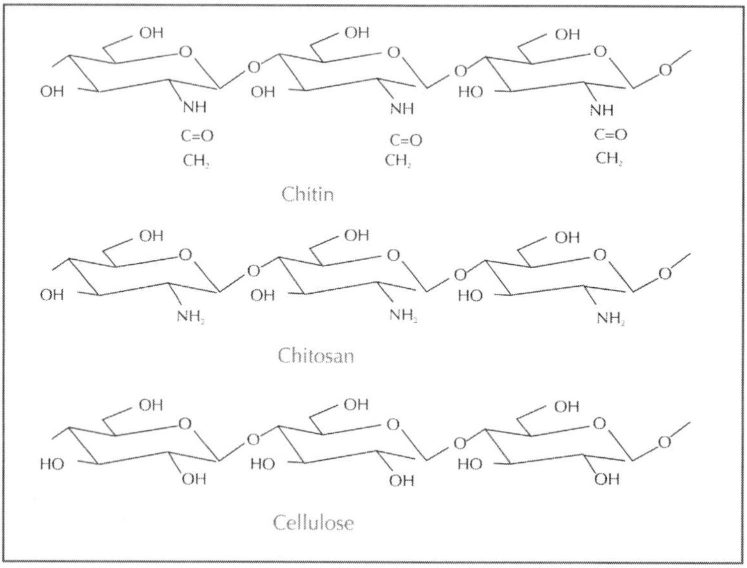

FIGURE 15.1 Chemical structures of chitin, chitosan, and cellulose

Commonly, isolated of chitin from crustacean shell waste consists of three basic steps: demineralization (calcium carbonate and calcium phosphate separation), deproteinization (Protein separation), decolorization (removal of pigments) and deacetylation (remove acetyl groups). These three steps are the standard procedure for chitin production [7]. The subsequent conversion of chitin to chitosan (deacetylation) is generally achieved by treatment with concentrated sodium hydroxide solution (40–50%) at 100°C or higher temperature to remove some or all acetyl group from thechitin [8–9]. Earlier studies by several authors have proved that the physicochemical characteristics of chitosan affect its functional properties which also differ due to crustacean species and preparation methods [10–11].

Several methods have been developed and proposedby many researchers over the years for preparation of chitosan from different crustacean shell wastes. Some of these formed the basis of acid and base concentration for industrial productionof chitosan. But most of the reported processes werecarried out at 100° C or higher temperature with autoclaving [12–13]. Therefore, the specific objectives of this work were to develop an optimum condition for production of chitosan from fish scales [14–15].

15.2 MATERIALS AND METHOD

15.2.1 RAW MATERIALS AND CHEMICALS

Fish Scales were taken from the local fish market for this project. Acid (HCL, 36.5g/mol) 1 percent solution of HCL was prepared for demineralization. Sodium hydroxide (NaOH, 40g/mol) 0.5 N solution of NaOH was prepared for Deproteinization. pH meter and distilled water were also used in the experimental process. All chemicals used were of laboratory grade.

15.2.2 EXPERIMENTAL PROCEDURE FOR THE PRODUCTION OF CHITIN AND CHITOSAN

Preparation of chitosan from fish scales was performed using the general method comprising of demineralisation, decolourisation and deacetylation. Raw fish scales were washed thoroughly with water, dried in oven and soaked in 1percent HCL solution for 36 hrs. It was then washed dried in oven and kept in 2N NaOH solution for 36 hrs for demineralisation. Fish scales were then kept in Potassium permanganate solution (having composition 1g of $KMnO_4$ in 100ml water) for 1 hr, followed by keeping it in Oxalic acid (having composition 1g of Oxalic acid in 100ml water) for the process of decolourization of the experimental sample.

These processes resulted in chitin as the product which was further treated with 50 percent w/v NaOH for the process of deacetylation resulting in chitosan as the end product (Figure 15.2).

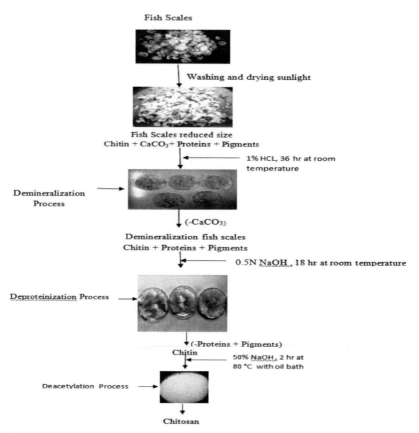

FIGURE 15.2 Overall Processes for preparation of chitin from fish scales

15.2.3 DEGREE OF DEACETYLATION

The DA of chitin/chitosan is the most important parameter that influences their various properties including biological, physicochemical and mechanical properties. The effectiveness and behaviours of chitin/chitosan and its derivatives has been found to be dependent on the DA. The expansion and stiffness of the macromolecular chain conformation and the tendency of the macromolecule chains to aggregate depend strongly on the DA. The determination of the DA for the two copolymers is essential for studying their chemical structures, properties, and structure-properties relationships. Knowledge on the DA is very important to maximize chitosan applications. The DA is known as many properties and applications can be predicated [16].

15.3 CHARACTERIZATION OF CHITIN AND CHITOSAN

15.3.1 FOURIER-TRANSFORM INFRARED SPECTROSCOPY ANALYSIS

Infrared spectra of KBr chitin or chitosan mixtures were obtained over the frequency range of 400 to 4,000 cm^{-1} at resolution of 4 cm^{-1} using Prestige-21, FTIR Spectrometer Shimadzu. The sample was thoroughly mixed with KBr, and the dried mixture was then pressed to result in a homogeneous sample disk [17].

15.3.2 X-RAY DIFFRACTION (XRD) ANALYSIS

The X-ray diffraction pattern of chitosan was recorded at room temperature. The data were collected in the 2θ range 2–70° with a step size of 0.02° and a counting time of 5 s/ step [18].

15.4 RESULTS AND DISCUSSION

15.4.1 DEGREE OF DEACETYLATION

Chitosan (0.2 gm) was dissolved in 20 ml 0.1 M hydrochloric acid and 25 ml of deionized water. After 30 min of continuous stirring additional 25 ml of deionized water was added and stirring was continued for next 30 min. When chitosan was completely dissolved solution was titrated with 0.1 M sodium hydroxide solution. Degree of deacetylation (DA) was calculated using formula.

$$DA (\%) = 2.03 \times (V_2-V_1)/ [m+0.0042 (V_2-V_1)]$$
$$= 98$$

15.4.2 FTIR

FTIR spectroscopic analysis of chitosan is shown in Figure 15.3. The absorption bands of experimentally prepared chitosan were identical to those of standard chitins. Different stretching vibration bands were observed in the range 3,425–2,881 cm^{-1} related to ν (N-H) in ν (NH$_2$) assoc. in primary amines (Dilyana, 2010). The band at 3,425–3,422 cm^{-1} could be assigned to ν (N-H), ν(O-H) and ν(NH$_2$) which present in chitosan in different amounts among which NH$_2$ groups being the least. The presence of methyl group in NHCOCH$_3$, methylene group in CH$_2$OH and methyne group in pyranose ring was proved by the corresponding stretching vibrations of these groups in the range 2,921-2,879cm^{-1}(Figure 15.3). The band at 1,597 cm^{-1} has a larger intensity than at 1,655 cm^{-1}, which suggests, effective deacetylation. When chitin deacetylation occurs, the band observed at 1,656 cm^{-1} decrease, while a growth at 1,597cm^{-1} occurs, indicating the prevalence of NH$_2$ groups. When the same spectrum is observed, in which the band from 1,500–1,700 cm^{-1} is stressed, indicated that there was an intensification of the peak at 1,597 and a decrease at 1,656 cm^{-1}, that suggests the occurrence of deacetylation [19].

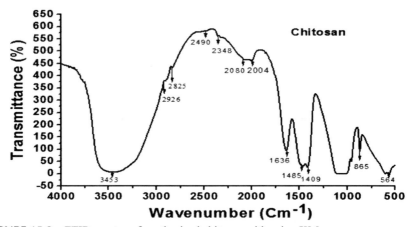

FIGURE 15.3 FTIR spectra of synthesized chitosan with using KMno$_4$

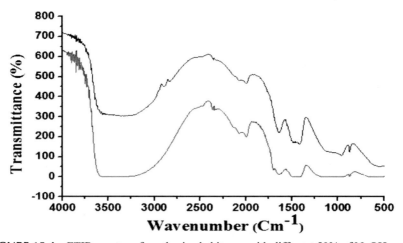

FIGURE 15.4 FTIR spectra of synthesized chitosan with different 50% of NaOH

15.4.3 XRD

The X-ray diffraction patterns of the obtained chitin and the corresponding hydrolyzed chitosan are present in (Figure 15.4). All chitin samples exhibited strong reflections at 2θ value of 9–10° and minor reflections at higher 2θ values. Generally, the sharpness of the bands is higher in the chitin samples than in their chitosan analogue with slight decrease in the crystalline percent [20].

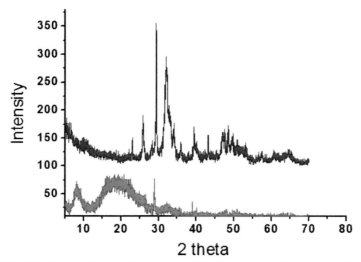

FIGURE 15.5 Comparison of X-ray power diffractograms of chitosan and fish scales

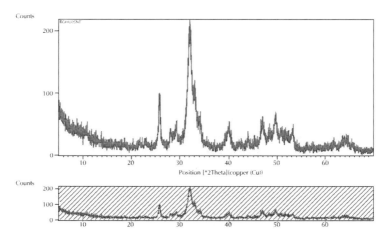

FIGURE 15.6 X-Ray diffraction pattern, 50 percent DDA chitosan diffractogram

The XRD pattern of thefish scales sample is shown in Figure15.4. Five crystalline reflections were observed in the 2θ range 10 to 80°. The peak intensity was 6.46, 24.42, 33.58, 37.42, 53.3, and 56.75 at 2θ. Chitosan sample crystalline reflections were observed in the 2θ range 10 to 80°. The peak intensity was 14.45, 20.02, 26.70, 37.32, and 54.32, respectively [20]. Figure 15.4 shows that fish scales are amorphous region in structure and Figure 15.5 chitosan is crystalline.

15.5 CONCLUSION

The chitin obtained from thewaste fishery scales hasbeen used in the variety of application especially, for waste water treatment. Chitosan is prepared from-Rohu fishery scales by chemical treatment withvarious concentration like 1 percent HCl and 0.5N NaOH. In deacetylation process15 to 50 percent NaOH concentration has been varied. However by FTIR analysis it is identified that 50 NaOH concentration is more effective to find out all the functional groups in chitosan macromolecules than other NaOH concentration. XRD analysis reveals that fishery scale is amorphous but chitosan is crystalline in structure.

KEYWORDS

- **Chitosan**
- **Demineralization**
- **Deacetylation**
- **FITR**
- **XRD**

REFERENCES

1. Austin, P. R.; Brine, C. J.; Castle, J. E.; et al; Chitin: New facets of research. *Science.* **1981**, *212*, 749–753.
2. Entsar, S.; Abdou, and Khaled, S. A.; Nagy and Maher Z. Elsabee; Extraction and characterization of chitin and chitosan from local sources. *Bioresour. Technol.* **2008**, *99*, 1359–1367.
3. Knorr, D.; Use of chitinous polymers in food –A challenge for food research and development. *Food technol-chicago.* **1984**, *38*, 85–97.
4. Allan, C.R.; and Hadwiger, L.A.; The fungicidal effect of chitosan on fungi of varying cell wall composition. *Exp. Mycol.* **1979**, *3*, 285–287.

5. No, H. K.; and Meyers, S. P.; Utilization of crawfish processing waste as carotenoids, chitin and chitosan souces. *J. Korean Soc. Food Nutr.* **1992**, *21*, 319–326.
6. Nessa, F.; Shah Md. Masum et al; A process for the preparation of chitin and chitosan from prawn shell waste. *J. Sci. Ind . Res.* **2010**, 45, 323–330.
7. No, H.K.; and Meyers, S.P.; Crawfish chitosan as a coagulant in recovery of organic compounds from seafood prcessing streams. *J.Agric. Food Chem.* **1989**, *37*, 580–583.
8. No, H. K.; and Meyers, S. P.; Preparation and characterization of chitin and chitosan—A review. *J. Aquat. Food Prod. Technol.* **1995**, 4, 27–52.
9. Galed G, Diaz E. and Heras A; Conditions of N-deactylation on chitosan production from alpha chitin.*Nat. Prod. Commun.* **2008**, 3, 543–50.
10. Cho Y. I. H. K. and Meyers S. P; Physicochemical characteristics chitin and chitosan production. *J. Agric. Food Chem.* **1998**, 46, 3839–3843.
11. No, H.K.; Lee, K.S.; and Meyers, S.P.; Correlation between physicochemical characteristics and binding capacities of chitosan products. *J. Food Sci.* **2000**, *65*, 1134–1137.
12. Horton, D.; and Lineback, D. R. N.; Deacetylation chitosan from chitin. In; Whistler RL, Wolfson ML, Eds. Methods in carbohydrate chemistry. New York: Academic Press. **1965**, 403.
13. Domard, A.; and Rinadudo, M.; Preparation and characterization of fully deactylated chitosan. *J. Biol.Macromol*, **1983**, 5, 49–52.
14. Prashanth, R.; and Tharanathan, R.; Chitin/chitosan: modification and their unlimited application potential—an overview. *Trends.Food. Sci. Tech.* **2007**, 18,117–131.
15. Galed, G.; Diaz, E. and Heras, A.; Conditions of N-deactylation on chitosan production from alpha chitin. *Nat. Prod. Commun.* **2008**, 3, 543–550.
16. Mohammad R. Kassai; Various methods for determination of the degree of N-Acetylation of chitin and chitosan: A Review .J. Agric Food Chem. 2009, 57, 1667–1676.
17. Mohamed, A.; Radwan, Samia A. A.; Farrag, Mahmoud, M.; and Abu et al; Extraction, characterization and nematicidal activity of chitin and chitosan derived from shrimp shell waste. *Biol. Fertil.Soils.* **2012**, *48*, 463–468.
18. Assaad, S.; Najwa, M.; Nadhem, S. et al; Chitin and chitosan extracted from chitosan waste using fish proteases Aided process: Efficiency of chitosan in the treatment of unhairing effluents. *J. Polym Environ.* **2013**.
19. Dilyana, Z.; Synthesis and characterization of chitosan from marine source in black sea. **2010**, *49*.
20. Zaku, S. G.; Emmanuel,S. A.; and Aguzue, O. C.; and Thomas, S. A.; Extraction and characterization of chitin; a functional biopolymer obtained from scales of common crap fish (cyprinuscarpioi): A lesser known source. *Afr. J. Food.Sci.* **2011**, *5*, 478–483.

CHAPTER 16

STUDY OF MIXING IN SHEAR THINNING FLUID USING CFD SIMULATION

AKHILESH KHAPRE* and BASUDEB MUNSHI

Department of Chemical Engineering, National Institute of Technology Rourkela, Odisha, India; *E-mail: akhilesh_khapre@yahoo.co.in

CONTENTS

16.1 Introduction 158
16.2 Problem Statement 159
16.3 CFD Methodology 160
16.4 Results and Discussion 162
16.5 Conclusions 168
Keywords 168
References 168

16.1 INTRODUCTION

Mixing operation in which the rheological properties of the medium change over the course of the mixing process are common in the chemical industries [1]. A standard stirred tank which equipped with the impeller is commonly used as mixing system. Also, the baffles are mounted on the tank wall to avoid circulatory motion of the fluid and to destroy surface vortices. Viscosity is one of the most important rheological properties that have great significance in the process [2]. For low viscosity non-Newtonian fluids, the high speed blade turbines such flat blade turbines, pitched blade turbines and hydrofoils etc. are commonly used for mixing system. The selection of good impeller is depended on the mixing duty requirement like the operating speed, discharge characteristics and the power draw. But for high viscosity fluids mixing, the close clearance impellers like helical ribbon impeller are preferred [3]. Hence, the mixing tank performance is depending on an appropriate adjustment of the reactor hardware and operating parameters like reactor and impeller shapes, aspect ratio of the reactor vessel, degree of baffling, etc., which provide effective means to control the performance of stirred reactors [4].

Many researchers have studied mixing phenomena in stirred tank using CFD by varying geometrical parameters and impeller shapes. Montante et al. [5] have done CFD simulation to calculate mixing time in stirred tank agitated with 45° pitched blade turbines for fluids characterised by either Newtonian or non-Newtonian rheological behaviour. Javed et al. [6] has carried out CFD modelling of turbulent batch mixing of an inert tracer with Newtonian liquid in a baffled vessel stirred by a Rushton turbine impeller. They have injected a tracer from the top surface of the vessel and mixing times are measured at 32 different locations in the vessel. Ochieng et al. [7] has investigated the effect of the impeller clearance position on the velocity flow field and mixing using CFD simulation. They have found that the standard double loop flow pattern is reduced to a single loop due to which axial velocity is increased and mixing time is decreased. They have also used draft tube with low clearance impeller, and the mixing time is decreased by 50 percent. Visuri et al. [8] have analysed mixing time by using novel digital imaging technique in stirred tank. Rushton turbine, EKATO Phasejet and EKATO Combijet impellers are used with three different xanthan gum solutions with three different mixing intensities. CFD simulations are carried out to study the effect of impeller geometry and effect of viscosity on mixing. Shekhar and Jayanti [9] have carried out number of CFD simulation of mixing of pseudoplastic with helical ribbon impellers to study the relation between shear rate and the rotational speed of impeller. They have shown that the shear rate near the impeller region varies linear with the rotation speed of impeller.

From the foregoing discussion, it can be concluded that no simulation work have been carried out to study the effect of baffles off wall mounting position on mixing behavior. Hence, the objective of this study is to investigate the effect of a distance of baffles mounting position from wall on mixing of non-Newtonian shear thinning fluid in stirred tank using CFD simulation. The predicted velocity profiles calculated for baffle mounted on the tank wall are compared with available literature data [10]. The CFD simulation is used to predict mixing time for different baffle mounting positions.

16.2 PROBLEM STATEMENT

A flat bottom standard stirred tank with four baffles installed on the tank wall is considered for simulation of non-Newtonian fluid as shown in Figure 16.1. The diameter of tank (T) is 0.627 m and the fluid is filled up to level H which is equal to tank diameter, T. A Six blades Rushton turbine impeller with diameter, $D = T/3$, is placed at an off bottom clearance of $C_i = D$. The width of baffle is, $W_b = T/10$. To simulate the effect of baffles position, the baffles are mounted at distance 1/10th and 1/5th of baffle width from the tank wall. The used shear thinning fluid is Carboxymethylcellulose (CMC). The flow index (n) of CMC is 0.85 and consistency index (K) is 0.0132 kg s^{n-2} m^{-1}. The simulation is carried out at 180 rpm of the impeller which correspond to Reynolds number of 16900 and simulated velocity profiles are compared with the data from Venneker et al. [10]. The mixing time is measured at different points which are placed axially in between two baffles near to tank wall.

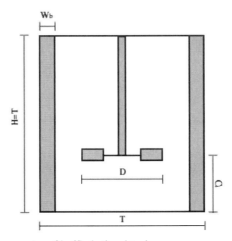

FIGURE 16.1 The geometry of baffled stirred tank.

16.3 CFD METHODOLOGY

16.3.1 GOVERNING EQUATIONS

The CFD simulation is performed using commercial CFD code, ANSYS-FLUENT 13. The realizable k-ε model is used to model turbulence flow and the Reynolds averaged Navier-Stokes equations are solved to obtain a velocity field inside stirred tank. The continuity equation for an incompressible fluid can be written as [11]:

$$\frac{\partial \rho}{\partial t} + \nabla \cdot (\rho \vec{v}) = 0 \qquad (16.1)$$

where, $\vec{v}$ is the velocity vector.

The conservation of momentum equation is given by [11],

$$\frac{\partial}{\partial t}(\rho \vec{v}) + \nabla \cdot (\rho \vec{v} \vec{v}) = -\nabla p + \nabla \cdot (\bar{\bar{\tau}}) + \rho \vec{g} + \vec{F} \qquad (16.2)$$

where, p is the static pressure and $\bar{\bar{\tau}}$ is the stress tensor. $\vec{g}$ and $\vec{F}$ are the gravitational body force and external body forces respectively. The stress tensor $\bar{\bar{\tau}}$ is given by

$$\bar{\bar{\tau}} = \mu \left[\left(\nabla \vec{v} + \nabla \vec{v}^T \right) - \frac{2}{3} \nabla \cdot \vec{v} I \right] \qquad (16.3)$$

where, μ is the molecular viscosity, I is the unit tensor.

The turbulence kinetic energy, k, and its rate of dissipation, ε, are calculated by the equations [11],

$$\frac{\partial}{\partial t}(\rho k) + \nabla \cdot (\rho k \vec{v}) = \nabla \cdot \left[\left(\mu + \frac{\mu_t}{\sigma_k} \right) \nabla \cdot k \right] + G_k - \rho \varepsilon \qquad (16.4)$$

and

$$\frac{\partial}{\partial t}(\rho \varepsilon) + \nabla \cdot (\rho \varepsilon \vec{v}) = \nabla \cdot \left[\left(\mu + \frac{\mu_t}{\sigma_\varepsilon} \right) \nabla \cdot \varepsilon \right] + C_{1\varepsilon} \frac{\varepsilon}{k} G_k - C_{2\varepsilon} \rho \frac{\varepsilon^2}{k} \qquad (16.5)$$

where, G_k represents generation of turbulent kinetic energy and μ_t is the turbulent or eddy viscosity computed by combining k and ε as follows:

$$\mu_t = \rho C_\mu k^2 / \varepsilon \qquad (16.6)$$

where, C_μ is a constant.

The model constant $C_{1\varepsilon}$, $C_{2\varepsilon}$, C_μ, σk, and σ_ε have the following default values:

$$C_{1\varepsilon} = 1.44, C_{2\varepsilon} = 1.92, C_\mu = 0.09, \sigma_k = 1.0, \sigma_\varepsilon = 1.2$$

For non-Newtonian fluid, the power law is used to model viscosity, is given as

$$\eta = K\dot{\gamma}^{n-1} \qquad (16.7)$$

where, η is effective viscosity, K is consistency index, n is flow index and $\dot{\gamma}$ is average share stress.

The Reynolds number for shear thinning fluid is calculated using the Metzner-Otto method [12],

$$Re = \frac{\rho N^{2-n} D^2}{K k_s^{n-1}} \qquad (16.8)$$

where k_s is Metzner-Otto constant with $k_s = 11.5$.

16.3.2 SIMULATION PROCEDURE

The stirred tank domain is discretized into a number of grid points where, Reynolds averaged Navier-Stokes equation is solved. The computational domain is meshed with approximately 800000 fine unstructured tetrahedral cells. A very refined mesh near the impeller region is done to capture the flow dynamics. While the inflation layer meshes, at the tank wall, is made to capture the boundary layer.

The transient Sliding mesh (SM) approach is used to model the impeller-baffles interaction. For this approach, the tank is divided into two domains. First is the rotating domain which engulfed the impeller and rotated with the same rotation of the impeller. The other is stationary domain which included baffles and remaining part of the tank, and is remained stationary. The both domains exchange values of flow parameters such as pressure, momentum etc., through common face known as interface. On the interface, the sliding mesh algorithm is applied which have taken into account for the relative motion between two domains [4].

The impeller is rotated (N) at 180 rpm which correspond to Reynolds number of 16900. The rotating zone around the impeller is also set to rotation at 180 rpm. The impeller blades and the shaft, which is outside rotating domain, are modelled as rotating walls. The baffles and the tank wall are kept stationary. The no-slip boundary condition is applied on the baffles and the tank wall. On the top wall of the tank, a symmetry boundary condition is applied. The second order transient solver is selected in fluent solver with second order upwind difference discretization scheme for all transport equations. A pressure-correction method, SIMPLE (Semi-Implicit Method for Pressure Linked Equation), is used to couple pressure and momentum equations. The realizable k-ε turbulence model is used for the turbulence prediction with initial condition of the turbulent intensity 10 percent. A turbulent Power Law is used to model a shear thinning non-Newtonian fluid behavior in the tank. For simulation, the time step is taken as 5×10^{-3} second and there are 30 iterations per time step which is enough for the simulation to converge at each time step. The simulation is run for 30 seconds to mix the tracer evenly. The root mean square residuals criteria for all discretized transport equations are set to 10^{-3}.

16.4 RESULTS AND DISCUSSION

16.4.1 FLOW FIELD

Let's consider three configuration of stirred tank, configuration 1 is a tank with baffles on wall, configuration 2 and 3 is tank with baffles mounted at distance of 1/10th and 1/5th of baffles width from wall respectively. The velocity components data are taken at impeller disk level, that is $H = 0.209$ m, along the non-dimensional radial distance, r/R, where r is a radial distance and R is a radius of the impeller. The predicted velocity components are made normalized with respect to the impeller blade tip velocity, $V_{tip} = \pi ND$.

From Figure 16.2, it observes that the predicted velocity components, in configuration 1, show overall reasonable agreement with literature data [10]. But near the impeller region, a simulation under predicts the value of all three velocity components. This might be due to under-prediction of turbulent intensity by k-ε turbulence model. The axial velocity component correctly predicts the change of flow direction that is the fluid movement in a downward direction near about $r/R = 2.4$. Thus, the change in direction of axial velocity plays an important role in determining a mixing time.

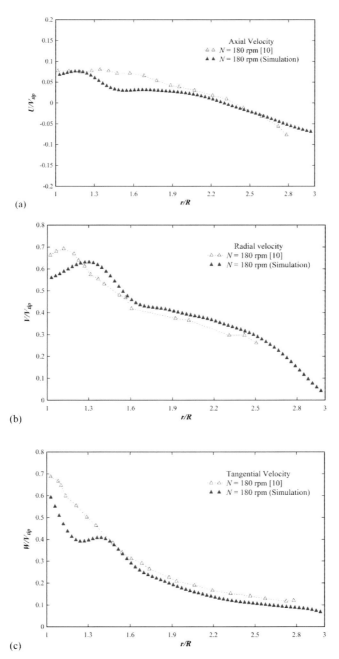

FIGURE 16.2 Comparison of predicted velocity components of configuration 1 with literature data [10]; (a) axial velocity, (b) radial velocity and (c) tangential velocity.

Figure 16.3 shows, the comparison of the radial profiles of axial velocity at all three stirred tank configurations. The predicted axial velocity at all axial position, for configuration 2, shows the higher value than the other two tank configurations near the tank wall. This increase in axial velocity in configuration 2 may help to minimize the local mixing time.

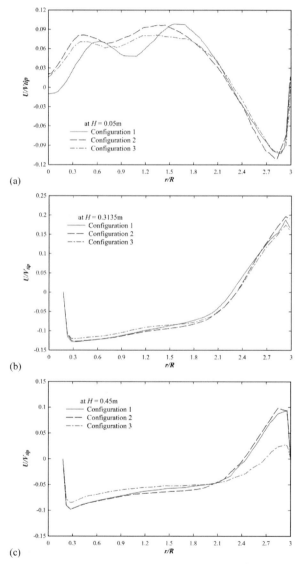

FIGURE 16.3 Comparison of radial profile of axial velocity at different axial position for three tank configurations.

16.4.2 MIXING TIME

In batch stirred tank, the mixing time or blending time is used as a key parameter for characterizing turbulent mixing. It is the time taken by a tracer between an instant when a tracer is released into the tank and when it reaches a specified degree of homogeneity. Experimentally, the mixing time is estimated by giving an input of tracer at a certain location in the tank and monitoring the change in concentration with time at some other location [13]. The mixing time, at the particular location, is considered as the time required reaching the concentration of tracer within 95–99 percent of the final steady state concentration at that location. A unit mass fraction of tracer concentration at zero time is taken at the center of the bottom surface of the tank. The concentrations at the remaining cells are initialized as zero.

Figure 16.4 shows, the CFD simulation predicted response curve of a tracer concentration at different axial point's positions on a plane midway between two baffles for different baffles mounting position inside a stirred tank. The dimensionless mixing time, N_t, is defined as $N_t = N \cdot t$ and C_0 and C is instantaneous and steady state concentration of a trace respectively. The dimensionless mixing time for 95 percent level of homogeneity is determined for all stirred tank configurations. From a tracer response profile, it is clear that the configuration 1 requires more recirculation loop than configurations 2 and 3. The height of tracer response curve for configuration 1 is larger than other two configurations. Thus, the rate of breakup and dispersal of a tracer in configuration 1 is less than other two configurations. It also observes that as a move away from the bottom to top of the tank, the mixing time is increases. This is due to less movement of fluid in the upper region of the tank.

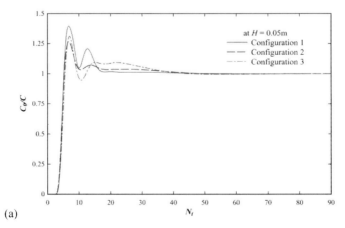

(a)

FIGURE 16.4 *(Continued)*

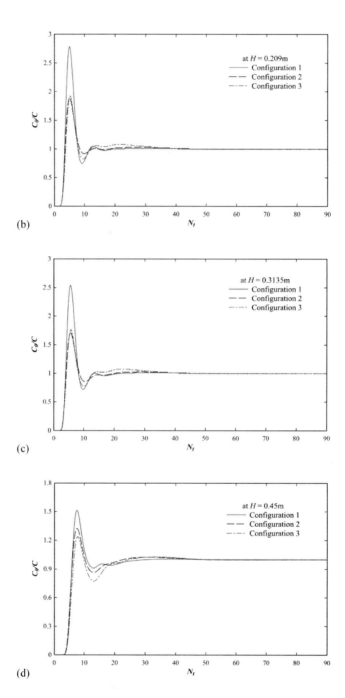

FIGURE 16.4 *(Continued)*

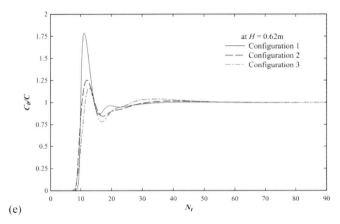

(e)

FIGURE 16.4 CFD simulation predicted tracer response curve at 5 different axial positions near tank wall.

TABLE 16.1 Mixing time for 95 percent homogenization

Position of points	Mixing Time, N_t		
	Configuration 1	**Configuration 2**	**Configuration 3**
$H = 0.05$m	15.84	16.02	31.065
$H = 0.209$m	11.235	11.13	29.685
$H = 0.3135$m	12.27	13.125	29.820
$H = 0.45$m	20.49	17.175	18.6
$H = 0.62$m	23.73	25.56	23.085

Table 16.1 shows, the local minimum mixing time required for all three tank configurations to achieve 95% of homogenization. At position H equal to 0.05, 0.209 and 0.3135 m, the local minimum mixing time for configuration 1 and 2 is almost equal. But for configuration 3, the mixing time is nearly double than configuration 1 and 2. This may be due to a trap of tracer in lower circulation loop. At position $H = 0.45$ m, the mixing time for configuration 2 and 3 requires 16% and 9% less time than configuration 1. This may be due to increasing in axial velocity and quickly breakup and dispersal of tracer near the tank wall. While at position $H = 0.62$ m, the configuration 3 requires less time to reach the top of the tank. The average mixing time for configuration 1, 2 and 3 is 16.713, 16.602 and 26.451 respectively. The average local mixing time for configuration 2 requires less time, by 0.1, than the configuration 1 which is very small in

magnitude than configuration 1. Thus, the choice of the selection of the baffles mounting position in tank purely depends on the amount of the power consumption by impeller to mix the fluid in stirred tank.

16.5 CONCLUSIONS

The CFD simulations for stirred tanks with different baffles mounting position configurations are done at fully turbulent flow regime. For validation purpose, the predicted velocity components of configuration 1, that is baffles mounted on the wall, are compared with literature data and found in good agreement with it. The predicted axial velocity in configuration 2 is more than other two tank configurations near the wall. The minimum mixing time is determined at five different axial positions near the wall. At position $H = 0.42$ m, the configuration 2 and 3 required 16 percent and 9 percent less time than configuration 1 respectively and to reach the top of tank configuration 3 required less mixing time. The configuration 2 required less average local mixing time, $N_t = 16.602$.

KEYWORDS

- **CFD**
- **Mixing time**
- **Shear thinning fluid**
- **Stirred tank**
- **Turbulence**

REFERENCES

1. Tatterson, G. B.; Fluid Mixing and Gas Dispersion in Agitated Tanks. McGraw-Hill, New York; **1991**.
2. Tanguy, P.; Heniche, M.; Rivera., C.; Devals., C.; and Takenaka, K.; Recent developments in CFD applied to viscous and non-Newtonian mixing in agitated vessels. Paper presented at 5th International Conference on CFD in the Process Industries, CSIRO, Melbourne, Australia, **2006**.
3. Tanguy, P. A.; Thibault, F.; Fuente, E. B. L.; Espinosa-Solares, T.; and Tecante, A.; Mixing performance induced by coaxial flat blade-helical ribbon impellers rotating at different speeds. *Chem. Eng. Sci.* **1997**, *52*, 1733–1741.
4. Joshi, J.; Nere, N.; Rane, C.; Murthy, B.; Mathpati, C.; Patwardhan, A.; and Ranade, V.; CFD simulation of stirred tanks: comparison of turbulence models. Part I: radial flow impellers. *Canad. J. Chem. Eng.* **2011**, *89*, 23–82.

5. Montante, G.; Moštek, M.; Jahoda, M.; and Magelli, F.; CFD simulations and experimental alidation of homogenisation curves and mixing time in stirred Newtonian and pseudoplastic liquids. *Chem. Eng. Sci.* **2005**, *60*, 2427–2437.
6. Javed, K. H.; Mahmud, T.; and Zhu, J. M.; Numerical simulation of turbulent batch mixing in a vessel agitated by a Rushton turbine. *Chem. Eng. Process.* **2006**, *45*, 99–112.
7. Ochieng, A.; Onyango, M. S.; Kumar, A.; Kiriamiti, K.; and Musonge, P.; Mixing in a tank stirred by a Rushton turbine at a low clearance. *Chem. Eng. Process.* **2008**, *47*, 842–851.
8. Visuri, O.; Moilanen, P.; and Alopaeus, V.; Comparative studies of local mixing times in water and viscous shear-thinning fluid for three impeller types. 13th European Conference on Mixing, London. **2009**, 14–17.
9. Shekhar, S. M.; and Jayanti, S.; Mixing of pseudoplastic fluids using helical ribbon impellers. *Am. Inst. Chem. Eng. J.* **2003**, *49*, 2768–2772.
10. Venneker, B.; Derksen, J.; and Van den Akker, H. E. A.; Turbulent flow of shear-thinning liquids in stirred tanks - The effects of Reynolds number and flow index. *Chem. Eng. Res. Des.* **2010**, *88*, 827–843.
11. Ansys Fluent 13, Theory Guide, Ansys Inc., U.S.A., **2011**.
12. Metzner, A. B.; and Otto, R. E.; Agitation of non-Newtonian fluids. *Am. Inst. Chem. Eng. J.* **1957**, *3*, 3–10.
13. Zhou, G.; Shi, L.; and Yu, P.; CFD study of mixing process in Rushton turbine stirred tanks. 3rd International Conference on CFD in the Minerals and Process Industries, CSIRO, Melbourne, Australia, 2003.

CHAPTER 17

NATURAL CONVECTION HEAT TRANSFER ENHANCEMENT IN SHELL AND COIL HEAT EXCHANGER USING CUO/WATER NANOFLUID

T. SRINIVAS and A. VENU VINOD*

Department of Chemical Engineering, National Institute of Technology, Warangal-506004 India; *E-mail: avv@nitw.ac.in

CONTENTS

17.1 Introduction 172
17.2 Materials and Methods 173
17.3 Results and Discussion 176
17.4 Conclusions 182
Keywords 183
References 183

17.1 INTRODUCTION

With the advantages of high efficiency in heat transfer and compactness in structure, helical coils are extensively used in engineering applications such as heat exchangers, steam generators and chemical reactors. For these reasons, heat transfer from fluids flowing in helical pipes has become a research subject. As fluid flows through a curved pipe, secondary flows occur in planes normal to the main flow by the action of centrifugal force [1], resulting in a high increase in friction factors and heat transfer coefficients compared with that in an equivalent straight pipe [2].

Rajasekharan et al. [3] have reported higher friction factors and heat-transfer coefficients for non-Newtonian fluids in coiled pipes compared to straight pipes. The addition of nanometer-sized solid metal or metal oxide particles to the base fluids shows an increment in the thermal conductivity of resultant fluids. Nanofluids are important because they can be used in numerous applications involving heat transfer, and other applications such as cooling is one of the most important technical challenges facing many diverse industries, including microelectronics, transportation, solid-state lighting, detergency and manufacturing [4],[5].

Xin and Ebadian [6] reported an experimental study on laminar natural convection heat transfer from helicoidal pipes in air in vertical and horizontal orientations. They developed two correlations for the overall average Nusselt number for the coils oriented vertically or horizontally. Taherian and Allen [7] investigated the natural convection heat transfer on shell-and-coil heat exchanger. In their experiments, the effects of tube diameter, coil diameter, coil surface and shell diameter on the shell-side heat transfer coefficient of shell-and-coil natural convection heat exchanger were studied. Nusselt number was correlated with the Rayleigh number based on the hydraulic diameter of the heat exchanger and the heat-flux on the shell-side. Ali [8] reported experimental investigation of laminar and transition free convection heat transfer from the outer surface of helical pipes with a finite pitch oriented vertically in a 57 percent glycerol-water solution. The solution had a Prandtl number range of 28–36. Prabhanjan et al. [9] presented results of an experimental investigation of natural convection heat transfer from helical coiled tubes in water. They used different characteristic lengths to correlate the outside Reynolds numbers to the Rayleigh number.

Ho et al. [10] experimentally investigated natural convection heat transfer of a nanofluid in vertical square enclosures with different sizes. They reported that usingby nanofluid containing particle fraction of 0.1 vol.%, heat transfer enhancement of around 18 percent compared with that of water was obtained.

In the present work, helical coil and nanofluids have been used for studying natural convection, and to show such a combination of helical coil and nanofluid has a unique practicality in enhancing the heat transfer in natural convection. The study is carried out at two concentrations of nanofluid, Rayleigh numbers, and the Nusselt number.

17.2 MATERIALS AND METHODS

17.2.1 EXPERIMENTAL SETUP

The schematic diagram of the experimental set-up is shown in Figure17.1.

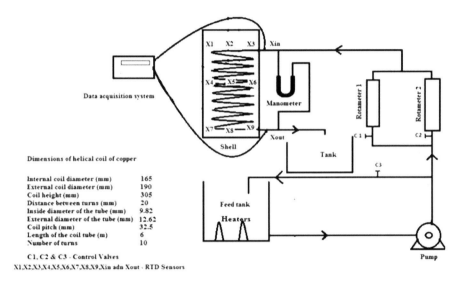

FIGURE 17.1 Schematic diagram of the Experimental set-up.

Helical coil is made of copper tube and has 10 turns. The length of the coil is 6 m. The inner diameter and outer diameter of coil tube i are 0.00982 m (9.82 mm) r and 0.01262 m (12.62 mm) respectively. The tube pitch is 0.0325 m (32.5 mm) and pitch circle diameter (PCD) of the coil is 0.165 m (165 mm). The coil is housed in a stainless steel cylindrical shell. The height and diameter of the shell are 0.42 m (420 mm) and 0.275 m (275 mm) respectively. Two electrical heaters (3 kW each) are used to heat the coil-side liquid in feed tank. Coil side temperature is maintained using a temperature controller. The shell is insulated with glass wool. Temperature measurements were made using PT-100 type RTD sensor (Arrow Instruments, INDIA). Nine sensors are placed at different locations in the shell to measure the shell-side fluid temperature. Hot water from feed tank is pumped through the coil using a magnetic drive pump (0.5 HP, Taha, INDIA). Flow rate of the water through the coil is measured using rotameter (range: 0.5 to 5 lpm; made of glass, SS-316 float, accuracy ± 2%, Make: CVG Technocrafts India). The set-up is provided with a data acquisition system (accuracy: 0.3%; Make: Ace Instruments, INDIA, model: Al-800D) to record all temperatures (inlet and outlet temperatures of water, shell-side fluid temperature). As water flows through the coil, heat is transferred from the coil-side fluid to fluid in the shell.

17.2.2 PREPARATION OF NANOFLUID

CuO/waternanofluid (particle size 40 nm; Sisco Research Laboratories Pvt Ltd. India) is prepared by dispersing required amount of nanopowder in ultra-pure water (Millipore). The SEM image of the nanopowder is shown in Figure17.2.

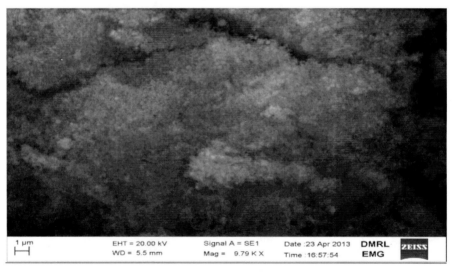

FIGURE 17.2 SEM image of CuO nanopowder.

The nanofluid is subjected to ultrasonic pulses using an Ultrasonicator (Hielscher, UP200H, Germany) for 3 h to get the uniform dispersion and stable suspension. The nanofluid with CuO nanoparticles of concentrations 0.3 percentand 0.6 percent by weight are prepared.

17.2.3 EXPERIMENTAL PROCESS

The shell of the exchanger was filled with ultra-pure water, and the initial water temperature was set at 25°C. Feed tank temperature (40°C) was maintained constant using temperature controller. Magnetic drive pump was switched on, and the flow rate of water through the helical coil was set at 0.5 lpm using rotameter. The experiment was allowed to run till the steady state was established (as indicated by the constant outlet temperature of the coil-side fluid). The procedure was repeated up to 5 lpm in the increments of 0.5 lpm. The experiments were repeated for feed water temperatures of 50 and 60°C, and for the nanofluid of different concentrations on shell-side.

17.2.4 EXPERIMENTAL STUDIES

The effect of the following on heat transfer has been investigated.
1. Effect of concentration of CuO nanoparticles in nanofluid (0%, 0.3% and 0.6% wt.).
2. Effect of coil-side fluid flow rate.
3. Effect of coil-side fluid temperature (40, 50 and 60°C).

2.5 Estimation of thermo-physical properties of CuO/water nanofluid:

The density is calculated from Pak and Cho [11] using the following equation:

$$\rho_{NF} = (1-\varphi)\rho_{BF} + \varphi\rho_{NP} \quad (17.1)$$

The specific heat capacity of nanofluid is estimated by using the equation given by Xuan and Roetzel [12]:

$$(\rho C_P)_{NF} = (1-\varphi)(\rho C_P)_{BF} + \varphi(\rho C_P)_{NP} \quad (17.2)$$

Einstein [13] model has been proposed to correlate viscosity data:

$$\mu_{NF} = \mu_{BF}(1+2.5\varphi) \quad (17.3)$$

The effective thermal conductivity is estimated by using the model proposed by Maxwell[14],

$$k_{NF} = \left[\frac{k_{NP} + 2k_{BF} + \varphi(k_{NP} - k_{BF})}{k_{NP} + 2k_{BF} - \varphi(k_{NP} - k_{BF})}\right]k_{BF} \quad (17.4)$$

17.2.6 DEAN NUMBER

Dean number is calculated using the following equation:

$$De = \text{Re}(d_i / 2R_C)^{0.5} \tag{17.5}$$

$$\text{Re} = d_i v \rho / \mu \tag{17.6}$$

where, R_c is curvature radius of the coil, di is inner diameter of the coil tube, v velocity of the fluid through the coil. Density and viscosity (μ) of the fluid were evaluated at the average of inlet and outlet temperatures of the coil-side fluid.. In the present study, laminar flow is obtained for flow range 0.5 –2.5 lpm (340 <De < 1,900) and turbulent flow for 3–5 lpm (2100 <De < 4,824).

17.2.7 RAYLEIGH NUMBER AND NUSSELT NUMBER

Rayleigh number is calculated using the following equation

$$Ra = \left(\frac{g \beta_{NF} \Delta T L^3}{\eta_{NF} \alpha_{NF}} \right) \tag{17.7}$$

$\Delta T = 0.5 * (T_{Cin} + T_{Cout}) - T_S$ where, T_S is shell-side bulk fluid temperature, L is length of coil tube[15].

Prabhanjan et al.[9] suggested the following relation to determine the Nusselt number for the range of Rayleigh number is 2×10^6 to 3×10^9.

$$Nu = 2.0487 * (Ra^{0.17868}) \tag{17.8}$$

17.3 RESULTS AND DISCUSSION

Natural convective heat transfer in a shell and coil heat exchanger has been studied using ultra-pure water, 0.3% and 0.6% weight concentration of CuO/waternanofluid. The effects of concentration of CuO nanopowder, feed temperature of coil-side fluidand coil-side fluid flow rate (Dean number) on heat transfer have been studied. As no agitation has been employed on the shell side, heat transfer from coil-side to shell side takes place by natural convection. The temperature of the surface of the coil has been measured at three different locations along the along the length of the coil. The temperature difference (driving force) between coil surface and nanofluid (shell-side) has been calculated at three different locations, and the average driving force has been used in the calculation of Rayleigh number.

17.3.1 STEADY STATE TIME

In any continuous operation, start-up time (time required to reach steady state) is crucial. In the present work, the time taken to reach steady state (as indicated by the constant outlet temperature of the coil-side fluid) has been experimentally determined, for various conditions.

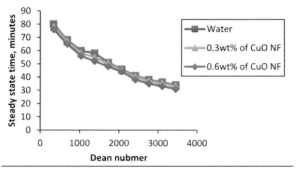

FIGURE 17.3 Variation of time for steady state with Dean number atdifferent CuO nanofluid concentrations (Feed temperature at 40°C)

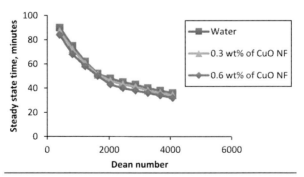

FIGURE 17.4 Variation of time for steady state with Dean number at different CuO nanofluid concentrations (Feed temperature at 50°C)

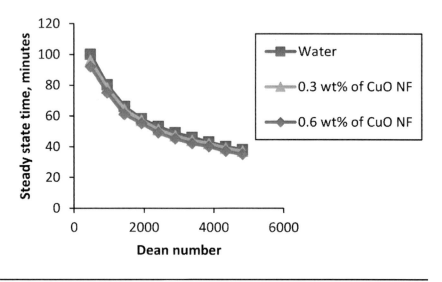

FIGURE 17.5 Variation of time for steady state with Dean number at different CuO nanofluid concentrations (Feed temperature at 60°C)

Figures 17.3, 17.4, and 17.5 present the variation of time required to attain steady state with Dean number at different concentrations of CuO/waternanofluid (0.3–0.6 percent weight). It can be observed from the results, that the steady state time has been found to be decreasing with increase in concentration of CuO/waternanofluid. This can be attributed to better heat transfer due to the use of nanofluid. Maximum reduction in time to reach steady state, when nanofluid is used instead of water, is 10.3 percent corresponding to 0.6% wt nanofluid concentration, 40°C and Dean number of 1383.

17.3.2 RAYLEIGH NUMBER VS NUSSELT NUMBER IN UNSTEADY STATE CONDITION

Rayleigh number and Nusselt number havebeen calculated using equations7 and8 respectively for the conditions prevailing after 15 and 30 min of the start of the experiment. These times correspond to unsteady state heat transfer condition. The results arepresented in Figures 17.6, 17.7, 17.8. 17.9, 17.10, 17.11.

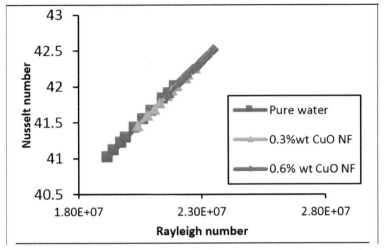

FIGURE 17.6 Variation of *Nu* with *Ra* at feed temperature at 40°C and different CuO nanofluid concentrations (at 15 mins from the start)

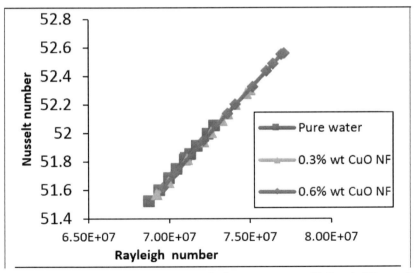

FIGURE 17.7 Variation of *Nu* with *Ra* at feed temperature at 50°C and different CuO nanofluid concentrations (at 15 min from the start)

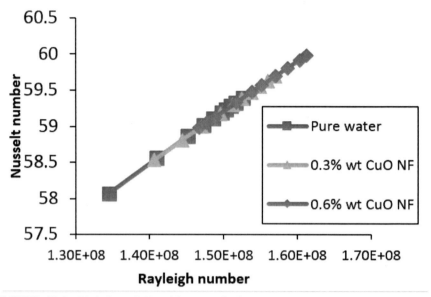

FIGURE 17.8 Variation of Nu with Ra at feed temperature at 60°C and different CuO nanofluid concentrations (at 15 min from the start)

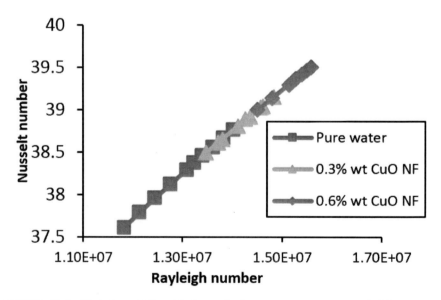

FIGURE 17.9 Variation of Nu with Ra at feed temperature at 40°C and different CuO nanofluid concentrations (at 30 min from the start)

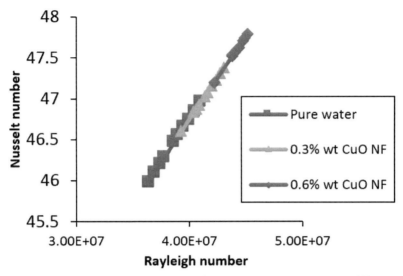

FIGURE 17.10 Variation of *Nu* with *Ra* at feed temperature at 50°C and different CuO nanofluid concentrations (at 30 min from the start)

Nomenclature

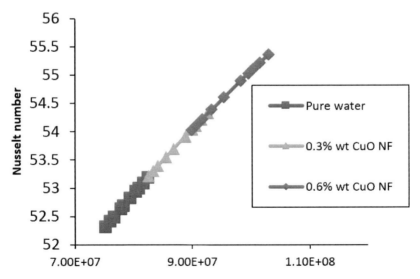

FIGURE 17.11 Variation of *Nu* with *Ra* at feed temperature at 60°C and different CuO nanofluid concentrations (at 30 min from the start)

From the resultsit can be observed that with increase in Rayleigh number, Nusselt number has been found to be increasing. Similar trend was reported previously [16, 17]. Lower values of Nusselt number have been obtained at 30 min of operation compared to those at 15 min. This is due to less driving force (ΔT) available at 30 min of operation. Corresponding to flow rate of 1 lpm and 0.6% wt. nanofluid concentration highest increase of *Nu* is 2.25 percent and 3.58 percent at 15 and 30 min respectively.

17.4 CONCLUSIONS

The effect of CuO/waternanofluid concentration, feed (coil-side) temperature and coil-side fluid flow rate (in terms of Dean number) on natural convection heat transfer has been studied in the shell and helical coil heat exchanger. It was observed that on increasing the nanofluid concentration and Dean number, the heat transfer rate increased, leading to steady state was reached in shorter time. Use of nanofluid resulted in higher Rayleigh number and Nusselt number compared to pure water.

NOMENCLATURE

A	area of heat transfer (m^2)
C_p	specific heat (kJ kg^{-1} K^{-1})
d	diameter (m)
De	Dean number
g	gravity (m/s^2)
k	thermal conductivity (W m^{-1} K^{-1})
L	length of coil tube
Nu	Nusselt number
Ra	Rayleigh number
R_c	curvature radius of the coil (m)
Re	Reynolds number
T	temperature (K)
v	velocity (m / s)

GREEK LETTERS

ΔT	temperature difference (K)
μ	viscosity (kg m^{-1} s^{-1})
r	density (kg m^{-3})
f	volume fraction

β	thermal expansion coefficient (K^{-1})
α	thermal diffusivity (m^2/s)
h	kinematic viscosity

SUBSCRIPT

BF	basefluid
Cin	Feed inlet temperature
Cout	Feed outlet temperature
crit	critical
i	inside
NF	nanofluid
NP	nanoparticle
o	outside
s	shell

KEYWORDS

- **Dean number**
- **Helical coil**
- **Nanofluid**
- **Natural convection**
- **Nusselt number**

REFERENCES

1. Chen, H.; and Zhang, B.; Fluid flow and mixed convection heat transfer in a rotating curved pipe. *Int. J. Therm. Sci.* **2003**, *42*, 1047–1059.
2. Prabhanjan, D. G.; Rennie, T. J.; and Raghavan, G. S. V.; Comparison of heat transfer rates between a straight tube heat exchanger and a helically coiled heat exchanger. Pergamon Press, *Int. commun. Heat. Mass. Transf.* **2002**, *29*(2), 185–191.
3. Rajasekharan, S.; Kurair, V. G.; and Kuloor, N. R.; Heat transfer to non-Newtonian fluids in coiled pipes in laminar flow. Pergamon Press. *Int. J. Heat. Mass. Transf.* **1970**, *13*, 1583–1594.
4. Wang, X.; Xu, X.; and Choi, S.U.S.; Thermal conductivity of Nanoparticle-fluid mixture. *J. Thermo-phys. Heat. Transf.* **1999**, *13*(4), 474–480.
5. Murshed, S. M. S.; Leong, K.C.; and Yang, C.; Enhanced thermal conductivity of TiO$_2$-water based nanofluids. *Int. J. Ther. Sci.* **2005**, *44*(4), 367–373.
6. Xin, R. C.; and Ebadian, M.A.; Natural Convection Heat Transfer from Helicoidal Pipes. *J. Therm. Phys. Heat. Transf.* **1996**, *12*(2), 297–302.

7. Taherian, H.; and Peter L. A.; Experimental study of natural convection shell-and-coil heat exchanger. *Am. Soc. Mech. Eng.* **1998**, *357*(2), ASME Proceeding of the 7th. AIAA/ASME.
8. Ali, M. E.; (2004) Free convection heat transfer from the outer sur- face of vertically oriented helical coils in glycerol-water solution.*Heat. Mass. Transfer.40*(8), 615–620.
9. Prabhanjan, D. G.; Rennie, T. J.; and Raghavan, G. S. V.; Natural Convection Heat Transfer from Helical Coiled Tubes. *Int. J. Ther. Sci*, *43*, 359–365.
10. Ho, C. J.; Liu, W. K.; Chang, Y. S.; and Lin, C. C.; Natural convection heat transfer of alumina-water nanofluid in vertical square enclosures: An experimental study. *Int. J. Therm. Sciences.* **2010**, *49*, 1345–1353.
11. Pak, B. C. and Cho, Y. I.; Hydrodynamic and heat transfer study of dispersed fluids with submicron metallic oxide particles, Experimental Heat Transfer. *A J. Ther. Ener. Gener. Transp., Stor., Conver.* **1998**, *11*(2), 151–170.
12. Xuan, Y.; and Roetzel, W.; Conceptions for heat transfer correlation of nanofluids. *Int. J. Heat Mass. Transf.* **2000**, *43*(19), 3701–3707.
13. Einstein, A.; *Annalen der Physik.* **1906**, *19*, 289–306.
14. Maxwell, J. C.; Electricity and magnetism. Oxford, UK: Clarendon.; **1873**.
15. Ali, M. E.; Natural convection heat transfer from vertical helical coils in oil. *Heat. Transf. Eng.* 27(3), 79- 85.
16. Abu-Nada, E.; and Oztop, H.F.; Effects of inclination angle on natural convection in enclosures filled with Cu–water nanofluid. *Int. J. of Heat and Fluid Flow, 30*, 669–678.
17. Ho, C. J.; Liu, W.K.;Chang, Y. S.; and Lin, C. C.; Natural convection heat transfer of alumina-water nanofluid in vertical square enclosures: An experimental study. *Int. J. Therm. Sci.* **2010**, *49*, 1345–1353.

CHAPTER 18

HOMOLOGY MODELING OF L-ASPARAGINASE ENZYME FROM *ENTEROBACTOR AEROGENES* KCTC2190

SATISH BABU RAJULAPATI and RAJESWARA REDDY ERVA*

Department of Biotechnology, National Institute of Technology, Warangal - 506004, India; *E-mail: rajeshreddy.bio@gmail.com

CONTENTS

18.1 Introduction 186
18.2 Materials and Methods 186
18.3 Results and Discussion 188
18.4 Conclusion 193
Keywords 193
References 193

18.1 INTRODUCTION

L-Asparaginase (EC 3.5.1.1) a wide spread enzyme can be found in many microbes like *Aerobacter, Bacillus, Pseudomonas, Serratia, Xanthomonas, Photobacterium* [1], *Streptomyces* [2], *Proteus* [3], *Vibrio* [4], and *Aspergillus* [5], etc., The medically important enzyme hydrolyze L-asparagine (an essential amino acid) to aspartic acid and ammonia. For protein synthesis in many types of cancer cells L-asparagine is essential and they are deprived of L-asparagine in presence of L-Asparaginase, thus resulting in cytotoxicity of tumor cells. All L-Asparaginase do not possess antitumor properties which seem to be related to the affinity of the enzyme for the substrate and factors affecting the clearance rate from the system [6]. Particularly against acute lymphoblastic leukemia L-Asparaginase from *E. coli* and *Erwinia carotovora* possess antitumor activity [7]. But, administering such enzyme protein for long duration generally produces the corresponding antibody in the tissues, resulting in anaphylactic shock and may neutralize the drug effect. So, there is a need for the search of new serologically different L-Asparaginase with similar therapeutic effect is highly desired. The present study is aimed at homology modeling of 3-D Structure of L-Asparaginase from *Enterobacter Aerogenes KCTC2190* using Swiss-Model [8–10]. The constructed model was further validated by Ramachandran plot. The model was again refined and subjected to energy minimization using CHIRON server and evaluated for quality assessment. Chiron performs rapid energy minimization of protein molecules using discrete molecular dynamics with an all-atom representation for each residue in the protein.

18.2 MATERIALS AND METHODS

18.2.1 SEQUENCE RETRIEVAL AND PROTEIN STRUCTURE PREDICTION

3D-Structure of L-Asparaginase was performed using Swiss-Model server. The amino acid sequence for L-Asparaginase enzyme was retrieved from the results of GENSCAN for the nucleic acid sequence of asnA gene from *Enterobactor Aerogenes* KCTC2190 (NCBI accession number: NC_015663). Swiss-Model is a fully automated protein structure homology-modeling server. The Swiss-Model is a web-based integrated service dedicated to protein structure homology modeling. Building a homology model comprises four main steps: identification of structural template(s), alignment of target sequence and template structure(s), model building, and model quality evaluation. These steps can be repeated until a satisfying modeling result is achieved. The backbone conformation of the modeled structure was calculated by analyzing the phi (ø) and psi (ψ) torsion angles using Rampage [12].

18.2.2 MODEL ASSESSMENT AND MOLECULAR DYNAMICS SIMULATIONS

Model assessment and molecular dynamics simulations were done by using the ANOLEA, QMEAN and DFire & GROMOS respectively.

The atomic empirical mean force potential ANOLEA [12] is used to assess packing quality of the models. The program performs energy calculations on a protein chain, evaluating the "Non-Local Environment" (NLE) of each heavy atom in the molecule. The y-axis of the plot represents the energy for each amino acid of the protein chain. Negative energy values (in green) represent favorable energy environment whereas positive values (in red) unfavorable energy environment for a given amino acid.

QMEAN is a composite scoring function for both the estimation of the global quality of the entire model as well as for the local per-residue analysis of different regions within a model. The global QMEAN6 scoring function is a linear combination of six structural descriptors using statistical potentials. The local geometry is analyzed by a torsion angle potential over three consecutive amino acids. Two distance-dependent interaction potentials are used to assess long-range interactions: the first is a residue-level implementation based on C-beta atoms only and the second an all-atom potential which is able to capture more details of the model. A solvation potential investigates the burial status of the residues. QMEAN6 is a reliability score for the whole model which can be used in order to compare and rank alternative models of the same target. The quality estimate ranges between 0 and 1 with higher values for better models. In addition to the raw scores, Z-scores of the QMEAN composite score as well as all terms are provided relating the quality estimates to scores obtained for high-resolution reference structures solved experimentally by X-ray crystallography [13]. The QMEAN Z-score represents the measure of the absolute quality of a model by providing an estimate of the degree of nativeness of the structural features observed in a model and by describing the likelihood that a given model is of comparable quality to experimental structures. Models of low quality are expected to have strongly negative QMEAN Z-scores (i.e. the model's QMEAN score is several standard deviations lower than expected for experimental structures of similar size). The analysis of the Z-scores of individual terms may help identifying the geometrical features responsible for an observed negative QMEAN Z-score.

GROMOS [14] is a general-purpose molecular dynamics computer simulation package for the study of bio molecular systems and can be applied to the analysis of conformations obtained by experiment or by computer simulation. The y-axis of the plot represents the energy for each amino acid of the protein chain. Negative energy values (in green) represent favorable energy environment whereas, positive values (in red) unfavorable energy environment for a given amino acid.

DFire [15] is an all-atom statistical potential based on a distance-scaled finite ideal-gas reference state. DFire is used to assess nonbonded atomic interactions in the protein model. Pseudo energy for the entire model is provided which reflects the quality of the model and can be used for ranking alternative predictions of the same target. A lower energy indicates that a model is closer to the native conformation. Then it was energy minimized using CHIRON tool (The energy after minimization is 300.007 Kcal/mol).

18.2.3 PROTEIN ACTIVE SITE PREDICTION

SCF Bio (Super Computing facility for Bioinformatics and Computational Biology IIT Delhi) [16] server was used to predict the structure-based protein Active Sites.

23.3 RESULTS AND DISCUSSION

23.3.1 SEQUENCE RETRIEVAL AND PROTEIN STRUCTURE PREDICTION

The Swiss-Model is a web-based integrated service dedicated to protein structure homology modeling. The server searched for structural homolog for L-Asparaginase using Gapped-BLAST, and the domains with structural homologs follow a template model (2p2d). The final structure prediction was done using 2p2d as template model (Figure 18.1).

The modeled L-Asparaginase enzyme was further subjected for optimization and validation. Geometric evaluations of the modeled 3D structure was performed using Rampage to get Ramachandran plot (Figure 18.2).

Homology Modeling of L-Asparaginase Enzyme

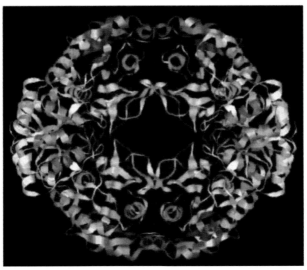

FIGURE 18.1 Modeled structure of L-Asparaginase

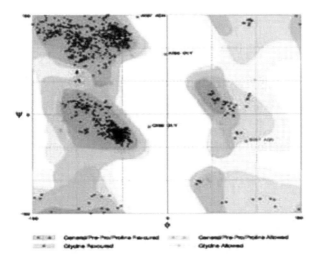

FIGURE 18.2 Ramachandran plot for model

The plot shows that 97.0 percent of residues were found in the favored and 2.7 percent allowed regions and 0.3 percent were in the outlier region. Presence of 99.7 percent of residues in favored and allowed regions confirms our model as a relatively good one.

18.3.2 MODEL ASSESSMENT AND MOLECULAR DYNAMICS SIMULATIONS

Molecular dynamics aimed to explain protein structure and function problems, such as, structural stability, folding, and conformational flexibility which includes, study of bio molecular systems by GROMOS. The model assessment is done by assessing packing quality of the models by ANOLEA, monitoring the local quality of the modeled enzyme by QMEAN, and the assessment of non-bonded atomic interactions in the model by DFire. As Models of low quality are expected to have strongly negative QMEAN Z-scores, results of the modeled enzyme support our modeled structure as a relatively good model as the QMEAN Z-score is 0.09.

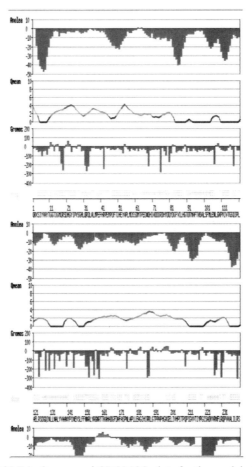

FIGURE 18.3 ANOLEA, Qmean and GROMOS plots for the modeled enzyme

As the DFire energy is very low (i.e., −2091.25), this indicates that a model is closer to the native conformation. The y-axis of the GROMOS plot represents the energy for each amino acid of the protein chain. Much of the portion in the plot is in green color (Negative energy values) representing us the favorable energy environment for a given amino acid and the same is the case even with ANOLEA, QMEAN plots (Figure 18.3).

18.3.3 PREDICTION OF ACTIVE SITES

Active site prediction in the modeled L-Asparaginase enzyme was done by SCF Bio (Super Computing facility for bioinformatics & Computational Biology IIT Delhi) "Active Site Prediction of Protein server" computes the cavities in a given protein. All together the server detected 79 putative active site cavities along with their amino acid residues in the modeled enzyme (Table 18.1).

TABLE 18.1 Active Sites present in the modeled L-Asparaginase

Cavity No.	Aminoacids	Cavity No.	Aminoacids	Cavity No.	Aminoacids
1	RLHIGVFEPQAYDSNT	28	EPRHLYQASM	55	LREQMIPF
2	AERYNGFLIQTDSH	29	DFHVIYQETMAGPS	56	FATRVYLINPE
3	VPTEMAHFDKLSYNGQCIR	30	GHNIPEQALRVDYSF	57	GDKSQMYTFEAPHI
4	DHYGRNAMEITQFVPKSLC	31	SVEFARYQLHNPIGK	58	NAKRFTYSPL
5	GNMHTALYKDQVIRPFESC	32	YLISGNVPFERTAKH	59	IGYPLRF
6	GNFMHTKDSIACQYRPEV	33	ILFGPQENAVDSYTK	60	DALRGIQVSKP
7	QIMEDNFGPLARSVYH	34	DTFLKRGQNSMVHPC	61	LTGRNAYPSF
8	YIHREGVLQFADSNM	35	KHGNPERAYLTIVS	62	KGYLPDRVAIF
9	YINLMREGPFASQD	36	SYKDTAPFEVMILHQ	63	PRNGTALIEY
10	NTSPHFCAMRQVELYGD	37	DTFYAVIHQELRGS	64	YIPSAFNLQVGM

TABLE 18.1 *(Continued)*

Cavity No.	Aminoacids	Cavity No.	Aminoacids	Cavity No.	Aminoacids
11	FMSRTENKVGDA-LIPHYCQ	38	TASNFVLQYIRG-PHEK	65	ANDLGIR-SKQVP
12	ITQPWDEFYVMG-SANL	39	ALVGTINQSMRFD	66	GNKPEI-RAYLV
13	QINDSPGEMY-ATFH	40	TKQPEVDWSNG-MYAIF	67	SFGHIDN-LTVA
14	IHYTMGDL-PASKQVN	41	VFMGLANSIQDR	68	SPHNG-ALRDTK
15	QRFIDSGMAVNT-LYKP	42	EHYITVDF	69	AVLTN-FGIMDS
16	VTFRCYGQMAD-HESILPNK	43	SRIELGYHN	70	THAIDFR
17	GNTMLSPV-CAERKQFHYID	44	TASWVEPFMYKIR-GQDCN	71	QIFPYSVLH
18	REPYQLMVIHASF	45	LPTVEFAMKRQ-GYHCDSI	72	ATIYVM-HLSGD
19	ELQAPMRSHIVY	46	YVPIRQLTKMS-GNDA	73	VFALN
20	KANYPVEIMRLH-FGSQ	47	QAFLMEIHVTRD-PKYS	74	RLIVFTNY
21	HLQAEMPRGIDT-SY	48	PNIAEMGYTSDV-LHQK	75	HEQITDFP-SRV
22	YAITPMQNVDKS-GLFEH	49	GMAQNSITVLDP	76	DYSPMTHI
23	TYINPSDMVL-GAHE	50	VGINSLAEPRTY-DKH	77	PILVNTRY
24	HAREGIFNQST-DLY	51	LRYDHVIAGT	78	AQVEIRL
25	PEDWTSMGYAIN-FQL	52	TSHMDGNPKLVRF	79	VKGLI
26	GKENILH-PYRAVFS	53	ELHAQNYIRF		
27	QMRLGAIYHFE-VTKPDS	54	NQFRGMISAVL		

18.4 CONCLUSION

The three-dimensional model of L-Asparaginase enzyme has been generated using Swiss-Model server. The model generated by using the template 2p2d proved to be the best model generated as compared to the other templates. The modeled L-Asparaginase enzyme was further validated using Rampage by calculating the Ramachandran plot. The plot (Figure-18.2) of our model shows that 97.0 percent of residues were found in the favored and 2.7 percent allowed regions and 0.3 percent were in the outlier region. The low DFire energy, positive QMEAN Z-Scores, plots of ANOLEA & GROMOS indicate that, as expected, the 3D structural model of L-Asparaginase enzyme represents a native conformation. In future the present study would aid in detailed molecular mechanism of L-Asparaginase enzyme in cancer therapy.

KEYWORDS

- L-Asparaginase
- Optimization
- Rampage
- Swiss-Model
- Validation

REFERENCES

1. Peterson, R. E. and Ciegler, A.; L-Asparaginase Production by Various Bacteria. *Appl. Microbiol.* **1969**, 929–930.
2. DeJong, P. J. et al.; L-Asparaginase production by Streptomyces griseus. *Appl. Environ. Microbiology.* 1972, 1163–1164.
3. Tosa, T.; Sano, R.; Yamamoto, K.; Nakamura, M.; Ando, K.; and Chibata, I.; L-Asparaginase from *Proteus vulgaris. Environ. Microbiol.* **1971**, 387–392.
4. Kafkewitz, D.; and Goodman, D.; *Appl. Environ. Microbiol.* **1974**, 206–209.
5. Sarquis MIdeMoura, Oliveira, E. M. M.; Santos, A. S.; and daCosta, G. L.; L-Asparaginase Production by he Rumen Anaerobe. *Vibrio succinogenes. Memórias do Instituto Oswaldo Cru.* **2004**, 489–492.
6. Cornea, C. P.; Lupescu, I.; Vatafu, I.; Caraiani, T.; Savoiu, V. G.; Campeanu, G. h.; Grebenisan, I.; and Negulescu, Gh.P.; Constantinescu D.; Production of l-Asparaginase II by recombinant Escherichia coli cells. *Roumanian. Biotechnol. Lett.* **2002**, 717–722.
7. Mashburn, L. T. and Wriston, J. C. (Jr.); Production of L-Asparaginase II by Recombinant *Escherichia coli* cells. *Arch. Biochem. Biophys.* **1964**, 450–452.

8. Arnold, K.; Bordoli, L.; Kopp, J.; and Schwede, T.; The SWISS-MODEL workspace: a web-based environment for protein structure homology modeling. *Bioinformatics.* **2006**, 195–201.
9. Schwede, T.; Kopp, J.; Guex, N.; and Peitsch, M. C.; et al.; SWISS-MODEL: an automated protein homology-modeling server. *Nucl. Acid. Res.* **2003**, 3381–3385.
10. Guex, N.; and Peitsch, M. C.; SWISS-MODEL and the Swiss-Pdb Viewer: an environment for comparative protein modeling. *Electrophoresis.* **1997,** 2714–2723.
11. Lovell, S. C.; Davis, I. W.; Arendall III, W. B.; de Bakker, P. I. W.; Word, J. M.; Prisant, M. G.; Richardson J. S.; and Richardson, D. C.; Structure validation by Cα geometry: ф, ψ and Cβ deviation. *Proteins.* **2002**, 437–450.
12. Melo, F.; and Feytmans, E.; Assessing protein structures with a non-local atomic interaction energy. *J. Mol. Biol.* **1998**, *277*(5), 1141–1152.
13. Benkert, P.; Biasini, M.; Schwede. T.; Toward the estimation of the absolute quality of individual protein structure Models. *Bioinformatics.* **2011**, 343–350.
14. Van Gunsteren, W. F.; Billeter, S. R.; Biomolecular Simulation: The GROMOS96 manual and user guide. *Zürich: VdF Hochschulverlag ETHZ.* **1996**.
15. Zhou, H.; and Zhou, Y.; Distance-scaled, finite ideal-gas reference state improves structure-derived potentials of mean force for structure selection and stability prediction. *Prot. Sci.* **2002**, 2714–2726.
16. Tanya Singh, D.; Biswas, B.; Jayaram.; AADS-An Automated Active Site Identification, Docking, and Scoring Protocol for Protein Targets Based on Physicochemical Descriptors. *J. Chem. Inf. Model.* **2011**, 2515–2527.
17. Ramachandran, S.; Kota, P.; Ding, F.; and Dokholyan.; Structure, Function and Bioinformatics. *Proteins.* **2011**, *79*: 261–270.

CHAPTER 19

MIXING OF BINARY MIXTURE IN A SPOUT-FLUID BED

B. SUJAN KUMAR[1] and A. VENU VINOD[2*]

[1,2]Department of Chemical Engineering, National Institute of Technology, Warangal, India; *E-mail: avv122@yahoo.com

CONTENTS

19.1 Introduction 196
19.2 Experimentation 196
19.3 Results and Discussion 198
19.4 Conclusions 204
Keywords 205
Nomenclature 205
References 205

19.1 INTRODUCTION

In many industrial applications of dense gas–solid fluidized beds, mixing and segregation phenomena play a very important role [1]. The hydrodynamic behavior of binary fluidized beds is strongly influenced by the difference inproperties of the respective particles, especially in density and size [2]. However, in industrial solids mixing, it is often required to mix particles differing widely in physical properties viz., size, density and/or shape.Rice et al. [3], Hoffmann et al. [4], Wu et al. [5], Sahoo et al. [6] and Bokkers et al. [7] investigated the mixing and segregation of the narrowsize distribution of binary particlesin fluidized beds.

Combining the features of both fluidized and spouted beds Chatterjee [8] proposed a novel gas-solid contactor known as spout fluidized bed or spout-fluid bed to overcome the limitations and drawbacks of individual systems. For a size-variant, equal-density system of particles, mixing index hasbeen proposed by Fan et al. [9], using two more system parameters such as, the bed aspect ratio and the ratio of the height of theparticles layer in the bed (from where sample is drawn) to column diameter. The model proposed by Fan et al. has been modified by Sahoo et al. [6] as follows Eq. (19.1) for static bed and Eq. (19.2) for fluidized bed condition). Equation (19.3) is used for spout-fluidization condition to calculate the mixing index [10].

$$I_M = 0.7554 * \left(\frac{\overline{d_p}}{d_F}\right)^{0.144} * \left(\frac{h_B}{D_C}\right)^{-0.1322} * \left(\frac{H_S}{D_C}\right)^{0.1596} * \left(\frac{U}{U-U_F}\right)^{0.3111} \quad (19.1)$$

$$I_M = 0.3725 * \left(\frac{\overline{d_p}}{d_F}\right)^{0.3679} * \left(\frac{h_B}{D_C}\right)^{-0.4864} * \left(\frac{H_S}{D_C}\right)^{0.8258} * \left(\frac{U}{U-U_F}\right)^{0.5084} \quad (19.2)$$

$$I_M = \frac{X^*}{\overline{X}_{bed}} \quad (19.3)$$

19.2 EXPERIMENTATION

A mixture of glass beads of density 2,600 kg/m³ having particle sizes of 0.0017 m and 0.00075 m is fluidized in a Perspex column (Figure 19.1) of 0.094 m internal diameter, 0.1 m outer diameter and 1.217 m height.

Mixing of Binary Mixture in a Spout-Fluid Bed

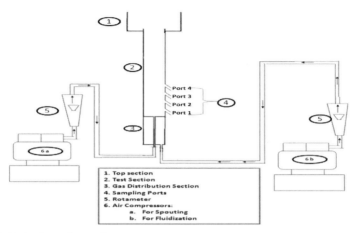

FIGURE 19.1 Experimental setup

The above-mentioned two particle sizes have been mixed in the weight ratio of 50:50 (total 1,000 g) with different initial particle bed arrangements (well-mixed, segregated). The static bed heights for particle sizes 0.0017 m and 0.00075 m are 0.04 m and 0.05 m respectively. The gas distributor section is attached to the test section, at the bottom and is designed in such a manner that uniformly distributed gas enters the test section. The cylindrical gas distributor section is made of Perspex and is 0.302 m in height, 0.094 m in internal diameter and has an outer diameter of 0.1 m. A perforated plate (distributor) made of SS and covered with a 50 mesh (BSS) stainless steel screen with 112 holes of 0.002 m diameter intriangular pitch of 0.008 m (for obtaining fluidization), is placed between test and distributor sectionslocated just below the screen.The gas distributor plate has a spout inlet diameter of 0.012 m located at the centre of the plate with spout armlength 0.076 m. The top section of the column is a cylinder with 0.18 m internal diameter and 0.202 m height. The top section allows gas to escape and is used to collect particles carried over when high velocities are used. The three sections are connected to each other with flange type arrangement.

Three types of experiments were carried out to determine the mixing index across the column with different initial particle bed arrangements (well-mixed, segregated). In segregated particle bed arrangement, layer wise (small over large and large over small) arrangement (Figure19.2) was employed with (i) only spouting, S (allowing only spout gasand no background gas (for fluidization)), (ii) only fluidization, F (background gas supply, no spout gas), and (iii) spout-fluidization, S-F was considered.

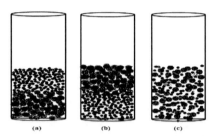

FIGURE 19.2 Schematic representation of different particle bed arrangements: (a) small-large (b) large-small (c) well-mixed

The samples drawn at different heights for both the static and fluidized bed conditions have been analyzed for the distribution of particle sizes across the columnat different axial positions (which are equispaced along the length of the test section with 0.025 m distance) for the bed arrangements mentioned above. One study by Roy and Sahoo[6] used horizontal ports for withdrawing samples from the column. For easy flow of particles from the column, in the present study, ports were provided at an angle of 45° to the column.

19.3 RESULTS AND DISCUSSION

All the experiments were conducted for three different flow conditions (S, F and S-F) witha gas velocity of 20.9 m/s with different particle bed arrangements. Mixing index was calculated using Eqs. (19.1) and (19.2) for only spouting and only fluidization conditions and Eq. (19.3) is used for spout-fluidization.Figures 19.3, 19.4, and 19.5 indicate the bed mixing index as a function of sampling time at first port.

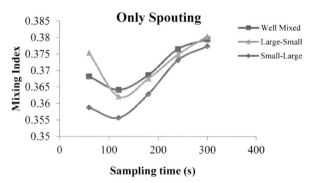

FIGURE 19.3 Mixing Index as a function of time for different bed arrangements (Only Spouting)

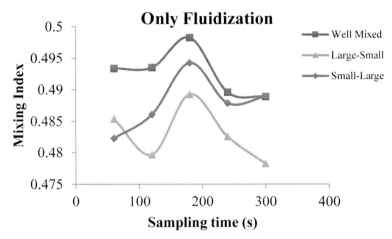

FIGURE 19.4 Mixing Index as a function of time for different bed arrangements (Only Fluidization)

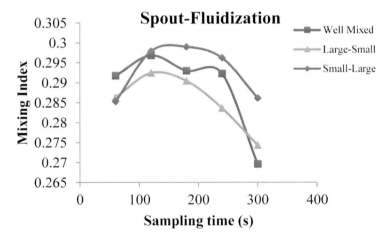

FIGURE 19.5 Mixing Index as a function of time for different bed arrangements (Spout-Fluidization)

In case of only spouting (Figure 19.3), mixing index decreases and then increases as time proceeds for all the three bed arrangements. Mixing index is found to be highest for well-mixed condition, because the samples are already well-mixed. Large-small arrangement is having high mixing index as compared to small-large bed arrangement, because the bottom layer of small particles fluidizes and mixes well with the upper layer of particles as the upper layer quickly interchanges the position because of their weight.In case of only fluidization (Figure 19.4), mixing index fluctuates with time.It is high in case of well-mixed condition. Large-Small arrangement haslow mixing index as compared to small-large bed arrangement, as the bottom layer of large particles fluidizes and mixes well with the upper layer of particles and settles down as time proceeds because of their weight.In case of spout-fluidization (Figure19.5), mixing index increases and then starts decreasing as time proceeds.The mixing index is high in case of small-large condition at higher sampling time and large for well-mixed condition at lower sampling time. In this case, large-small arrangement shows low mixing index, as the bottom layer of large particles fluidizes and mixes well with the upper layer of particles and settles down as time proceeds because of their weight. The small-large bed arrangement shows high mixing index value as time proceeds because the particles in the layers start interchanging their positions where as in the other two bed arrangements particles starts segregating.

Figure 19.6 indicates the mixing index as a function of sampling time. This is plotted for the samples collected at first port for well-mixed bed arrangement. Similar profiles were obtained in the case of small-large and large-small bed arrangements. Mixing index is high for the case of only fluidization, low for the case of spout-fluidization for all bed arrangements.This is due to the segregation of particles, which was observed in case of spout-fluidization at any sampling time. Segregation was less (better mixing) in only fluidization condition when compared to other flow conditions.

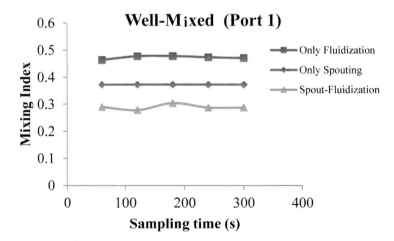

FIGURE 19.6 Mixing Index as a function of time for different flow conditions for well-mixed bed arrangement

Figures 19.7, 19.8, and 19.9 shows axial variation of mixing index as a function of sampling time for only spouting condition. Mixing index fluctuates as time proceeds at all sampling locations. Mixing index is high at port 3 at smaller times.

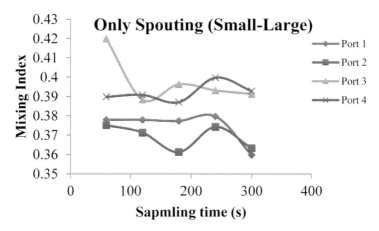

FIGURE 19.7 Mixing Index as a function of time for Only Spouting conditions for Small-Large bed arrangement at different Axial positions

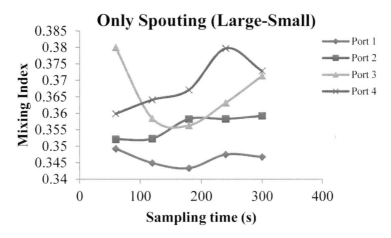

FIGURE 19.8 Mixing Index as a function of time for Only Spouting conditions for large-small bed arrangement at different axial positions

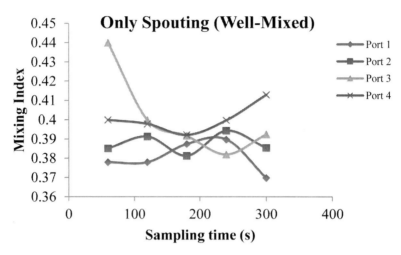

FIGURE 19.9 Mixing index as a function of time for only spouting conditions for well-mixed bed arrangement at different axial positions

The Figures 19.7 and 19.8 have been plotted for the case of layer-wise (small-large or large-small) bed arrangement. In this arrangement, the interface of the two layers is close to the port 3, because of which the sample from the port would contain particles of both sizes, giving rise to better mixing. Ports 1 and 2 shows same index at higher sampling time. In case of well-mixed condition (Figure 19.9), mixing index is found to be very less as compared to other two bed arrangements (Figure 19.7 and 19.8), this is mainly because the samples are already well-mixed and the particles started segregating. Similar behavior was observed in case of only fluidization.

Figure 19.10 shows mixing behavior for spout-fluidization condition. This was plotted for two different axiallocations (port 1 and 2). In case of large-small bed arrangement, mixing index is increasing at port 2 but it is fluctuating at port 1 as time proceeds. This is mainly because of the reason that at port 1 the particle density is high and particles start segregating as time proceeds. VA

Figures 19.11, 19.12, and 19.13 indicate mixing index as a function of axial position for static bed condition (after the experiment) for different experimentation times (5 min (Figure 19.11), 3 min (Figure 12) and 1 min (Figure 19.13). The bed is divided into three equal parts and the particles are collected using respective collection ports provided. For all bed arrangements, it is observed that the mixing index is high for only fluidization condition for long time experimentation and for shorter time of experimentation the spout-fluidization gives good mixing.

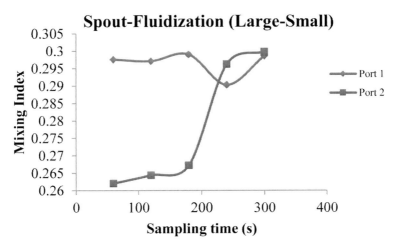

FIGURE 19.10 Mixing index as a function of time for spout-fluidization conditions for large-small bed arrangement at different axial positions

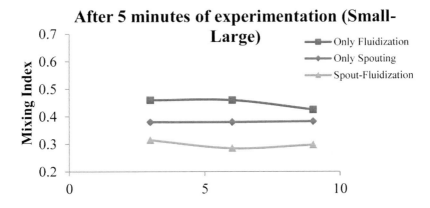

Axial sapmling location from bottom of the bed (cm)

FIGURE 19.11 Mixing index as a function of axial location for different flow conditions for small-large bed arrangement (5 min of experimentation)

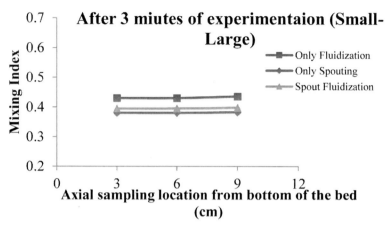

FIGURE 19.12 Mixing index as a function of axial location for different flow conditions for small-large bed arrangement (3 min of experimentation)

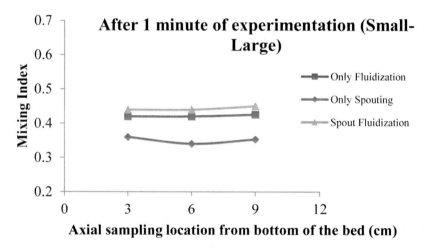

FIGURE 19.13 Mixing index as a function of axial location for different flow conditions for small-large bed arrangement (1 min of experimentation)

19.4 CONCLUSIONS

As the time proceeds mixing index increases in case of only fluidization condition where as it is high at smaller times for spout-fluidization condition. The well-mixed bed arrangement shows high mixing index in case of only spouting and only fluidization conditions. Only spouting condition with large-small bed arrangement and in only fluidization condition with small-large bed arrangement gave high mixing index values. For spout-fluidization condition, small-large bed arrangement showed higher mixing index values.For all bed arrangements in static bed condition, it is observed that the mixing index is high for only fluidization condition for longer time of experimentation whereas for shorter time of experimentation the spout-fluidization gives good mixing.

KEYWORDS

- **Binary Mixture**
- **Fluidization**
- **Mixing Index**
- **Spout-Fluid Bed**
- **Spouting**

NOMENCLATURE

d_F : Diameter of the flotsam particle [m]
: Average particle size of the mixture [m]
D_c : Diameter of the column [m]
h_B : Height of particles layer in the bed from the distributor [m]
H_s : Initial static bed height [m]
I_M : Mixing index, dimensionless
U : Superficial velocity of the fluidizing medium [m/s]
U_F : Minimum fluidization velocity of the flotsam particles [m/s]
X^* : Percentage of jetsam particle in any layer
: Percentage of jetsam particle in the bed

REFERENCES

1. Zhang, Y.; Jin, B.; and Zhong, W.; Experimental investigation on mixing and segregation behavior of biomass particle in fluidized bed. *Chem. Eng. Proc.* **2009**, *48*, 745–754.
2. Rice, R. W.; and Brainovich, J. F.; Mixing/segregation in two- and three-dimensional fluidized beds: binary systems of equidensity spherical particles.*AIChE. J.* **1986**, *32*, 7–16.
3. Hoffmann, A. C.; and Romp, E. J.; Segregation in a fluidized powder of continuous size distribution. *Powd. Technol.* **1991**, *66*, 119–126.
4. Wu, S. Y.; and Baeyens, J.; Segregation by size difference in gas fluidized beds. *Powd. Technol.* **1998**, *98*, 139–150.

5. Marzocchella, P.; Salatino, V. D.; and Pastena, L. Lirer; Transient fluidization and segregation of binary mixtures of particles.*AIChE.J.***2000**, *46*, 2175–2182.
6. Sahoo, A.; and Roy, G. K.; Mixing characteristic of homogeneous binary mixture of regular particles in a gas–solid fluidized bed.*Powd. Technol.* **2005**, *159*, 150–154.
7. Chatterjee, A.; Spout-fluid bed technique. *Ind. Eng. Chem. Proc. Des. and Dev.* **1970**, *9*, 340–341.
8. Bokkers, G.A.; Sint Annaland, M.; van. Kuipers, J. A. M.; Mixing and segregation in a bidisperse gas–solid fluidised bed: a numerical and experimental study. *Powd. Technol.* **2004**, *140*, 176–186.
9. Fan, L.T.; Chen, Y.; and Lai, F.S.; *Powd. Technol.* **1990**, *61*, p 255.
10. Naimer, N.; Chiba,T.; and Nienow, A.W.; *Chem. Eng. Sci.* **1982,** *37*, 1047.

CHAPTER 20

EFFECT OF VARIOUS PARAMETERS ON CONTINUOUS FLUIDIZED BED DRYING OF SOLIDS

G. SRINIVAS[1] and Y. PYDI SETTY[2*]

[1,2]Department of Chemical Engineering, National Institute of Technology Warangal, Warangal, India-506004; *E-mail: psetty@nitw.ac.in

CONTENTS

20.1 Introduction 208
20.2 Experimental Setup and Procedure 208
20.3 Results and Discussion.. 209
20.4 Conclusions 213
Keywords 214
References... 214

20.1 INTRODUCTION

Drying, generally occurs due to heating of moist material and with the surrounding air humidity being less compared to its saturation level. Heat can be supplied to the material to be dried by conduction, convection, and radiation [1]. Most of the drying equipments fall in the category of the conduction and convection type. In convection type, the heating medium, usually air or the flue gas of combustion comes in direct contact with the wet material. In conduction type, heat is transmitted indirectly by contact of the wet material with a hot surface. In radiation, heat is transmitted directly and solely from a heated body to the wet material by radiation of heat.

The present study involves fluidized bed drying in convection and conduction type mainly. Heat to the flowing air is supplied through the wall surface of the fluidized bed dryer. This heat can be transferred to the wet material by convection and conduction effectively across and along a flowing air medium [1].

Continuous fluidized bed drying is the most difficult operation in processing of large amounts of solids in all types of process industries. Continuous drying behavior has a many influencing parameters such as, air velocity, temperature, initial moisture content of solids and solids flow rate. It is necessary to investigate the minimum operating conditions to get a desired uniform product with respect to various parameters. The minimum fluidization velocity has been found from equation suggested by Kunni and Levenspiel [2] for uniformly sized particles of size 2.6 mm which belongs to Geldart group D according to Geldart classification [3]. In the present study air at above minimum fluidization velocity ranging from 2.13 to 2.98 m/s with varying initial moisture content from 4 to 10 percent at three different temperatures ranging from 40 to 60°C at three different solids flow rates ranging from 4 to 12 kg/h.

20.2 EXPERIMENTAL SETUP AND PROCEDURE

A schematic experimental setup is shown in Figure 20.1. Air at room temperature has been collected in a compressor and with the help of rotameter air at known velocity has been passed in to the fluidized bed drying zone through calming section.

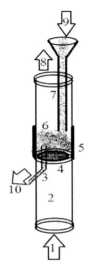

FIGURE 20.1 Experimental setup.

1. Air inlet
2. Calming section
3. Downcomer
4. Flange with distributor plate
5. Heater with insultion
6. Fluidized bed
7. Feed supporting pipe
8. Air outlet
9. Solids inlet
10. Solids outlet

Known solids flow rate from the top of fluidized bed zone has been initiated after initiating the air velocity from the bottom of the column. The solids flow rate has been monitored wit vibratory feeder DR100 (Retsch India). Temperature has been measured at different locations in the fluidized bed using temperature sensors of PT100. The heat to the drying zone has been supplied from the hot inner surface of the column. The drying zone has been wound with electrical heater (Strip) and the surface outside the heater has been thoroughly insulated with glass wool and fiber glass tape. A downcomer has been provided in the fluidized bed drying zone to draw continuously solids from the drying zone. Aluminum funnel support has been provided at the top of the fluidized bed column and belowof the vibratory feeder tofacilitate flow of solids into the fluidized bed drying zone without any disturbances. The solids feed rate has been controlled by vibratory feeder (DR100).

20.3 RESULTS AND DISCUSSION

In general the drying curve will exhibits the different drying rate regimes such as, warm up period, constant drying rate period and falling rate period one and two.Figure20.2 presents the drying rate with respect to moisture content with time at air velocity of 2.13 m/s with initial moisture content of 5 percent at for 12 kg/h solids feed flow rate and the temperature used is 40°C. From the Figure 20.2 it can observed clearly that three different drying rate regimes namely constant drying rate period and falling rate period one and falling rate period two have been exhibited. The solids have been introduced after attaining the required wall temperature and hence there is no warm up period in the present study.

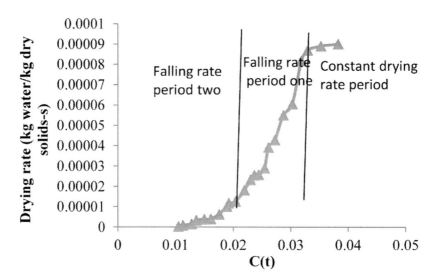

FIGURE 20.2 Different drying rate regimes in drying rate curve.

20.3.1 EFFECT OF AIR VELOCITY

Experiments have been conducted at three different air velocities ranging from 50 to 70 kg/h for uniformly sized solids of size 2.6 mm with initial moisture content of 5 percent. Figure 20.3 presents the results of drying rate variation with varying air velocity. It has been observed from the experimental results that increasing the air velocity the drying rate has been found to be increasing and the equilibrium moisture constant has been found to be decreasing slightly. It has been also observed that from the experimental results that with increase in air velocity the critical moisture content has been found to be decreasing slightly and the drying rate under constant drying rate period has been found to be increasing with increase in air velocity.

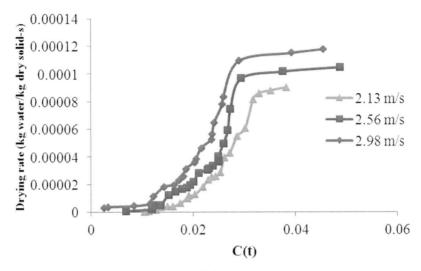

FIGURE 20.3 Effect of air velocity on drying rate.

20.3.2 EFFECT OF TEMPERATURE

Experiments have been carried out at three different temperatures ranging from 40 to 60°C at air velocity of 2.13 m/s with initial moisture content of 5 percent and with solids flow rate of 15 kg/h. Figure 20.4 presents the results of drying rate at different temperatures. It has been observed from the results that increase in temperature the drying rate has been found to be increasing and the outlet solids equilibrium moisture content has been found to be decreasing. It has been also observed from the results that with increase in temperature, the critical moisture content has been found to be approximately same for all temperatures and the drying rate under constant drying rate period has been found to be increasing with increase in temperature.

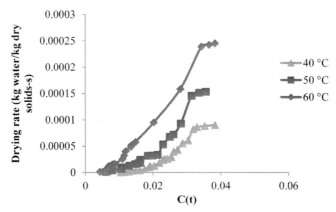

FIGURE 20.4 Effect of temperature on drying rate.

20.3.3 EFFECT OF SOLIDS FLOW RATE

Experiments have been conducted at three different solids flow rate ranging from 5 to 15 kg/h with initial moisture content of 5 percent at an temperature of 40°C with air velocity of 50 kg/h. Figure 20.5 presents the results of drying rate with varying inlet feed solids flow rate. It has been observed from the results that increase in solids flow rate, the drying rate has been found to be decreasing and the equilibrium moisture content has been found to be increasing in the outlet solids. It has been also observed from the results that the critical moisture content has been found to be approximately same for all inlet solids feed flow rates and the drying rate under constant drying rate period has been found to be increasing with decrease in inlet solids feed rate.

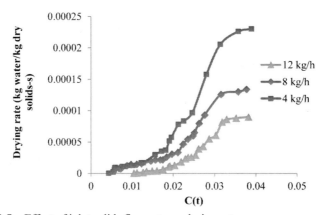

FIGURE 20.5 Effect of inlet solids flow rate on drying rate.

20.3.4 EFFECT OF INITIAL MOISTURE CONTENT

Experiments have been carried out at four different initial moisture content of solids ranging from 4 to 10 percent with an air velocity of 2.13 m/s at temperature of 40°C for 4 kg/h solids feed flow rate. Figure 20.6 presents the results at different initial moisture content of solids. It has been found from the experimental results that increase in initial moisture content of solids the drying rate, has been found to be decreasing and equilibrium moisture of outlet solids has been found to be increasing. It has been also observed from the experimental results that with increase in initial moisture content of solids, the critical moisture content has been found to be increasing.

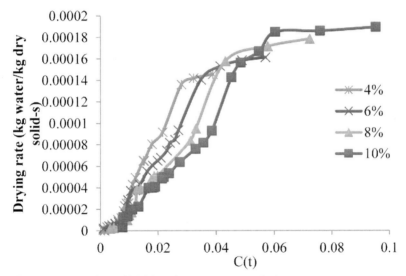

FIGURE 20.6 Effect of initial moisture content on drying rate.

20.4 CONCLUSIONS

Experiments have been carried out in a continuous fluidized bed with varying air velocity, temperature, initial moisture content, and solids feed rate. From the results it has been observed that have been exhibited. Three drying regimes such as, constant and falling rate one and two periods. It has been observed that with increase in air velocity the drying rate has been found to be increasing and the equilibrium moisture content has been found to be decreasing. The critical moisture content has been found to decrease slightly with increase in air velocity. It has been observed from the results that with increase in temperature the drying rate has been found to be increasing and the equilibrium moisture content has been found to be decreasing and the critical moisture content has been found to be constant. It has been observed from the results that with increase in inlet solids feed rate the drying rate has been found to decreasing and equilibrium moisture content in the outlet solids has been found to be increasing and it has been also observed that the critical moisture content has been found to be nearly constant with increase in inlet solids feed rate. It has been observed from the results that increase in initial moisture content of solids the drying rate has been found to be decreasing and the equilibrium moisture content in outlet solids has been found to be increasing and it has been also observed that with increase in initial moisture content of solids the critical moisture content has been found to be increasing. The present results were found to be in line with literature reported results [4, 5, 6].

KEYWORDS

- **Continuous fluidized bed drying**
- **Drying behavior**
- **Equilibrium moisture content**
- **Geldart group D particles**

REFERENCES

1. William-Gardner, A.; Industrial Drying. London: Leonard Hill Books; **1971**.
2. Kunii, D.; and Levenspiel, O.; Fluidization Engineering. USA: Butterworth-Heinemann; **1991**.
3. Geldart, G.; Types of gas fluidization. *Powder Technol.* **1973**, *7,*285–292.
4. Lenin Babu, K.; and Pydi Setty, Y.;Drying of solids in a continuous fluidized bed. *J. Energy, Heat Mass Trans.* **2001**, *23,* 135–151.
5. Satish, S.; and Pydi Setty, Y.; Modeling of continuous fluidized bed dryer using artificial neural networks. *Int. Commun. Heat Mass Trans.* **2005**, *32,* 539–547.
6. Srinivasa Kannan, C.; Thomas, P. P.; and Varma, Y. B. G.; Drying of solids in fluidized beds. *Ind. Eng. Chem. Res.* **1995**,*34,* 3060–3077.

CHAPTER 21

STUDY OF FLOW BEHAVIOR IN MICROCHANNELS

S. ILAIAH[1], USHA VIRENDRA[2], N. ANITHA[3], C. E. ALEMAYEHU[4], and T. SANKARSHANA[5*]

[1,3,4,5*]University College of Technology, Osmania University, Hyderabad, India
[2]Indian Institute of Chemical Technology, Hyderabad, India
*E-mail: tsankarshana@yahoo.com

CONTENTS

21.1 Introduction 216
21.2 Experimental 217
21.3 Results and Discussion 219
21.4 Conclusions 223
Keywords 224
References 224

21.1 INTRODUCTION

The microdevices utilized in engineering for process intensification and are becoming effective platforms for liquid–liquid or gas–liquid mass transfer [1, 2]. It is the science of designing, manufacturing, and formulating devices with an internal diameter in the range of μm, with volumetric flow rate in the range of μl/min. The equivalent hydraulic diameters of the microfluidic devices are up to a few hundreds of micrometers which can provide large surface-to-volume ratio and short mass transfer distance [3]. Microfluidic technology has also demonstrated its advantages such as, lightweight, compact, inherent miniaturization and portability, intensification of heat and mass transfer [4] and less reaction time, high throughput and low consumption of reagents, and stable laminar flow, even at high shear rates. It is reported that, the mass transfer coefficient in microscale dispersion extraction process can be 10–100 times larger than that in conventional extraction columns [5]. Mass transfer process can be completed in several seconds [6]. Due to the small diameters of the channels and the high specific surface area, high mass transfer performance can be achieved in the microchannels.

21.1.1 SINGLE-PHASE FLOW

Several researchers devoted growing effects to improve understanding of physical phenomena occurring in flows at small length-scale geometries, with characteristic lengths varying from 1to 1,000μm, and to develop their technological applications. Bayrktar and Pidugu [7] established velocities below 100 mm/s which are characteristics of fluid flows at such scales and defined the driving forces as pressure gradients, applied electrical fields, capillary forces and free surface phenomena, like gradients in the interfacial tension as in marangoni flows. Campbell and Kandilkar [8] studied the effects of entrance conditions on microchannel and minichannel pressure gradient and laminar to turbulent transition in a circular pipe. They observed, fully developed, friction factor times Reynolds number had an increasing value with increasing Reynolds number, even for low Reynolds numbers, for both tubes. Peng and Peterson [9] studied microchannels with hydraulic diameters of 133–367 μm. They showed that the transition to turbulence occurs at Reynolds numbers as low as 200–700. Furthermore, they showed that this holds truly only for liquid flow in microchannels with identical microchannel dimensions. Barajas and Panton [10] determined the flow patterns visually in 1.6 mm horizontal channels of four different contact angles for an air-water system. Different flow patterns, such as wavy-stratified, plug slug, annular, bubbly, rivulet, etc. were observed in their experiments for closed channels.

21.1.2 TWO-PHASE FLOW

Two-phase microfluidic systems have been attracting increasing research interest for applications such as separations, mixing, encapsulation and chemical and biological analysis among others. Serizawa et al. [11] investigated the visualization of the two-phase flow pattern in circular microchannels. The flowing mixture of air and water in a circular microchannel of 50μm in diameter were conducted experimentally. Two-phase flow patterns obtained from both air-water and steam-water flows were quite similar and their detailed structures were described. The study confirmed that the surface wettability had a significant effect on the two-phase flow Patterns in very small channels. There have been a number of studies involving the application of μ-PIV on gas–liquid flows [12, 13], but only limited ones for liquid-liquid systems [14]. The objective of this study is to characterize the behavior of single-phase liquid flow in straight, U-shaped and Y-shaped microchannels and also two-phase liquid-liquid flow in Y-shaped microchannels.

21.2 EXPERIMENTAL

Particle Image Velocimetry (PIV) is a nonintrusive optical technique used to measure the kinematic parameters of fluid flows. With this technique, the flow velocity is determined by measuring the displacement of a collection of seeding particles between two recorded images separated by a known period of time. A schematic illustration of typical μ-PIV system is shown in Figure 21.1. The system consists of a CCD camera, a microscope equipped with fluorescent filters, an experimental light source, and appropriate optics such as optical fibers, beam expanders, etc. The fluid inside the transparent glass microchannel was seeded with nine red fluorescent particles of size 1μm, which were illuminated by the light source and imaged through the microscope objective onto CCD array of the camera.

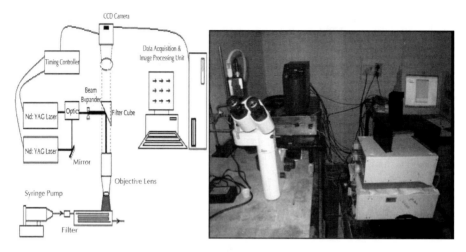

FIGURE 21.1 (a) Schematic diagram of Micro PIV setup and (b) experimental set up for µ-PIV.

21.2.1 SINGLE PHASE

Channel was arranged in such a way that the flow was due to gravity. The water was seeded with fluorescent (dye) particles. The images were captured and analyzed through Dantec Dynamicsµ-PIV system. The flow was modeled using Fluent and Gambit software applying the basic laminar flow and energy equations.Experiments were conducted in Straight, U-shaped and Y-shaped microchannels. The fluid used in all the experiments was deionized water. Respective microchannel was mounted on the working table overthe objective lens.

21.2.2 TWO-PHASE FLOW

Extraction of acetic acid from water using ethyl acetate as solvent was performed in microchannels. In order to determine the fluid flow behavior in the organic (ethylacetate) slugs, the extraction operation was performed in 0.73mm microchannel operating with solvent and feed flow rates of 0.04 ml/min each usingPostnova PN1610 high precision syringe dosing pumpsand the fluid flow behavior inside the microchannel was determined using micro-PIV. In this operation the laser was allowed to fall on the microchannel, which was taken in a beaker containing glycerol solution to match the refractive index of glass to getclarity in the images. The flow rates of both the liquids were maintained and operated at 0.04 ml/min. After some timeuniform slugs formed was seen clearly on the computer screen. During the flow, several images were recorded by the camera provided for determining the flow sense of the liquid inside the channel and the time gap between each image was 0.25 s. The image contents were further analyzed using cross-correlation and adaptive correlation techniques.

21.3 RESULTS AND DISCUSSION

21.3.1 SINGLE-PHASE FLOW

Based on the cross-section obtained with µ-PIV a similar geometry was generated using Gambit with the mesh generator code used in the present CFD study. The numerical simulations were performed with Fluent and the flow field was modeled as steady and laminar. In all the Figures 21.1, 21.2, and 21.3 obtained using CFD, the color indicator indicates the intensity of the velocity.The velocity increases from blue to red and zero to maximum respectively. In these three different kinds of microchannels the velocity profile obtained at the outletappears to be parabolic. From the Figures 21.3.1(a), (b) and 21.3.2 (a), (b), it can be seen that the velocity is maximum at the center of the microchannel forboth the experimentally captured profile and the profile obtained by CFD and is almost zeroat the walls of the microchannels. In Figures 21.3.3(a), (b), it can be seen that the mixing of two fluids occurs at the junction of the two legs of 'Y' and expands as a single fluid in the straight channel.

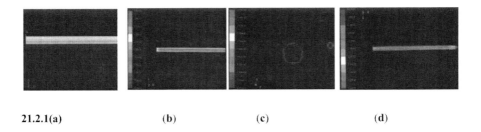

21.2.1(a) (b) (c) (d)

FIGURE 21.2 *(Continued)*

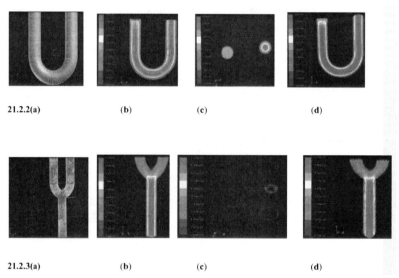

FIGURE 21.2 21.1, 21.2.2, and 21.2.3 Velocity profile for water flowing in straight, U and Y-shaped micro channels: (**a**) Gambit geometry, (**b**) contour display of velocity profiles, (**c**) contour display of velocity profiles at inlet and outlet, and (**d**) vector display of the velocity profile respectively.

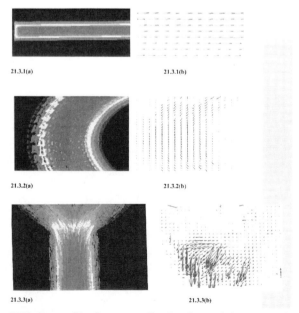

FIGURE 21.3 Velocity profile for water flowing in straight, U-, and Y-shaped micro channels using (**a**) CFD (**b**) μ-PIV instrument respectively.

21.3.2 TWO-PHASE FLOW

Flow patterns were observed by micro-PIV set up as mentioned in experimental section. Due to the hydrophobic wall properties, aqueous phase distributed as a film and also, the size of the ethyl acetate bubbles decreased. This increased the surface between the organic phase and the aqueous phase, which promotes faster mixing and higher extraction efficiencies. Not only high surface-to-volume ratio in microchannels but also the internal circulation of fluid in the slug and diffusion of solute from aqueous phase to organic slug provide better mixing and hence higher mass transfer in microchannels. Another advantage of slug flow is the secondary flow pattern within the slug, which provides faster mixing due to surface renewal. After obtaining the flow sense image contents as seen in a Figures 21.4 (a), (b), a flow sense at a particular time was selected for the analysis. For the analysis by µ-PIV the flow sense image selected was at 8.5s and the image was gridded and then a particular portion of the slug was selected for analysis and it was analyzed using cross-correlation and adaptive correlation methodologies. The results of cross-correlation and adaptive correlations are shown in Figures 21.5 (a), (b) and Figures 21.6 (a), (b) respectively. From the cross correlation the direction of fluid flow is obtained inside the slug and it is observed that the direction of the vectors inside the channel is random. From the adaptive correlation the directions as well as the magnitude of velocity vectors corresponding to the fluid flow inside the slug were obtained and a great chaos was observed in the direction of vectors inside the slug. The volumetric mass transfer coefficient, $K_L a$, Reynolds number, N_{Re} Weber number, N_{We} and Capillary number, N_{Ca} were evaluated and are given in Table 21.1. The Solvent to Feed ratio, S/F was maintained equal to one.

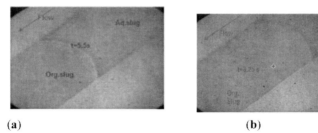

(a)　　　　　　　　　　(b)

FIGURE 21.4 Flow pattern of Micro- PIV for water-acetic acid-ethyl acetate system at different time intervals (**a**) 5.5 s and (**b**) 8.25 s (micro channel diameter =0.73 mm, flow rate of Ethyl acetate=0.04ml/min and 20% acetic acid percentage in the water=0.04 ml/min).

FIGURE 21.5 Vector map of flow using crosscorrelations using (a) 32 × 32 pixel size and (b) 64 × 64 pixel size.

FIGURE 21.6 Vector map using Adaptive Correlation for (a) 32 × 32 pixel size and (b) 64 × 64 pixel size.

From Table 21.1 it is evident that the increase in $K_L a$ with N_{Re} is the consequence of the increase in the surface renewal velocity and increase in the interfacial mass transfer area due to interface disturbance of the two immiscible phases, as the mass transfer distance becomes short, the mass transfer process is intensified. For a particular channel as N_{Re} increases the percentage extraction has decreased. This may be due to increase in N_{Re} and so slug linear velocity, which in turn causes less residence time in microchannels. $K_L a$ increased with increase in the N_{We} in the three different microchannels. Interfacial forces have more influence than the inertial forces. It can also be seen that this effect of interfacial forces is more in the small channel diameters than in the larger diameters. Same reason can be assigned to this as above. For a particular channel as N_{We} increases, percentage extraction decreases whereas microchannels with constant volume, a decrease in diameter gave higher N_{We}, which in turn gave higher percentage extraction when compared to channels with higher diameters. This may due to interfacial forces have more influence than the inertial forces. It can also be seen that this effect of interfacial forces is more in the small channel diameters than in the larger diameters. The shape of the interface between the immiscible liquids was controlled by a competition between the viscous forces and the local interfacial tension. For a particular channel a decrease in diameter gave higher N_{Ca} numbers which in turn gave higher K_La values. For the value of N_{Ca} between 0.0001 to 0.0006, the N_{Re} value varies between 1 to 8. This implies that interfacial forces have more influence than viscous forces. It can also be seen that this effect of interfacial forces is more in the small channel diameters than in the larger diameters, that is interfacial forces have more influence on flow, than viscous forces. Though the value of N_{Re} is less than 10 in slug flow in microchannels, from µ-PIV it was observed that there is a random motion of vectors inside the slugs. From the results of µ-PIV it is observed that because of chaosin the movement of vectors inside the organic slug and leading to higher internal recirculation of liquid inside the slug, and hence every time may be a new surface is being created resulting in the diffusion of acetic acid from the aqueous phase in to the organic slug and hence high mass transfer and percentage extraction in microchannels were obtained.

TABLE 21.1 Dimensionless analysis: 20% acetic acid, $V_m = 0.1708$ cm^3, $C_{or,i} = 0$

D (mm)	Q_{aq} (ml/min)	Q_{or} (ml/min)	S/F ratio	N_{Re}	new	N_{Ca}	N_{Re}/N_{Ca}	K_La (sec^{-1})	% E
1.8	0.08	0.08	1	3.05	2.76×10^{-4}	9.04×10^{-5}	33816.37	8.98×10^{-3}	57.14
	0.06	0.06	1	2.29	1.55×10^{-4}	6.78×10^{-5}	33820.05	7.74×10^{-3}	64.28
	0.04	0.04	1	1.52	6.91×10^{-5}	4.52×10^{-5}	34955.7	5.6×10^{-3}	68.57
0.9	0.08	0.08	1	6.11	2.21×10^{-3}	3.61×10^{-4}	16936.28	9.49×10^{-3}	60.0
	0.06	0.06	1	4.58	1.24×10^{-3}	2.71×10^{-4}	16918.81	9.21×10^{-3}	72.85
	0.04	0.04	1	3.05	5.53×10^{-4}	1.81×10^{-4}	16889.5	6.97×10^{-3}	78.57
0.73	0.08	0.08	1	7.53	4.14×10^{-3}	5.49×10^{-4}	13728.59	1.08×10^{-2}	66.66
	0.06	0.06	1	5.65	2.33×10^{-3}	4.12×10^{-4}	13720.87	1.04×10^{-2}	78.57
	0.04	0.04	1	3.76	1.04×10^{-3}	2.75×10^{-4}	13705.4	7.75×10^{-3}	82.85

21.4 CONCLUSIONS

From the CFD simulations, velocity profile at outlet of the microchannels appears to be parabolic and was maximum at the center and almost zero at the walls of the respective microchannels. The CFD simulations were in good agreement with experimental results obtained through µ-PIV for single-phase flow in all the microchannels. From the two-phase flow results obtained in µ-PIV, it was observed that, because of chaos in the movement of vectors inside the organic slug and lead to higher internal recirculation and every time a new surface was being created for the diffusion of acetic acid from the aqueous phase in to the organic slug and hence high mass transfer and percentage extraction in the microchannels. Interfacial forces play major role than the inertial and viscous forces.

KEYWORDS

- CFD
- Microfluidics
- μ-PIV
- Single-phase flow
- Two-phaseflow

REFERENCES

1. Benz, K.; et al. Utilization of micromixers for extraction processes. *Chem. Eng. Technol.* **2001**, *24*, 11–17.
2. Kashid, M. N.; Renken, A.; and Kiwi-Minsker, L.; Gas-liquid and liquid-liquid mass transfer in microstructured reactors. *Chem. Eng. Sci.* **2011**, *66*, 3876–389.
3. Hudson, S. D.; Poiseuille flow and drop circulation in micro channels.*Rheol. Acta.***2010**, *49*, 237–243.
4. Burns, J.R.;and Ramshaw, C.; The intensification of rapid reactions in multiphase systems using slug flow in capillaries.*Lab Chip.***2001**, *1*, 10–15.
5. Tan, J.; Du, L.; Lu, Y. C.; Xu, J. H.; and Luo, G. S.; Development of a gas-liquid microstructured system for oxidation of hydrogenated 2-ethyltetrahydroanthraquinone.*Chem. Eng. J.***2008**, *102*, 302–306.
6. Kashid, M. N.; Harshe, Y. M.; and Agar, D. W.; Liquid-liquid slug flow in a capillary: an alternative to suspended drop or film contactors,joint 6th International Symposium on Catalysis in Multiphase Reactors/5th International Symposium on Multifunctional Reactors (CAMURE6/ISMR-5-).*Amer. Chem. Soc.* Pune, India; **2007**, 8420–843039.
7. Bayraktar, T.; and Srikanth Pidugu, B.; Characterization of liquid flows in micro fluidic systems. *Int. J. Heat Mass Trans.* **2006**, *49*, 8159–824.
8. Campbell, L. A.; and Kandlikar, S. G.; Effect of entrance condition on frictional losses and transition to turbulence in micro channel flows. In: Proceedings of the Second International Conference on Micro Channels and Mini Channels. Rochester, NY; **2004**, ICMM,2004–2339.
9. Peng, X. F.; and Peterson, G. P.; Convective heat transfer flow friction for water flow in micro channels structures. *Int. J. Heat Mass Trans.***1996**, *39*, 2599–2608.
10. Barajas, A. M.; and Panton, R. L.; The effect of contact angle on two-phase flow in capillary tubes. *Int. J. Multiphase Flow.* **1993**, *19*, 337–346.
11. Seriazawa, A.; Feng, Z.; and Kawara, Z.; Two-phase flow in micro channels. *Exp. Therm. FluidSci.* **2002**, *26*, 703–714.

12. Thulasidas, T. C.; Abraham, M. A.; and Cerro, R. L.; Flow patterns in liquid slugs during bubble-train flow inside capillaries. *Chem. Eng. Sci.* **1997,** *52, 17,*2947–2962.
13. Malsch, D.; et al. μPIV-analysis of Taylor flow in micro channels. *Chem. Eng. J.* **2008,** *135, 1,*S166–S172.
14. Kashid, M. N.; et al. Internal circulation within the liquid slugs of a liquid-liquid slug-flow capillary micro reactor. *Ind. Eng. Chem. Res.* **2005,** *44,*5003–5010.

CHAPTER 22

SOLID DISSOLUTION OF CINNAMIC AND BENZOIC ACIDS IN AGITATED VESSEL WITH AND WITHOUT CHEMICAL REACTION

DR. B. SARATH BABU[1*] and MR. P. SREEDHAR[2]

[1]Assistant Professor, Department of Chemical Engineering, S. V. University College of Engineering, Tirupati; *E-mail: bsarathbabu75@gmail.com

[2]Research Scholar, Department of Chemical Engineering, S. V. University College of Engineering, Tirupati

CONTENTS

22.1 Introduction	228
22.2 Experimental Work	229
22.3 Results and Discussions	231
22.5 Conclusions	236
Keywords	237
References	237

22.1 INTRODUCTION

22.1.1 MOLECULAR DIFFUSION IN LIQUIDS

Molecular diffusion is concerned with movement of individual molecules through a substance by virtue of their thermal energy, the kinetic theory of gases gives an idea about visualizing molecular diffusion and what exactly happens and indeed the success of this theory is quantitatively helpful in describing diffusional phenomena due to which it is widely accepted. Thus the phenomenon of molecular diffusion leads to completely uniform concentration of substances in a solution that may initially be in uniform concentration, however it is necessary to differentiate between molecular diffusion and more rapid mixing as molecular diffusion is a slow process, rapid mixing can be done by mechanical stirring and convective fluid movement. Mechanical stirring generally results in rapid movement of large eddies of fluid which is a characteristic of turbulent motion.

Rates are described in terms of molar flux or mol/ (area) (time), where area is measured generally in a direction normal to diffusion. The diffusivity or diffusion coefficient, D_{AB} of a constituent A in solution B, which is a measure of diffusive mobility, is then defined as ratio of its flux J_A to its concentration gradient.

$$J_A = -D_{AB} \frac{\partial C_A}{\partial Z}$$

$$= -C\, D_{AB} \frac{\partial X_A}{\partial Z}$$

which is known as Fick's first law written for Z-direction. The negative sign emphasizes that the diffusion occurs in the direction of a drop in direction. The diffusivity is a characteristic of a constituent and its environment (Temperature, Pressure, Concentration, Water in liquid, gas or solid solution, and nature of other constituents).

22.1.2 AGITATION OF LIQUIDS

Agitation may be defined as induced motion of material in a specified way, usually in circular pattern inside some sort of container.

22.1.3 AGITATION EQUIPMENT

Liquids are often agitated in some kind of vessel, usually cylindrical in form and with a vertical axis, the top of vessel may be open to air or closed, the liquid depth is nearly equal to diameter of tank. An impeller is mounted on a one hung shaft, driven by a motor. Impeller creates a flow pattern in system causing liquid to circulate through the vessel and return to the impeller, type of flow in agitated vessel depends on type of impeller, characteristics of fluid, the size and proportions of the tanks, baffles, and agitator.

22.1.4 SUSPENSION OF SOLID PARTICLES

Particles of solids are suspended in liquids for many purposes, to produce a homogeneous mixture. Agitation generally creates fluid flow in horizontal flow as well as upward and downward flow. To keep the solids suspended in tank, much higher velocities are required. When solids are suspended in agitated tank, there are several ways to define condition of suspension, they are: (i) Nearly complete suspension (ii) complete particle motion (iii) complete suspension or complete off-bottom suspension.

22.1.5 CINNAMIC ACID ($C_6H_5CHCHCOOH$)

Cinnamic acid is a white crystalline organic acid, which is slightly soluble in water. Saturation solubility is of 0.004 k-mol/m [3].It is obtained from oil of Cinnamon or Balsams such as Storax. Cinnamic acid is used in flavors, synthetic indigo and certain pharmaceuticals, though its primary use is in the manufacturing of the methyl, ethyl, and benzyl esters for the perfume industry. Cinnamic acid has a honey-like odor. Cinnamic acid is freely soluble in benzene, diethyl ether, acetone and it is insoluble in hexane.

Cinnamic acid is also a kind of self-inhibitor produced by fungal spore to prevent germination.

22.2 EXPERIMENTAL WORK

The procedures employed in the present work are detailed here in four parts: (1) Preparation of Cinnamic acid Moulds,(2) mass transfer without chemical reaction, (3) mass transfer with chemical reaction, and(4) range of variables studied.

22.2.1 PREPARATION OF CINNAMIC ACID MOULDS

The solute used was pure "Cinnamic acid" and liquid used was distilled water. Since, Cinnamic acid is generally obtained in powdered form; it is taken into a steel kettle and heated to a temperature around 130°C, where it melts. The molten Cinnamic acid is poured into a brass mold specially designed, a brass shaft held in center of mold. The shaft held the solidified solute in form of a 3.9 and 2.8 cm long smooth and strong cylinder. All the cast cylinders so made are freed from surface dust, washed with water and dried in desiccators before being used for the actual run, care is taken so as to prevent inhalation of Cinnamic acid vapors which can cause health hazards and also care is taken so that Cinnamic acid moulds are free from moisture.

22.2.2 MASS TRANSFER WITHOUT CHEMICAL REACTION

Cinnamic acid moulds are mounted tightly in center of vessel which was filled with measured volume of distilled water. The rotation of cylinder was then started at a preset speed; 5 ml samples of solution were collected after every 10 min and titrated against standardized Sodium hydroxide. The experiment is continued until there is a constant titer value. Initial and final length, diameter is determined before and after the experiment. Diameter measurements of each cylinder were taken at four different positions, the mean of these were used in calculations. The experiment is repeated with different speeds, 200, 300 and 400 rpm respectively with two different mold sizes.

22.2.3 MASS TRANSFER WITH CHEMICAL REACTION

A known amount of sodium hydroxide solution is taken in cylindrical vessel. To this solution 3–4 drops of "phenolphthalein" indicator is added so that the color of solution turns pink, the solution is thoroughly mixed before starting the experiment, then rotation of Cinnamic acid laden stirrer is continued until there is a color change in solution that is, pink to colorless, time taken for run is noted. The solvent used was standardized Sodium Hydroxide at concentration 0.01. Initial and final diameter and length of cylinder are taken before and after the experiment. The experiment is run at speeds of 200, 300, and 400 rpm respectively. Care is taken so that the mold is supported at the center of vessel and it does not touch the bottom of vessel and is properly dipped in solution.

22.2.4 RANGE OF VARIABLES STUDIED

The range of variables studied is as follows.
   With and Without Chemical Reaction
   (1) Three different speeds
      (i) 200 rpm (ii) 300 rpm (iii) 400 rpm
   Concentration of NaOH 0.01 N
   (2) Two different lengths and diameters

(i) 4.2 and 2.8 cm (ii) 2.5 and 3.8 cm

Experimental set up

22.3 RESULTS AND DISCUSSIONS

The experimental study undertaken in the present work involves mass transfer without chemical reaction and mass transfer with chemical reaction. Specifically dissolution of Cinnamic acid from two different size moulds at three different speeds into water and also the dissolution of Cinnamic acid at uniform speed into two different concentrations of water-sodium hydroxide system are examined.

The information collected in mass transfer without chemical reaction is all presented in table numbering.

22.3.1 MASS TRANSFER WITHOUT CHEMICAL REACTION

Mass transfer studies in absence of chemical reaction were undertaken using solid Cinnamic acid in different moulds into water and results are discussed below:

22.3.2 EFFECT OF SHAFT SPEED

Variation of Cinnamic acid concentration with time is shown in Figure 22.1 for small mould at three different stirring speeds

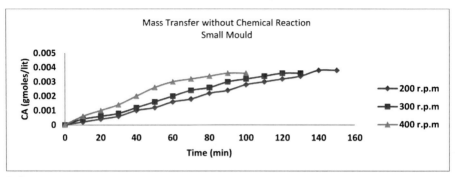

FIGURE 22.1 Variation of Cinnamic acid concentration with time for small mould at three different stirring speeds.

and in Figure 22.2 for big mould at three different stirring speeds.

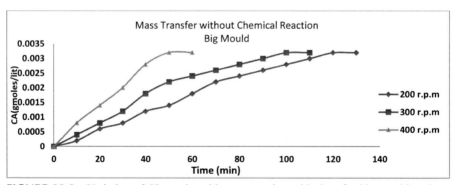

FIGURE 22.2 Variation of Cinnamic acid concentration with time for big mould at three different stirring speeds.

It is observed that the acid concentration increases with time for a fixed stirrer speed. The rate of mass transfer observed to be decreasing with increase in time; this is due to decrease in driving force due to increase in acid concentration, which in turn decreases the flux. In both cases, it is observed that the increase in stirring speed increases the rate of dissolution; this is due to decrease in mass transfer resistance as stirrer speed.

Figure 22.3 shows that the variation of Cinnamic acid concentration with time at constant speeds for two different sizes of mould.

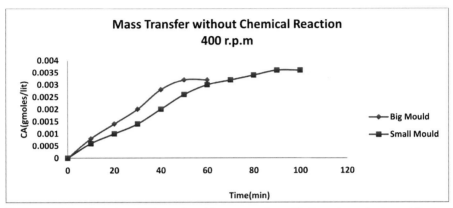

FIGURE 22.3 Variation of Cinnamic acid concentration with time at constant speed for two different sizes of mould.

It is observed that the time required for the dissolution of the big mould is less than the small mould. Further, observed that the concentration of the big mould is always greater than the smaller one. The reason for this is that the surface area is more in case of big mould.

22.3.3 ESTIMATION OF MASS TRANSFER COEFFICIENT

During solid dissolution the flux is proportional to driving force and proportionality constant given by mass transfer coefficient.

The molar flux is given by

$$N_A = \frac{1}{a}\frac{dCA}{dt}$$
$$= K_S \cdot \Delta C_A$$
$$= K_S(C^* - C_A) \quad (22.1)$$

where C^* = solubility of Cinnamic acid in water

Integrating Eq. (22.1) gives

$$\ln\left(\frac{C^*}{C^* - CA}\right) = K_S\, a.t \quad (22.2)$$

Equation (22.2) shows that ln ($\frac{C*}{C*-CA}$) versus time is a straight line and having a slope of $K_S\,a$, Figures 22.4, 22.5, 22.6, and 22.7 show the variation of ln ($\frac{C*}{C*-CA}$) with time for two types of molds at a stirring speeds of 300 and 400 rpm respectively, the data is well fitted by straight line and from the slope of these lines the mass transfer Coefficient "$K_S\,a$" is calculated.

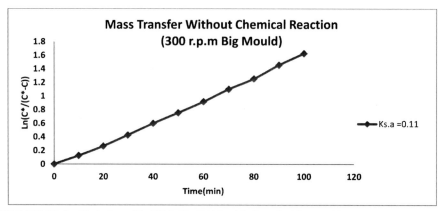

FIGURE 22.4 Variation of ln(C*/(C*-CA)) with time for big mould at 300 r.p.m.

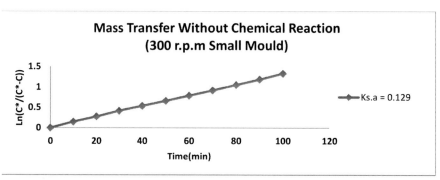

FIGURE 22.5 Variation of ln(C*/(C*-CA)) with time for small mould at 300 r.p.m.

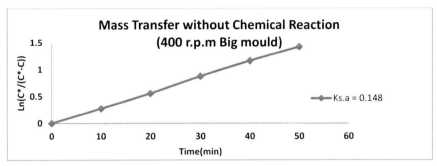

FIGURE 22.6 Variation of ln(C*/(C*-CA)) with time for big mould at 400 r.p.m.

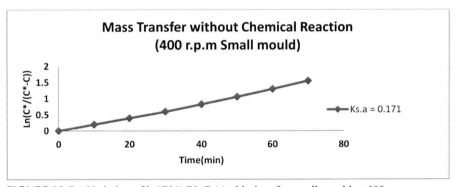

FIGURE 22.7 Variation of ln(C*/(C*-C_A)) with time for small mould at 400 r.p.m.

Further it is also observed that mass transfer coefficient decreases with increase in cylindrical mold diameter.

22.3.4 COMPARISON OF CINNAMIC ACID AND BENZOIC ACID CONCENTRATIONS AT 400 RPM (BIG MOLD)

We observed that, both Cinnamic acid and Benzoic acid dissolution rate increases with increase in time. And when we compare both it is seen in Figure 22.8 that benzoic acid dissolution is more than that of Cinnamic acid dissolution.

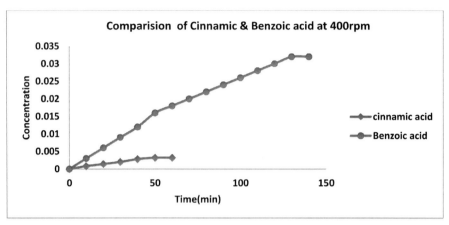

FIGURE 22.8 Comparison of $C_A V_s$ time between cinnamic and benzoic acid moulds (big) at 400 r.p.m

22.5 CONCLUSIONS

The research work covered the Reynolds number range 4,500–37,000. The results are expressed in terms of Sherwood number (Sh), Reynolds Number (Re) and diameter of mould to that of diameter of Vessel (d/D_t) and found that the coefficients on the RHS is 0.55, the power of Re is 0.49 and the power of d/D_t is -1.78. Increase in shaft speed increases the rate of dissolution. Increase in mould size decreases the time for dissolution. The mass transfer coefficient decreases with increase in mould diameter.

The correlation obtained is for mass transfer without Chemical Reaction

$$Sh = 0.55 Re^{0.49} (d/D_t)^{-1.78}$$

For Mass Transfer with Chemical Reaction, reaction factor (F_R) is the ratio of Sherwood number with chemical reaction to the Sherwood number without chemical reaction is calculated and is found to be independent of Reynolds Number.

The scope for future work can be, considering different temperatures and the work with different combinations of solids (Benzoic acid, Barium chloride, Naphthalene) liquids (Water, benzene, methanol) in cylindrical vessels fitted with a four bladed propeller attached to a vertical shaft.

KEYWORDS

- **Benzoic acid**
- **Cinnamic acid**
- **Reaction factor**
- **Reynolds number**
- **Sherwood number**

REFERENCES

1. Robert E. Treybal Third Edition; Mass Transfer Operations, McGraw-Hill International Editions, Chemical Engineering Series, 1981.
2. Coulson, J. M.; and Richardson, J. F.; Chemical Engineering, Volume 6, Third Edition, Butterworth Heinemann, an imprint of Elsevier, 2000.
3. Sherwood, T. K.; and Woertz, B. B.; *Trans. Am. Inst. Chem. Eng.*1939, *35,* 517.
4. Hixson, A.W.; and Crowell, J. H.; A survey of American Chemistry (86), *Ind. Eng. Chem.* 1931, 23:923, 1002, 1160,
5. Hixson, A.W.; and Baum, S. J.; Agitation: mass transfer coefficients in liquid- solid agitated systems, *Ind. Eng.Chem.*1941, *33,* 478–485, 1433.
6. Harriot, P.; Mass Transfer to particles, *A.I.Ch.E. J.* 1962, *8(1),* 93–102.
7. Dryden, C. E.; Strang, D. A.; and Wathrow, A. E.; Heat and Mass Transfer in packed beds, *Chemical Engineering Progress.*1953, *49,*191.:
8. Tripathi, G.; Singh, S. K.; and Upadhyay, S. N.; Mass Transfer rates at circular cylinders rotating about their axis in the vessel, *Ind. J. Tech.*1971, *9,* 237–241.
9. Baldi, G.; and Specchia, V.; Current Awareness in particle technology, *Chem. Eng. J.*1976, volume-9, *11,* 81–88,
10. Kirk and Othmer; *Encycl. Chem. Eng. Series.*
11. Delgado, J. M. P. Q.; ASM International *,JPEDW.* 2007, *28,*427–432.
12. Peter Sabev; and Hon Long Lam; *Chemical Engineering Transaction.* 2012, *29.*
13. Patil, V. K.; Joshi, J. B.; and Sharma, M. M.; JBJ group papers, Chem. Eng. Res. Dev. 1984, *62,* 254–274.
14. Steele, L. R., and Geankoplis, C. J. A.I.Ch.E. J., 5, 178 (1959).

PART II

ENVIRONMENTAL ENGINEERING, NANOTECHNOLOGY, AND MATERIAL ENGINEERING

CHAPTER 23

ENERGY AUDIT, DESIGNING, AND MANAGEMENT OF WALKING BEAM REHEATING FURNACE IN STEEL INDUSTRY

SHABANA SHAIK[1*], K. SUDHAKAR[2], and DEEPA MEGHAVATU[3]

[1,3]Depertment of Chemical Engineering, Andhra University College of Engineernig, Andhra University (A), Vishakapatnam; *E-mail: shabana89.chem@gmail.com

[2]Assistant General Manager, EMD Department, Vizag Steel Plant

CONTENTS

23.1 Introduction 242
23.2 Methodology 242
23.3 Data Collection 243
23.4 Results 252
Keywords 255
References 255

23.1 INTRODUCTION

India is the 10th largest Steel producer and 6th largest energy consumer in the world and annually consumes about 3 percent of the world's total energy [1]. Energy consumption and economic growth are interlinked. The energy consumption appears to be directly related to the level of living of the population and the degree of industrialization of the country [2]. In addition to global imbalances energy resources becoming scare. This made the all-round efforts to reduce energy by optimizing energy usage at all sectors and bring awareness of the importance of the conservation and development of new energy resources. It is the responsibility of the scientists, engineers, and technicians to locate the development and exploit these new sources and should have an intimate knowledge of the various energy forms, sources, conservation methods, conservation techniques along with their limitations, and inherent problems. Energy conservation programs often start from energy audits. Today the engineers must be concern with the 3 "E"s—Energy, Economy, and Ecology. Thus the modern energy engineers must try to develop systems that produce large enormities of energy at low cost with less impact on the environment. Energy audit co-ordinates the effect among various departments for conservation of energy, optimization of available energies and minimization of losses.

Walking beam reheating furnace is an improved modified version of the reheating furnace and is heated by gaseous fuels by capturing waste heat energy though recuperators by balanced draft control of continuous furnace and gases coming from LD Converter, Blast Furnace and Coke Oven gases [3]. That is a device to move the slab in the furnace by lifting it, moving it in forward direction lowering it down and then again returning to its original position. Reheating furnace comprises of three sections namely slab charging system, furnace proper and slab discharging [4]. The main function of reheating furnace is to raise the temperature from 600 to 1,400°C until it is plastic enough to be pressed or rolled to desired shapes and sizes. In walking beam reheating furnace excellent surface condition of the product is achieved and bottom firing of the entire length of the furnace, thereby ensuring a higher specific production rate per square meter of the hearth and the controlling of heating system is easy.

23.2 METHODOLOGY

The System Boundary for the purpose of carrying out the Thermal Energy Audit of the Reheating Furnace was first defined. The Furnace, the Air Recuperators and the Gas Recuperators, were considered as different system boundaries. Next, based on each of the system boundaries defined—the input and output enthalpies of all the three systems were considered. The input enthalpy was the heat input given by the fuel gas entering the furnace where combustion with air takes place. The output enthalpy were useful heat absorbed by the steel billets and the balance being heat loss through flue gases, skid cooling water, radiation, and conduction losses through walls and openings. The methodology contains preparation of heat balance, identification of loss streams, quantification of losses and reasons for losses and measures to minimize losses.

The efficiency of a furnace may be obtained from a heat balance [5], which is a comparison between the sum of the items that contribute to the heat supplied to the furnace and the sum of the items representing the heat leaving the furnace. By performing a heat balance it is possible to determine the following:

(a) The thermal energy efficiency of the process.
(b) The way in which the heat supplied is used.

By comparing the data with known standards based on experience or design data, any wastage of heat can be located and corrective measures taken to counteract this heat loss and showed in sankey diagram.

23.3 DATA COLLECTION

The TOSDIC control system is a Microprocessor based system for controlling [6] the various flow and temperature parameters of fuel gas, air, flue gas, cooling water, Oxygen content, rolling rate, etc. Contents of Oxygen at various points in flue duct were taken manually through a special arrangement and then measured using ORSAT apparatus. Temperatures of stock (bloom) and furnace openings were taken using noncontact Optical Pyrometers. The calculations of combustion and thermal equations derived at on MS EXCEL worksheets, which would help in representing graphical curves of various parametric relationships.

23.3.1 THERMAL ENERGY AUDIT—CALCULATION PROCEDURE

23.3.1.1 ANALYSIS OF MIXER GAS

Ratio proportion calculation [7]:
  Basis: 1 mol – 1 m^3/hr.
  B.F. calorific value = 700 kcal/nm^3
  Co gas calorific value = 4,200 kcal/nm^3
  Our required mixer gas ratio cv is 2,400 kcal/nm^3
  B.F. gas = 1 nm^3, co gas = x nm^3.
  B.F. gas × CV + co gas × CV = 2,400 (B.F. gas + co gas).
  $(1 \times 700) + (4{,}200 \times x) = 2{,}400\,(1+x)$

$700 + 4{,}200x = 2{,}400 + 2{,}400x.$
$x = 0.94$
B.F. gas: co gas:: 1:0.94.
Now, B.F. gas calorific value = 700 kcal/nm³
Co gas calorific value = 4,200 kcal/nm³
LD gas calorific value = 1,800 kcal/nm³
Our mixer gas calorific value = 2,400 kcal/nm³
(B.F. gas × cv) + (co gas × cv) + (LD gas × cv) = 2,400 (B.F. gas + co gas + LD gas).
$(700 \times A) + (4{,}200 \times B) + (1{,}800 \times C)/(A+B+C) = 2{,}400.$
A: B: C = 0.05:0.2:0.4 = 2,453.
Stoichiometric calculations (Table 23.1) [7, 8, 9]:

TABLE 23.1 Stoichiometric calculations of required O_2

S. No	Stoichiometric Reactions	Composition	Stoichiometric O_2	Basis Gives1	Gives2	1 nm³ of Flue Gas Required Composition	Required O_2
1.	$CO + 1/2 O_2 = CO_2$	1mole CO	0.5	1 mole of CO_2	—	0.2554	0.1277
2.	$H_2 + 1/2 O_2 = H_2O$	1mole H_2	0.5	1 mole of H_2O	—	0.1756	0.0878
3.	$CH_4 + 2O_2 = CO_2 + 2H_2O$	1mole CH_4	2.0	1 mole of CO_2	2 mole of H_2O	0.0817	0.1634
4.	$H_2S + 3/2 O_2 = H_2O + SO_2$	1mole H_2S	1.5	1 mole of H_2O	1 mole of SO_2	0.0011	0.0016
5.	$C_2H_4 + 3O_2 = 2CO_2 + 2H_2O$	1mole C_2H_4	3.0	2 moles of CO_2	2 moles of H_2O	0.006	0.018

23.3.1.2 THERMAL ENERGY AUDIT

The energy audit can be done in many ways like thermal energy audit, commercial energy audit, power energy audit, industrial energy audit etc. Here the thermal energy audit of a reheating furnace is done by using the below following procedure [3, 10].

Total heat input:
1. Heat consumption of the flue gas = volume of flue gas × calorific value
= [(m³/hr) × kcal/hr³/1,000]Mcal/hr

2. Heat of oxidation = [material oxidized × conversion factor]/1,000 Mcal/hr
(i) Conversion factor ⟹ ton=1,350,000 kcal
Material oxidized = (total weight of blooms) × (% oxidized)
Total weight of blooms = No. of blooms × conversion factor
(ii) Conversion factor ⟹ 1 bloom=2.78 ton.

Total heat output:
Heat loss by stock = total weight of blooms × specific heat of flue gas × (door discharge temp − charging temp)
= tons/hr × Mcal/ton × (°C− °C) [3, 11, 12].

1. Heat loss by waste flue gas = heat consumption of flue gas × specific heat of waste flue gas × (gas recuperator discharge temp − surrounding temp) Mcal/hr
= Mcal/hr × Mcal/hr (°C−°C)
2. Heat loss by water vapor = (mass of water vapor × [584 + 0.45 × (temp of flue gas at exit − surrounding temp)])/1,000 Mcal/hr
Mass of water vapor= kg/hr
Latent heat of vaporization = 584 kcal/kg
Specific heat of water vapor = 0.45 kcal/kg °C
3. Structural heat loss:
(i) Loss through skin: $Q = a \times (T_1 - T_2)^{5/4} + 4.88\ E\ [(T_1+273/100)^4 - (T_2+273/100)^4]$
a = factor regarding direction of the surface of natural convection
E = emissivity factor = 0.88 (emissive of external wall surface of the furnace)
T_1 = temperature of external wall surface of the furnace (°C)
T_2 = temperature of air around the furnace (°C)
a_1 = 2.8 = ceiling
a_2 = 2.2 = side walls
(a_2—sidewall1; a_3—sidewall2; a_4—discharge door; a_5—charging door)
a_6 = 1.5 = hearth
(ii) Heat loss by radiation = $c_o \times (T/100)$ × area of opening × time of opening
c_o = emissivity of black body = 4.96 kcal/hr
T = temperature of furnace
Area of opening = 13 × 0.7 m²
Time of opening = time of discharge/hr
During charging time= area × time of opening = 13 × 0.5 m²× time
During normal time = 13 × 0.25 m² × time
(iii) Heat loss through indirect cooling water = Mcal/hr

Mass of cooling water × specific heat of water × raise in temp × density of water vapor

Mass of cooling water = m³/hr
Specific heat of water = Mcal/kg °C
Raise in temperature = °C
Density of water = kg/m³

(iv) Heat loss through evaporative cooling system = (vol of feed water × change in enthalpy)/1,000 Mcal/hr

Volume of feed water = mass of feed water/1,000
Structural heat loss = loss through skin + loss through radiation + loss through cooling water + loss through ECS

(v) Moisture loss:
Moisture loss through air = [density of air × total air flow × specific heat × humidity factor × (temp before air recp − ambient temperature)]/1,000,000 Mcal/hr

Density of air = 1.16
Total air flow = m³/hr
Specific heat = 0.45
Humidity factor = 0.024

Moisture loss through fuel = density of fuel × total gas flow × specific heat (temp before air recuperator − surrounding temp)/1,000,000 Mcal/hr

Density of fuel = 0.904
Humidity factor = 0.06

Total output loss = heat loss by stock + heat loss by waste flue gas + heat loss by water vapor + structural heat loss + moisture loss.

Total heat account:

TABLE 23.2 Heat account model table [8].

S. No	Input Term	Heat (Mcal/hr)	Output Term	Heat (Mcal/hr)
1.	Heat consumption of flue gas		Heat loss by stock	
2.	Heat of radiation		Heat loss by waste flue gas	
3.			Heat loss by water vapor	
4.			Heat loss through skin	

TABLE 23.2 *(Continued)*

S. No	Input Term	Heat (Mcal/hr)	Output Term	Heat (Mcal/hr)
5.			Heat loss by radiation	
6.			Heat loss through moisture	
7.			Heat loss through ICS	
8.			Heat loss through ECS	
	Total		Total	

Energy audit results:
The total input given to the furnace = 173,990 Mcal/hr
The total output or losses from the furnace = 121,918 Mcal/hr
That is, efficiency = 45–55 percent

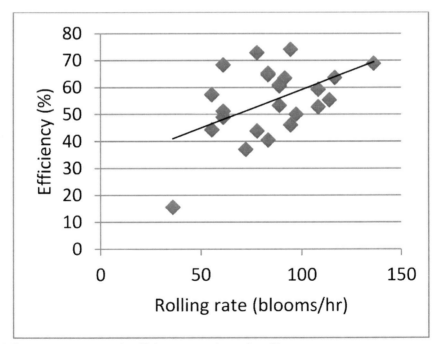

FIGURE 23.1 Change in efficiency w.r.t. change in rolling rate.

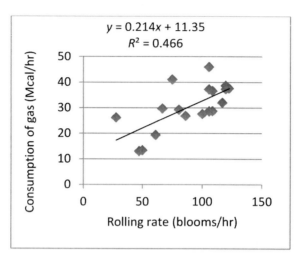

FIGURE 23.2 Consumption of fuel gas w.r.t. rolling rate

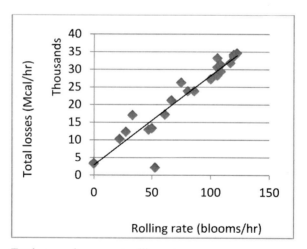

FIGURE 23.3 Total energy losses w.r.t rolling rate

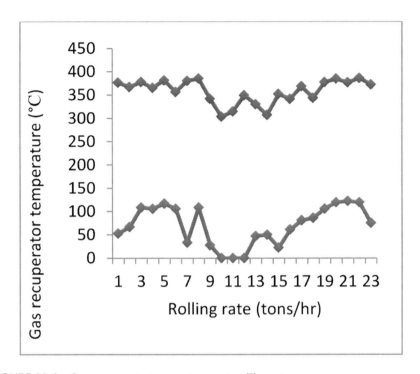

FIGURE 23.4 Gas recuperator temperature w.r.t. rolling rate

From the Figures 23.1–23.4, we can state that efficiency of the furnace increases when the rolling rate of beam is operated at high blooms, heat from the gas recuperator and air recuperator involve more in the combustion reaction therefore the more release of gas from the recuperators may lead to heat escape and waste of energy and more scale up and more losses. So, operation of the reheating furnace at optimum temperature leads to reduce wastage of energy and heating time and therefore obtains good quality products.

23.3.1 3DESIGN CALCULATIONS: THE DESIGN OF THE REHEATING FURNACE IS DONE BY REFERRING [13] AS SHOWN IN THE TABLE 23.3.

TABLE 3 Design Calculations

Parameter	Symbol	Value	Unit		Parameter	Symbol	Value	Unit		Parameter	Symbol	Value	Unit
Productivity (effictive)	G	50000	Kg/hr		Total metal emissivity	Em	0.8			tc-2	tc-2	532.5	deg C
Bottom tonnage capac	Ha	500	Kg/m2.hr		initial coeffcicient	Co	5.7			Heating time,sect	T(ph-3)		
The area of the bottor	Fa	100	m2		Coefficient,section 1	Cglm(ph-1)	3.795	W/m2K4		Gas temp	Tg	1125	deg C	1094.28
Furnace Dimensions					Coefficient of radient heat	αr(ph1)	93.834	W/m2K		metal temp	Tm(avg)	995	deg C	378.72
Height	h	5	m		heat conductivity	λ	48.4	W/m Deg C		metal temperatur	Tm(out)	1120	deg C	0.85
pre heating zone	ph	1.5	m		Specific heat	Cp	0.542	KJ/kg deg C		Eco₂	Eco₂	0.12	(from graph)	1173.57
welding zone	w	2.2	m		S	S	0.1	m		EH₂O	EH₂O	0.14	(from graph)
soaking zone	s	1.3	m		Bi(ph)	Bi(ph)	0.194			total gas emissivi	Eg	0.2712	
Breadth	B	13	m		L	L	4	m		coefficient	Cglm(ph-3)	4.360	
length	L	21.7	m		B	B	0.2	m		Coefficient of rad	αr(ph-3)	473.999	
Degree of developm	W				H	H	0.2	m		heat conductivity	λ	30.2	W/m Deg C
Wph	Wph	4	m		Density of bloom, m	m	7800	kg/m	2.17 Kg/m2 hr	Specific heat	Cp	0.637	KJ/kg deg C	637
Ww	Ww	4.35	m		Productivity (effictive are	G	1248			Bi(ph-3)	Bi(ph-3)	1.570	
Ws	Ws	3.9	m		specific heat of bloom	Cp	0.542	KJ/kg deg C	542 J/Kg deg C	Temp criterion o	θs	0.020	
Effective thickness of	Sef				preheating area of furnace	Fa	1.6			Fo	Fo	0.2	for the given Bi & TET
effcicieny		0.9			Time(ph-1)	T(ph-1)	0.486	hr		Density	ρ	7800	kg/m3
Sef(ph)	Sef(ph)	2.421			Heating time,section 2	T(ph-2)				Thernmal diffusiv	a	0.0219	m2/hr
Sef(w)	Sef(w)	3.387			Gas temp	Tg	1075	deg C	1348 Deg K	Time(ph-3)	T(ph-3)	0.091	hr (check)
Sef(s)	Sef(s)	2.127			metal temp	Tm(avg)	585	deg C	858 Deg K	TETA-c	θc	0.8	for the given Bi & TET
Pre heating time	T(ph)				metal temperature-outlet	Tm(out)	870	deg C		tc-3	tc-3	1121	deg C
Heating time,sectio	T(ph)1				Eco₂	Eco₂	0.13	(from graph)		T(ph)	T(ph)	1.258	hr
Gas temperature	Tg	890	deg C	1163 Deg K	EH₂O	EH₂O	0.165	(from graph)		welding zone			
metal temperature	Tm(avg)	165	deg C	438 Deg K	total gas emissivity,Eg	Eg	0.3082			Heating time, T(w)			
metal temperature-ink	Tm(in)	30	deg C	303 Deg K	coefficient	Cglm(ph-2)	3.605			Partial pressure o	Pco₂	0.146	Kgf/cm2	14.3
metal temperature-out	Tm(out)	300	deg C	573 Deg K	Coefficient of radiant heat	αr(ph-2)	203.073			Partial pressure o	PH₂O	0.138	Kgf/cm2	13.5
Partial pressure of CO	Pco₂	0.146	Kgf/cm2	14.3 K-N/m2	heat conductivity	λ	35	W/m Deg C		Sef(ph)*Pco2	Sef(ph)*P	0.494	KN/m
Partial pressure of H2	PH₂O	0.138	Kgf/cm2	13.5 K-N/m2	Specific heat	Cp	0.687	KJ/kg deg C	687 J/Kg deg C	Sef(ph)*PH2O	Sef(ph)*PH	0.467	KN/m
Sef (ph)*Pco2	Sef (ph)*P	34.616			Bi(ph-2)	Bi(ph-2)	0.580			Gas temp	Tg	1125	deg C	1398
Sef(ph)*PH2O	Sef(ph)*P	32.679			Temp criterion of metal s	θs	0.265			metal temp	Tm(avg)	1160	deg C	1433
Eco2	Eco2	0.15			Fo	Fo	1.6	for the given Bi & TETA-s(fron	metal temperatur	Tm(out)	1200	deg C	1473
Eh20	Eh20	0.19			Thernmal diffusivity	a	0.024	m2/hr		Eco₂	Eco₂	0.12	
beta		1.08			Time(ph-2)	T(ph-2)	0.680	hr (check)		EH₂O	EH₂O	0.16	
Total gas emissivity	Eg	0.3552			TETA-c	θc	0.7	for the given Bi & TETA-s(fron	beta	β	1.07		

total gas emissivity	Eg	0.291			billet density	ρ	7.8	
coefficient	Cglm(w)	4.350			mass of billet	m(b)	1.248	tons/hr
Coefficient of radiant h	αr-w	493.566			number of billets	n	111.38	
heat conductivity	λ	26.7	W/m Deg C		length of the furnace	L	22.28	m (for billets laid in one r)
Specific heat	Cp	0.69	KJ/kg deg C		Width of the furnace	B	12.75	m
Bi(W)	Bi(W)	1.849			Active bottom area	Fa	89.10	m2
Thermmal diffusivity	a	0.018	m2/hr		Total bottom area	Ft	284.01	m2
TETA-s	θs	-0.091			Preheating zone	Lph	13.25	m
Fo	Fo	0.6	for the given Bi &TETA-s(from graph fig-4	Welding zone	Lw	3.54	m
T(w)	T(w)	0.336	hr		Soaking zone		5.48	m
soaking zone					Total bottom tonnage capacity of		176	Kgf/m2.hr
Heating time, T(s)	T(s)							
TETA-c	θc	0.6	for the given Bi &TETA-s(from graph-fig-48,pg-104)					
tc(w)	tc(w)	1122.6	deg C					
delta t(b)	Δt(b)	275.4	deg k					
tolerable temp differen	Δt(e)	50	deg C	323 deg K				
TETA	θ	1.17						
viscosity	μ	0.5						
Fo	Fo	1.2	for the given Bi &TETA-s(from graph-fig-48,pg-104)					
tavg	tavg	1180.7	deg C					
heat conductivity	λ	29.5	W/m Deg C					
Specific heat	Cp	0.68	KJ/kg deg C					
Bi(S)	Bi(S)	0.68						
Thernmal diffusivity	a	0.023	m2/hr					
Time(s)	T(s)	0.521	hr					
total time, T (t(ph)	T	2.114	hr					
Determination of Furnace dimensions								
Specified productivity	p	50	tons/hr					
bloom weight	wt	2.78	tons/hr					
One bloom length	l	6	m					
No of spaces in the furnaces		3	spaces					
Space btw the bloom		0.25	m					
mass of metal contain	m	139	tons/hr					

23.3.1.4 ENERGY MANAGEMENT TECHNIQUES

1. From the analysis, it is observed that the major heat loss is through flue gases.
   (i) Oxygen enrichment of combustion air [14]:
   Air is mainly composed of 21 percent oxygen (by volume) and rest 79 percent of nitrogen. When combustion takes place, the oxygen combines with the carbon and hydrogen of the fuel and liberates heat. The inert gases of the air absorb heat from the combustion and carry it out of the furnace resulting in loss. It reduces flame temperature by absorbing heat and thus reducing the rate of heat transfer to the stock. The fuel input may be decreased when enriched air is used, to maintain the same production rate as obtained with more fuel using ordinary air. The use of pure oxygen reduces the stack losses of fuels. Use of oxygen enrich combustion air offers a method of achieving a measure of stack loss reduction when using oxygen to combust the high calorific value fuels with the technical disadvantages associated with the use of pure straight oxygen.
   (ii) Change in Calorific Value by Changing Mixture Gas Ratios and Gas Temperature:
   On increasing the calorific value of the fuel gas the temperature of the fuel gas increases simultaneously, this leads decrease in volume of the furnace and increase in productivity.
2. One of the most important factors influencing the efficient use of the fuels is their stack loss, that is, that proportion of the fuel energy which is carried out of the operating furnace as the sensible heat of the flue gases. The important factors governing this parameter are normally the temperature of the flue gases leaving the furnace and their oxygen content. So those are to be maintained in proportion.
3. The excess air required for the combustion of fuel in the furnace should be exactly of 10, more or less excess air should be strictly prohibited.
4. High furnace pressure causing leaking out of flames, damaging furnace parts. This could be due to air ingress from water seals and also there exists air inflation from the water blocks at the bottom of the furnace results in cooling the furnace and thereby increasing fuel consumption.
5. Hot charging is a method in which steels of high temperature are directly charged to a reheating furnace to conserve energy.

23.4 RESULTS

1. It is observed that increase in calorific value of mixed gas from 2,000 to 3,000 kcal/Ncum has impact on furnace heating time and productivity. By increasing the calorific value of gas flame temperature increases, therefore the heat transfer to steel material also increases rapidly. From Figures 23.5 and 23.6, it is observed that increasing calorific value from 2,000 to 2,878 kcal/Ncum reduces the heating time of the steel reheating from 2.11 to 1.85 hr. This result in increasing furnace efficiency by 18 tons/hr and the specific fuel consumption reduces significantly with higher rolling rate.

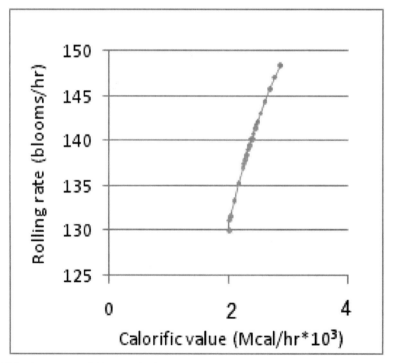

FIGURE 23.5 Rate of change of rolling rate with respect to change in calorific value

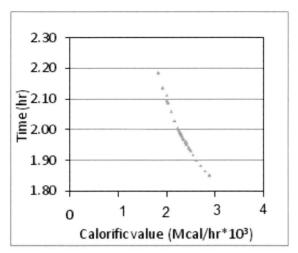

FIGURE 23.6 Rate of change of time with respect to increase in calorific value

1. It is also observed that charging hot steel blooms in to reheating furnace reduces steel reheating time. The steel reheating time reduces from 2.106 to 2.04 hr. This has increased the output of the furnace by 4 tons/hr and reduced specific fuel consumption marginally is shown in Figure 23.7.

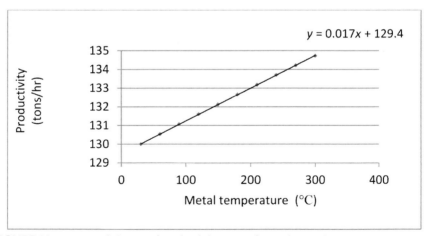

FIGURE 23.7 Rate of change of productivity w.r.t. change in metal temperature.

1. If we combine these two options and implement, there is greater possibility of reducing fuel consumption significantly up to 28 tons/hr.

2. On preheating air temperature before passing it into air recuperators we can reduce the flue gas consumption.
3. By increase in gas temperature in gas recuperator the flue gas consumption reduces.
4. By increasing flame temperature and by intake of preheated air to the recuperator, higher heat transfer rate can be obtained.
5. Usage of ceramic air recuperator by metallic air recuperator can increase the heat transfer rate more efficiently, as the ceramic air recuperator can hold temperatures beyond 650°C.
6. Sankey Diagram: The energy audit (thermal energy audit) is best show in sankey diagram (Figure 23.8) and well described in Table 23.4.

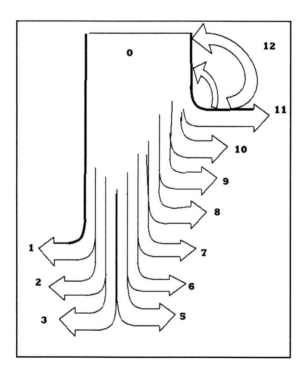

FIGURE 23.8 Sankey Diagram

TABLE 23.4 Heat component of sankey diagram (from Table 23.3)

Legend	Heat Output Component	M.Cal/t	%
0	Heat input	490	100
1	Through heating billets	195	39
2	Scale formation	4	1.0
3	Heat carried by moisture	17	4.5
4	Loss by Cooling water	35	9.5
5	Loss through wall-conduction	9	2.5
6	Loss through wall-convection	3	0.8
7	Loss through wall-radiation	3	0.7
8	Loss through discharge side	1	0.1
9	Loss through charging side	2	0.5
10	Carried by dry flue gas	96	26.1
11	Recovery from air recuperator		
12	Recovery from gas	Recup	

KEYWORDS

- **Thermal Energy audit**
- **Energy conservation**
- **Energy consumption**
- **Inflation**
- **Mixer gas**
- **Oxygen enrichment**

REFERENCES

1. Lynch, R.; An Energy Overview of India. Washington, DC: U.S. Department of Energy, Office of Fossil Energy; **2002**.
2. M.; J. Environ. Stud. Policy. **1998**, *1*(2),89–96.
3. Sharma, B. K.; Engineering Chemistry, 14th Edition, Goel Publishing House, Meerut **1960**.
4. Vishal, D; Energy Efficiency Office. A Practice Guide-77, Continuous Steel Reheating Furnace: Operation & Maintenance; **2010**.
5. Kern, D. Q.; Process Heat Transfer, Mc Graw-Hill, International Edition, ISBN 0-07-085353-3, **1965**.

6. Vishal, D; Energy Efficiency Office. A Practice Guide-76, Energy Efficient Operation of Continuous Steel Reheating Furnace: Specifications, Designing of Equipments, **2010**.
7. Gupta, O. P.; Elements of Fuels, Furnaces and Refractories, Khanna Publishers, **1998**.
8. Reedy, T. S.; Impact of Energy Expenditure on Total Cost. AGM, Steel Authority of India Limited.
9. Hougen, O. A.; Watson, K. M.; Ragatz, R. A.; Chemical Process Principle: Material & Energy Balances. CBS Publishers & Distributors, New Delhi, Part I, Second Edition, Reprinted in **2004**.
10. Rao, K. R.; Energy and Power Generation Handbook: Established and Emerging Technologies. New York: ASME Press; **2011**.
11. Saunders, E. A. S.; Fundamentals of Heat & Mass Transfer, 4th Edition, John Willey & Sons.
12. Incropera, F. P.; De Witt, D. P.; Fundamentals of Heat and Mass Transfer. 4th edition, willey, New York.
13. Krivandin, V.; Markov, B.; and Afanasyev, V. V. (trans); Design of Metallurgical Furnace.
14. Ebeling, C.; Axelsson, C. L.; and Coe, D.; Oxy-Fuel Applications for Steel Reheating Furnace (AISE Iron and Steel Exposition & Annual Convention; **1999**. Cleveland, OH).
15. Murphy, W. R.; McKay, G.; Energy Management. Butterworth-Heinemann, New Delhi, Reprinted 2001.

CHAPTER 24

SYNTHESIS AND CHARACTERIZATION OF PVDT/PAN HOLLOW FIBER BLEND MEMBRANES FOR PURIFICATION OF SURFACE WATER TREATMENT

K. PRANEETH[1,2], JAMES TARDIO[1], SURESH K. BHARGAVA[1], and S. SRIDHAR[2*]

[1]Royal Melbourne Institute of Technology (RMIT), School of Applied Sciences, Melbourne, VIC-3001, Australia

[2]Membrane Separations Group, Chemical Engineering Division, Indian Institute of Chemical Technology (CSIR-IICT), Hyderabad, India – 500007;
*E-mail: sridhar11in@yahoo.com

CONTENTS

24.1 Introduction 258
24.2 Experimental 259
24.3 Results and Discussions 262
24.4 Conclusions 267
Keywords 267
References 267

24.1 INTRODUCTION

The field of membrane separation technology is presently in a state of rapid growth and innovation. While new membrane processes are being conceived with remarkable frequency, existing processes are under constant improvement to enhance their economic competitiveness. Significant progress is currently being made in many aspects of membrane technology for the development of new membrane materials with higher selectivity and permeability [1].

A Hollow Fiber (HF) is a capillary having a diameter of less than 1mm, whose wall functions as a semi-permeable membrane. To produce wastewater of a quality superior to the minimum effluent disposal standards, the standard of membrane modules must be state of the art. Membrane technology pertaining to HFs belongs to the category of pressure driven filtration, mainly ultrafiltration (UF) and microfiltration (MF) [2]. HF membranes are commercially utilized by integrating them into compact packages/modules. Each module contains a bundle of numerous fine fibers capable of withstanding a pressure gradient across the fiber wall. HF modules can be operated with feed flowing in either the tube-side (lumen side) or shell-side and permeate may flow in either a countercurrent or co-current direction relative to the feed. Variation in fiber inner diameter and fiber properties such as permeability and selectivity are critical factors governing module performance [3]. PVDF is a semi-crystalline polymer suitable for synthesis of HF membrane with high chemical resistance, besides thermal and mechanical stability. However, the hydrophobicity of PVDF remains an issue and limits its application and thus, inducing hydrophilicity to PVDF membrane is an area of focus in UF. Blending is often used to alter polymer material properties. PAN is one of the well-known polymeric materials for making UF membranes and possesses considerable hydrophilicity with added advantages and was hence chosen in this study for blending with PVDF. Although both miscibility and morphology of PVDF/PAN blends have been studied earlier, the effect of blending with PAN on the separation properties of PVDF hollow fibers is not reported.

In the current study, a manual spinning machine was designed and fabricated indigenously along with the critical spinneret component used for extrusion. PVDF/PAN HF blend membranes were synthesized by the dry-wet spinning method and characterized by scanning electron microscopy (SEM), pure water flux and Bovine Serum Albumin (BSA) rejection. Effect of different parameters such as flow rate of polymer dope, bore liquid and polymer pump speed on specific macroscopic and microscopic properties of the hollow fibers was evaluated. These membranes were further fabricated into different modules, based on permeate collection mode, by equipping them with single or twin permeate outlets. Flux and turbidity rejection of the fabricated modules was studied for the clarification and disinfection of surface water.

24.2 EXPERIMENTAL

24.2.1 MATERIALS

Dimethyl formamide (DMF) procured from s.d. Fine Chemicals, Mumbai was used as the solvent for preparing the polymer blend solution. Poly(vinylidene fluoride) (PVDF) and Poly(acrylonitrile) (PAN) polymers were supplied by Permionics Membranes Pvt. Ltd., Vadodara, India. Water is usually chosen as the nonsolvent owing to economic reasons to bring the polymer solution close to the two-phase unstable region on the ternary phase diagram, such that rapid phase inversion of the polymer occurs when the dope is immersed in the quench bath. Deionized water for bore fluid was prepared in-house using the laboratory reverse osmosis system. Tap water was used for gelation of the hollow fiber membranes.

24.2.2 HOLLOW FIBER MEMBRANE SPINNING PROCESS

HF membranes were spun at room temperature (25–30°C) employing the dry-wet spinning technique. The spinning solution was prepared from 10 wt% PVDF + 10 wt% PAN in DMF by continuous stirring at approximately 60°C for about 12–15 h which ensured complete dissolution of the polymers. The prepared polymer dope was observed to be transparent and homogenous at room temperature and the solution was then degassed overnight. Further, the polymer solution was loaded into a reservoir and forced into the spinneret using pressurized nitrogen. The dope solution and the internal coagulant liquid were forced through a tube-in-orifice spinneret, in such a manner, that the polymer solution flowed through an outer ring nozzle while the coagulating fluid was fed through the inner tube [2]. Figure 24.1 reveals the methodology of hollow fiber spinning employed in this study. The polymer solution was directly extruded into a coagulation bath after an air gap of 13 cm. The fibers were collected in a take-up drum and immersed in ethanol solution for about 24 h to replace water in membrane pores with ethanol that possess lower surface tension [4, 5].

The dimensions of the annular spinneret opening, the polymer to bore volumetric flow rate ratio, and the draw ratio are the primary factors that determine the final fiber dimensions. The ultimate outer to inner fiber radii ratio is determined by the polymer to bore volumetric flow rate ratio. The membrane structure in terms of pore diameter and pore size distribution is determined by some factors like air gap length, quench air temperature, viscosity of dope and type of solvent used [6].

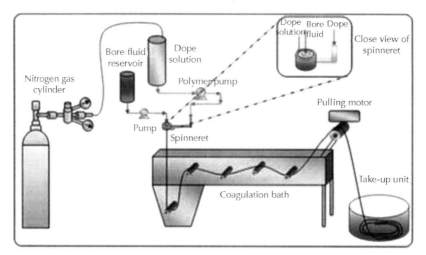

FIGURE 24.1 Schematic view of hollow fiber spinning machine process

24.2.3 MODULE FABRICATION AND FEATURES

The membrane module is the most vital part in a membrane separation system. Separation occurs in the membrane module in which the HF membranes are systematically packed to maximize the membrane area per unit volume, also known as packing density.

A bundle of HFs were introduced into the poly(vinyl chloride) (PVC) tube or housing of 2.54 cm diameter and 30.48 cm length. Either ends of the module were potted using epoxy resin. A nylon rod was used for making end caps with provision of openings for feed inlet and reject outlet flow. This forms a tube-side flow configuration as the permeate flows through the shell side. The PVC housing based HF membrane modules are shown in Figure 24.2(a). The internal diameter and wall thickness of the synthesized UF HF membrane were found to be 0.9 and 0.4 mm, respectively. The effective area of the membrane module was 0.07 m^2. Standard solutions of known molecular weights of Dextran were used to determine the molecular weight cut off (MWCO) of the membranes. The MWCO of 20 wt% PVDF/PAN HF membranes was found to be around 40 kDa.

24.2.4 EXPERIMENTAL SETUP AND PROCEDURE

A bench scale system was built to incorporate a manually fabricated HF membrane module made of PVDF/PAN blend. The feed input to the system was given through a feed tank of 5L capacity. The feed tank also had a provision for recycling of the reject. The feed passed through a prefilter before reaching the membrane module to prevent clogging of the HF membrane by sediments, silts, and sand present in the feed water. Feed then finally flowed through the HF membrane module to separate the pyrogenic bacteria, virus and turbidity as retentate with purified water coming out through the membrane pores as permeate. The required pressure was maintained by adjusting a needle valve arranged in the reject line. In this study, surface water was sourced from Uppal Cheruvu lake, Hyderabad, India. 5 L of feed containing 230 FAU turbidity, 92 mg/L suspended solids, 1.3×10^3 (MPN/100ml) *E. coli* and 7.2 pH was introduced into the UF system using a diaphragm pump at a moderate hydraulic pressure (0.2–2bar). The flow sheet of the experimental setup for surface water purification is shown in Figure 24.2 (b).

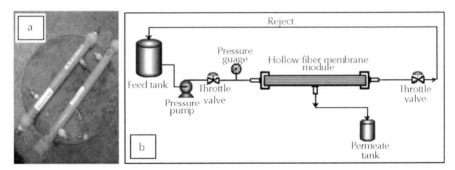

FIGURE 24.2 (a) Pictorial view of PVC hollow fiber membrane modules and (b) flow sheet of laboratory ultrafiltration setup used in surface water treatment

24.2.5 MEMBRANE CLEANING AND MAINTENANCE

At the end of the operation back flushing is carried out for 10 min with raw water to displace the impurity or solid particles deposited within the membrane pores or in the form of scales on the barrier surface . A 1.0 percent solution of citric acid prepared in 5L of deionized water was circulated for 10 min with complete recycle of permeate, every alternate week to restore flux and rejection properties. Backwashing was done on alternate days with tap water to remove pore clogging. To prevent biological fouling and irreversible destruction, the membrane was stored in 0.5 percent solution of sodium metabisulfite ($Na_2S_2O_5$) during shut down.

24.3 RESULTS AND DISCUSSIONS

24.3.1 SCANNING ELECTRON MICROSCOPY (SEM) STUDIES

The morphological characteristics of surface and cross-section of the HF membrane were explored using SEM. Figure 24.3 presents the cross-section and surface SEM images of PVDF/PAN HF membrane. Figure 24.3(a) shows a typical asymmetric structure of HF with a tight skin layer and a more porous substructure. The formation of porous substructure and voids near both inner and the outer edges of the membrane can be observed. This can be attributed to the penetration of bore fluid and external coagulant from the inner and outer surfaces of the membrane during the phase inversion process. The surface morphology of membrane exhibited by Figure 24.3(b) reveals the presence of fine pores that are much tighter and appear to be distributed uniformly across the membrane.

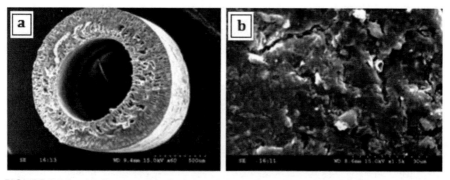

FIGURE 24.3 (a) Cross-sectional view and (b) surface view of PVDF/PAN membrane

24.3.2 EFFECTS OF SPINNING PARAMETERS ON HF MEMBRANE STRUCTURE

24.3.2.1 EFFECT OF POLYMER EXTRUSION RATE

Spinning speed determines the productivity of HF membrane manufacturing. Any variation in speed of extrusion directly affects the flow chemistry of the dope flowing through the spinneret and subsequently the structure of fiber and its separation behavior. The effect of dope extrusion rate on outer diameter and wall thickness of the HF membrane at constant bore flow rate (8 mL/min) and speed of pulling motor (30 rps) is graphically represented in Figure 24.4. Increase in dope extrusion rate resulted in enlargement of the outer diameter and wall thickness of the fibers due to a rise in shear rate, which in turn leads to decrease in viscosity of the polymer solution. Low viscosity of the polymer solution leads to a loose membrane structure bringing about greater wall thickness and outer fiber diameter.

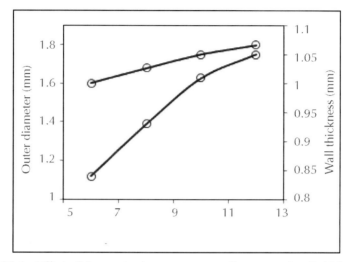

FIGURE 24.4 Effect of dope extrusion rate on outer diameter and wall thickness of HF membrane

24.3.2.2 EFFECT OF BORE FLUID FLOW RATE

During the dry/wet phase inversion process, the use of bore fluid as an internal coagulant not only influences the stress to open the HF but also offers the nonsolvent, water in bulk for exchange with the DMF solvent from the inner wall of the nascent fiber. Figure 24.5 (a) shows the effect of bore flow rate on fiber dimensions at constant dope extrusion rate of 270 rps and pulling motor speed of 30 rps. As expected, an increase in the bore flow rate increased liquid pressure in an axial direction. Therefore, the inner diameters of the HF would enlarge without any change in the outer diameter.

The bore fluid can be used to fine-tune the structure of the HF, which has an impact on the membrane permeability. The solvent–nonsolvent exchange at the interface between polymer solution and the nonsolvent begins after the nascent fiber leaves the spinneret and contacts the internal coagulant in the bore side of the fiber. An increase in internal coagulant flow velocity enhances the rate of solvent–nonsolvent exchange and further accelerates pore formation.

24.3.2.3 EFFECT OF PULLING MOTOR SPEED

Figure 24.5 (b) describes the effect of take-up speed on the outer diameter of fiber at constant dope extrusion rate (270 rps) and bore fluid flow rate (8 mL/min). The curve indicates that the outer diameter of the HF decreases with an increase in the take-up speed. This can be attributed to the orientation of the polymer molecules. The molecular chains in the HF are more oriented when the fiber take-up speed is increased but the volume between the polymer chains decreases, which results in HF with smaller diameters. However, it is observed that there is no considerable change in the wall thickness of the fiber. A higher take-up speed will tend to stretch the fibers, leading to more porous morphology and lower outer and inner fiber diameters.

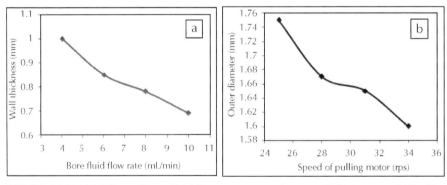

FIGURE 24.5 (a) Effect of bore fluid flow rate on membrane wall thickness and (b) effect of pulling motor speed on membrane outer diameter

24.3.3 TESTING HF MEMBRANE PERFORMANCE

Permeation characteristics of pure water and rejection of Bovine serum albumin (BSA) were studied with the indigenously developed HF membrane modules by varying the feed pressure from 0.2 to 1.5 bar at ambient temperature. Effect of applied pressure on pure water flux and BSA rejection for PVDF/PAN HF membranes is depicted in Figure 24.6 (a). At a pressure of 0.2 bar, the membrane exhibited a flux of 59.3 L/m² h. Increase in pressure to 1.5 bar further enhanced the flux to 218.5 L/m² h. Rejection of 95.6 percent of BSA was obtained at 0.2 bar pressure.

Effect of time on permeate flux at a constant pressure of 1 bar and feed turbidity of 230 FAU for 20 wt% PVDF/PAN HF membrane is shown in Figure 24.6 (b). Flux lowered with filtration time due to concentration polarization and gradual fouling of the membrane. High initial flux followed by a rapid decline is characteristic of operations carried out at constant trans-membrane pressure (TMP) gradient. High initial flux causes rapid deposition of rejected solute molecules which results in the buildup of a boundary layer at the membrane surface causing resistance to solvent, water. Flux declined from 77.2 to 71.8 L/m² h after 120 min of continuous operation at 1 bar pressure. However, initial flux was restored to 76.9 L/m² h after 10 min of backwashing.

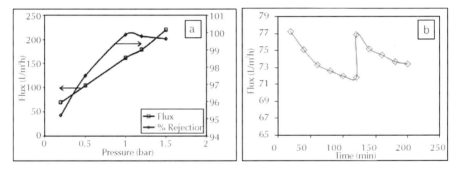

FIGURE 24.6 Pure water flux and BSA rejection of membranes at (a) varying feed pressure and (b) effect of operating time on flux at 1 bar pressure.

The effect of pressure on flux and turbidity rejection for surface water feed processed through PVDF/PAN HF membranes are described in Figure 24.7(a) and (b). At 0.5 bar, a flux of 46.3 L/m² h was obtained with a turbidity rejection of 95.6 percent. An enhancement in flux up to a value of 91.2 L/m² h was achieved with 99.2 percent rejection when the applied pressure was increased to 1.2 bar. Further rise in flux was observed with pressure but resulted in decreased turbidity rejection indicating that these HFs should preferably be operated at lower pressures (~1bar) to achieve optimum results. In addition, these membranes have shown an *E. coli* reduction of 5 log, falling in the desirable 4–6 log reduction range at 1 bar, which exhibits acceptable quality of water obtained in permeate.

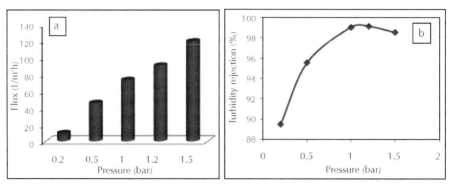

FIGURE 24.7 Effect of pressure on (a) flux and (b) turbidity rejection

A comparison of performance was made between 20 wt% PAN HF membranes and 20 wt% PVDF HF membranes with that of 20 wt% PVDF/PAN 1:1 blend HF membrane as depicted in Figure 24.8 (a) and (b). From these experimental observations, it is clear that blend HF membranes have shown peculiar behavior with respect to flux and turbidity rejection when compared to pristine PAN and PVDF HF membranes. The flux at 1 bar for PAN and PVDF HF membranes based on the individual polymers were determined to be 49.1 and 47.2 L/m^2 h, respectively. On the other hand, PVDF/PAN blend HF membrane exhibited much higher flux of 73.9 L/m^2 h with 99.1 percent turbidity rejection revealing that blending enabled utilization of the important properties of both the polymers.

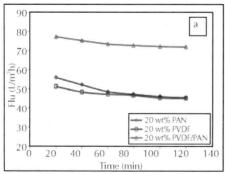
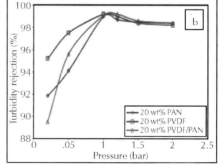

FIGURE 24.8 Comparison between PAN, PVDF, and PAN/PVDF blend membranes: (a) variation of flux with time and (b) turbidity rejection at 1 bar pressure

24.4 CONCLUSIONS

Hollow fiber ultrafiltration membranes provides an effective and economical platform for purification of surface water using physical barrier filtration through remove of pathogens and turbidity at low applied pressures. Most of the current water treatment technologies are based on chlorination, boiling, or adsorption which get affected by sudden spikes in turbidity or pathogen concentrations. The membrane filters out turbidity and disinfects the water in a single step and avoids presence of dead microbes in the treated water. Indigenous hollow fiber ultrafiltration blend membranes based on PVDF/PAN were synthesized using a dry/wet spinning process using forced convection in the dry air gap. Tap water was chosen as the external coagulant and deionized water was used as the bore fluid. Membrane modules were manually fabricated and their performance for purification of surface water was investigated. The membrane exhibited a turbidity rejection of 99.1 percent at a low hydraulic pressure of 1 bar with a high flux of 73.9 L/m^2 h. An overhead tank pressure (0.5–1 bar) could be sufficient to treat surface water at desirable flux of pure water without using electric power. The large surface area per unit volume and self-supporting structure makes hollow fiber systems very compact, reliable, and back washable.

KEYWORDS

- **Flux**
- **Hollow fiber membrane**
- **PVDF/PAN blend**
- **Rejection**
- **Surface water treatment**

REFERENCES

1. Guo, H.; Wyart, Y.; Perot, J.; Nauleau, F.; and Nauleau, P.;*Water Res*. **2010**, *44*, 41–57.
2. Ahmad, A. L.; Sarif, M.; and Ismail, S.;*Desalination.***2005**, *179*, 257–263.
3. Andrzej, C.; Cezary, W.; Konrad, D.; and Ewa, L.; *Biocybern. Biomed. Eng.* **2009**, *29*, 47–59.
4. Chung, T. S.; Qin, J. J.; and Gu, J.; *Chem. Eng. Sci.* **2000b**, *55*, 1077–1091.
5. Yu, D. G.; Choua, W. L.; and Yang, M. C.; *Sep. Purif. Technol.* **2006**, *52*, 380–387.
6. Khayet, M.; Feng, C. Y.; Khulbe, K. C.; and Matsuura, T.;*Polymer*. **2002**, *43*, 3879–3890.

CHAPTER 25

DEVELOPMENT OF EPOXIDE MATERIAL FROM VEGETABLE OIL

SRIKANTA DINDA*, NIHIT BANDARU, and RADHEV PALETI

Department of Chemical Engineering, BITS Pilani Hyderabad Campus, Hyderabad – 500078, India; *E-mail: srikantadinda@gmail.com

CONTENTS

25.1 Introduction 270
25.2 Experimental Details 271
25.3 Results and Discussion 272
25.4 Conclusions 278
Keywords 278
References 279

25.1 INTRODUCTION

Vegetable oils are sustainable and renewable resources that can be treated chemically or enzymatically to produce bio-based materials that can often act as replacement for materials derived from petroleum [1]. Epoxidation of vegetable oils is a commercially important reaction, because the epoxides obtained from these renewable raw materials have wide applications in different areas such as for making plasticizers and polymer stabilizers [2]. Epoxidation of soybean oil for the synthesis of plasticizers and polymer stabilizers has already been successfully implemented at plant scale [3–5]. Epoxide serves as a plasticizer and as a scavenger for hydrochloric acid liberated from PVC when the PVC undergoes heat treatment. Okieimen et al. [6] have studied the epoxidation of rubber seed oil by peroxyacetic acid (PAA) generated *in situ*. Industrially, various peroxyacids such as peroxyformic acid, peroxyacetic acid, and peroxybenzoicacid are used for the epoxidation reaction. Out of the above mentioned peroxyacids, peroxyacetic acid is widely used due to its easy availability, lower price, high epoxidation efficiency, and reasonable stability at ordinary temperatures. Dindaet al. [7] have studied the kinetics of epoxidation of cottonseed oil by peroxyacetic acid generated *in situ* from hydrogen peroxide. The epoxidation of vegetable oil can be carried out with *in situ* formed or preformed peroxyacid in presence of an acidic catalyst. The *in situ* process is widely used because the preformed concentrated organic peroxyacid is unstable. The acidic catalyst may be a homogeneous or heterogeneous one. The main drawback of a homogeneous catalyst is that, it helps the oxirane cleavage reaction. The use of heterogeneous acidic catalyst for the epoxidation of vegetable oils was found to minimize the side reactions and to improve selectivity. Goudet al. [8] have studied the epoxidation reaction of karanja (*pongamiaglabra*) oil catalyzed by acidic ion exchange resin. The epoxidation kinetics of anchovy oil with partially preformed PAA in the presence of a resin catalyst has been reported by Wisniak and Navarrete [9]. Petrovicet al. [10] and Jankovićand Sinadinovic–Fišer [11] have studied the reaction kinetics of epoxidation of soybean oil and the extent of side reactions with *in situ* formed PAA and peroxyformic acid and in the presence of an ion exchange resin as catalyst.

The cost effectiveness of the route greatly depends upon the local and cheap availability of the raw material. Both edible and nonedible vegetable oils such as rice–bran, cottonseed, groundnut, sunflower, rapeseed, coconut, linseed, castor, neem, karanja, and nahor oil are easily available in India. Epoxidation of edible oils may not be cost effective for the manufacture of epoxides. Hence, it would be always better to target large number of nonedible vegetable oils to study the epoxidation reactions. To the best of our knowledge, there is no literature available on the epoxidation of nahor oil. Therefore, the objective of the present work was to study the epoxidation of nahor oil (*Mesuaferrea Linn*) with *in situ* generated peroxyacetic acid, with a view of obtaining value-added products from locally available renewable natural resources.

25.2 EXPERIMENTAL DETAILS

25.2.1 MATERIALS

Glacial acetic acid (CH_3COOH) (99–100%), aqueous hydrogen peroxide (H_2O_2) (50%), were obtained from Merck (India) Limited. Acidic ion exchange resin Amberlite IR 120 (20–50 mesh) was procured from LobaChemie. All other chemicals for analytical purpose were obtained from Merck (India) Limited and Sd fine-chem Limited. Nahor trees are common in tropical parts of India, Sri Lanka, southern Nepal, Burma, Thailand, Indochina, the Philippines, Malaysia, and Sumatra, where it grows in evergreen forests, especially in river valleys. Nahor seed kernel contains about 70 wt% oil with an unpleasant pungent odor [12]. For the present study, refined nahor oil was obtained from a Mudar India Export.

25.2.2 EXPERIMENTALPROCEDURE

The reactions were carried out in a mechanically agitated batch reactor. The reactor assembly was immersed in a thermostatic water bath, with a temperature control ±1°C. A suitable amount of nahor oil was taken in the reactor. Calculated amount of acetic acid (AA) and AIER were added to the reactor and the mixture was stirred for about 15 min. Then, the required amount of 50 percentaqueous H_2O_2 was added dropwise at a rate such that the addition was completed in half an hour and the reaction was continued further for the desired time duration. Samples were withdrawn intermittentlyandthe collected samples were then extracted with ether, washed with water and then analyzed for iodine value, oxirane content and α–Glycol content.

25.2.3 ANALYTICALTECHNIQUES

Iodine value was determined according to Wijs method [13]. The percentage oxirane oxygen was determined by direct method using hydrobromic acid solution in glacial acetic acid [14]. α–Glycol content was determined by the method reported by May [15]. From the oxirane content values, the relative percentage conversion to oxirane was calculated using the following expression:

$$\text{Relative percentage conversion to oxirane} = (OO_{exp}/OO_{the}) \times 100 \quad (25.1)$$

where, OO_{exp} is the experimentally obtained and OO_{the} is the theoretically obtainable maximum oxirane oxygen.

The relative percentage conversion to α–glycol was calculated from the following expression:

$$\text{Relative percentage conversion to } \alpha\text{-glycol} = (G_{exp}/G_{the}) \times 100 \quad (25.2)$$

where, G_{exp} is the experimentally obtained and G_{the} is the theoretically obtainable maximum glycol.

25.3 RESULTS AND DISCUSSION

Epoxidation was carried out with the parameters varied in the following ranges: stirring speed 500–2000 rev/min.; temperature 30–75°C; H_2O_2–to–ethylenic unsaturation (EU) mole ratio 1.25–2.0; AA–to–EU mole ratio 0.3–0.6. The AIER used as catalyst for epoxidation was Amberlite IR 120. The catalyst loading was expressed as weight percentage with respect to "oil phase" and was 5–20 percent for Amberlite IR–120. The properties of oil, as experimentally determined were as follows: specific gravity 0.95 at 30°C; iodine value 90±1 (g I_2/100 g oil); saponification value 198±2 (mg KOH/g oil) and acid value 5±0.5 (mg KOH/g oil). Some of the experiments were repeated under identical conditions to find out the percentage deviation between two same set of experimental results and it was found that the deviation lies within ±4 percent.

25.3.1 EFFECT OF STIRRING SPEED

To investigate the effect of stirring speed on catalytic epoxidation rate under tri–phase conditions, 0.25 mole oil was treated with H_2O_2 (H_2O_2–to–EU mole ratio 2), AA–to–EU mole ratio 0.5, and 15 percent solid catalyst at 75±1°C. The experiments were performed in the range of stirring speeds 500–2000 rev/min. Figure 25.1 shows that, the oxirane formation rate increased with an increase of stirring speed up to 1500 rev/min and beyond 1500 rev/min epoxidation rate was not substantially affected by stirring speeds, and hence, it was assumed that the reaction was free from mass transfer resistance beyond 1500 rev/min under the given conditions. However, all subsequent experiments were performed at a stirring speed of 2000±100 rev/min to ensure that the reaction was kinetically controlled.

Development of Epoxide Material from Vegetable Oil

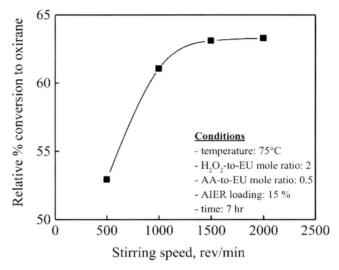

FIGURE 25.1 Effect of stirring speed on oxirane conversion.

25.3.2 EFFECT OF RESIN PARTICLE SIZE ON OXIRANE CONVERSION

To find the effect of particle size on the epoxidation rate (i.e., intraparticle diffusional resistance), the experiments were conducted by using two widely different sizes of resin particles under otherwise similar conditions. The particle sizes used were greater than 610 µm and less than 110 µm, and the results are shown in Table 25.1. The results show that both particle sizes gave nearly the same oxirane conversion even though the particle sizes differed nearly by a factor of over 5. Therefore, it can be concluded that the intraparticle diffusional resistance was practically absent under the experimental conditions considered for the present study. Hence, the commercial AIER (particle size ≈ 600 µm) was used for subsequent experimentation.

TABLE 25.1 Effect of resin particle size on oxirane conversion

Particle size (µm)	Parameter	Time (hr)		
		1	5	9
>610	Relative % conv. to OO	19.5	48.4	67.1
<110	Relative % conv. to OO	20.2	47.3	68.9

Conditions: temperature = 75°C; H_2O_2–to–EU mole ratio = 2; AA–to–EU mole ratio = 0.5; AIER loading = 15%; stirring speed = 2000±100 rev/min.

25.3.3 EFFECT OF TEMPERATURE

To study the effect of temperature on epoxidation rate, calculated amount of H_2O_2, CH_3COOH, and AIER Amberlite IR 120 were added with nahoroil under a stirring speed of around 2000 rev/min and the runs were taken at different temperatures in the range of 30–75°C. The conversions of oxirane oxygen at different temperatures are shown in Figure 25.2. The oxirane content increased monotonically with reaction time in the temperature range 30–60°C. But, at 75°C the extent of epoxidation initially increased and reached a maximum value, after which it decreased with the reaction time. It was found that as temperature increased, the epoxidation rate increased. The decrease of oxirane content after a certain time, at high temperature may be due to the hydrolysis of oxirane to α–glycol. The iodine value conversions, α–glycol value of the final product at the corresponding temperatures are shown in Table 25.2. These results suggested that an optimum level of epoxidation could be attained at moderate reaction temperature of about 75°C at which epoxide degradation would be minimal and the thermal stability of resin catalyst would be quite good.

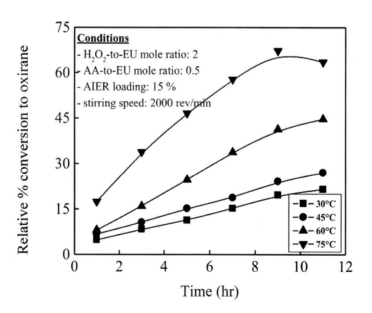

FIGURE 25.2 Effect of temperature on oxirane conversion.

TABLE 25.2 Effect of temperature on iodine value, and α–glycol conversion

Temperature (°C)	Parameters	Time (hr)			
		5	7	9	11
30	% conv. of IV	13.1	14.9	18.2	21.6
	relative % conv. to α–G				1.7
45	% conv. of IV	16.8	20.2	23.3	27.1
	relative % conv. to α–G			2.2	2.5
60	% conv. of IV	27.3	37.1	42.0	50.9
	relative % conv. to α–G			2.9	3.3
75	% conv. of IV	52.2	60.8	66.4	71.3
	relative % conv. to α–G			3.1	4.9

Conditions: H_2O_2–to–EU mole ratio = 2; AA–to–EU mole ratio = 0.5; AIER loading = 15%; stirring speed = 2000±100 rev/min.

25.3.4 EFFECT OF H_2O_2–TO–EU MOLE RATIO

The effect of H_2O_2–to–EU mole ratio on the conversion to oxirane and α–glycol was studied in the mole ratio range 1.25–2. Figure 25.3 shows that the epoxidation rate increased as the concentration of H_2O_2 in the system increased from 1.25 to 1.75 and after which it had no significant effect on extent of oxirane formation. Although the maximum conversion to oxirane was obtained for a mole ratio of 2.0, the stability of the oxirane ring was comparatively less at this high mole ratio. On the other hand, at low concentrations of H_2O_2, oxirane ring was quite stable. Almost negligible difference was observed in the final conversions attained for mole ratios of 1.75 and 2.0.

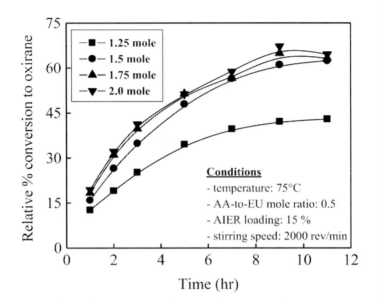

FIGURE 25.3 Effect of H_2O_2 concentration on oxirane conversion.

25.3.5 EFFECT OF AA–TO–EU MOLE RATIO

To investigate the effect of AA concentration on the epoxidation rate and glycol formation of nahor oil, the mole ratio was varied in the range 0.3–0.6. *In situ* epoxidation of nahor oil, catalyzed by AIER is a heterogeneous reaction process. The PAA is generated in the aqueous phase and must be transferred to the oil phase to effect epoxidation. The mass transfer process may be diffusion controlled. For a reaction mixture containing a fixed amount of double bonds, there may be a concentration of AA required for optimum epoxidation, beyond which oxirane cleavage may become important. Figure 25.4 shows that the extent of epoxidation increased as the concentration of AA in the system increased but at high mole ratio of AA, the stability of the oxirane ring decreased. It has also been observed that, the α–glycol content in the final product increased by 4 percent, as the AA mole ratio increased from 0.5 to 0.6. Therefore, to obtain the maximum oxirane, the optimum level of concentration of AA should be used where both the effects are optimized. Hence, within the experimental conditions, the most favorable concentration of AA appeared to be 0.5 mole ratio.

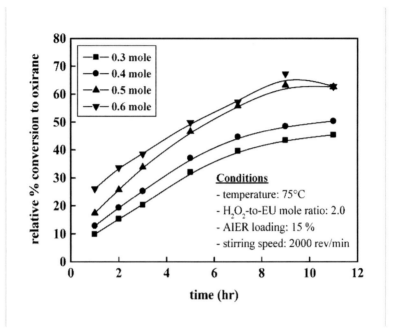

FIGURE 25.4 Effect of CH_3COOH concentration on oxirane conversion.

25.3.6 EFFECT OF CATALYST (AIER) LOADING

The active cites present on the surface of solid catalyst might play an increased contributory role with an increase in catalyst concentration. It was expected that an increase of catalyst concentration might increase the rate of *in situ* PAA formation. Hence, to investigate the effect of AIER catalyst concentration on the epoxidation reaction, concentration of catalyst was varied from 5 to 20 wt% of Amberlite IR–120. Figure 25.5 shows that, when the catalyst loading was increased from 5 to 15 wt% the oxirane conversion increased and after which its effect was insignificant specifically after 9 hr. Therefore, the catalyst loading for the subsequent experiments were fixed to 15 wt% of Amberlite IR–120.

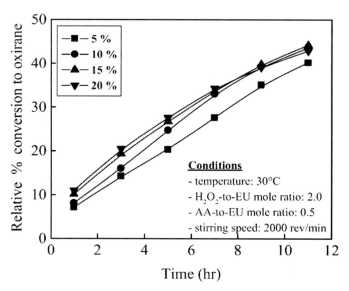

FIGURE 25.5 Effect of catalyst (AIER) loading on oxirane conversion.

25.4 CONCLUSIONS

The epoxidation of nahor oil using *in situ* generated peroxyacid could be carried out at moderate temperature of about 75°C. Higher temperature and higher acid concentrations reduced the reaction time needed to reach the maximum conversion to oxirane value; however, it simultaneously increased the extent of oxirane ring cleavage to glycols. The reaction was kinetically controlled beyond a stirring speed of about 1500 rev/min. These optimum conditions include a temperature range of 60–75°C, H_2O_2–to–EU mole ratio range 1.75–2.0, AA–to–EU mole ratio of about 0.5 and catalyst loading of about 15 percent. It was possible to obtain more than 67 percent (Fig. 25.2) relative conversion to oxirane at 75°C under the present experimental conditions.

KEYWORDS

- AIER catalyst
- Hydrogen peroxide
- *In situ* epoxidation
- Nahor oil
- Oxirane

REFERENCES

1. Gan, L. H.; Ooi, K. S.; Gan, L. M.; and Goh, S. H.; *J. Am. Oil Chem. Soc.* **1995**,*72*, 439–442.
2. Wallace, J. C.; Encyclopedia of Chemical Technology, 3rd ed., vol. 9. New York, NY: John Wiley & Sons; **1978**.
3. Rios, L. A.; Weckes, P.; Schuster, H.; and Hoelderich, W. F.; *J. Catal.* **2005**, *232*, 19–26.
4. Biermann, U.; Friedt, W.; Lang, S.; Luhs, W.; Machmuller, G.; Metzger, J. O.; Klaas, M. R.; Schafer, H. J.; and Schneider, M. P.; *Angew. Chem. Int. Ed.* **2000**, *39*, 2206–2224.
5. Klaas, M. R.; and Warwel, S.; *Ind. Crops Prod.* **1999**, *9*, 125–132.
6. Okieimen, F. E.; Bakare, O. I.; and Okieimen, C.O.; *Ind. Crops Prod.* **2002**, *15*, 139–144.
7. Dinda, S.; Patwardhan, A. V.; Goud, V. V.;and Pradhan, N. C.;*Bioresour. Technol.* **2008**, *99*, 3737–3744.
8. Goud, V. V.; Patwardhan, A. V.; Dinda, S.; and Pradhan, N. C.; *Eur. J. Lipid Sci. Technol.* **2007**, *109*, 575–584.
9. Wisniak, J.; and Navarrete, E.; *Ind. Eng. Chem. Prod. Res. Dev.* **1970**, *9*, 33–41.
10. Petrovic, Z. S.; Zlatanic, A.; Lava, C. C.; and Sinadinovic–Fišer, S.; *Eur. J. Lipid Sci. Technol.* **2002**, *104*, 293–299.
11. Janković, M.; and Sinadinovic–Fišer, S.; *J. Am. Oil Chem. Soc.* **2005**, *82*, 301–303.
12. Salunkhe, D. K.; Chavan, J. K.; Adsule, R. N.; and Kadam, S. S.; World Oilseeds: Chemistry, Technology and Utilization, 1st ed. New York, NY: Van Nostrand Reinhold; **1992**.
13. Siggia, S.; and Hanna, J. G.; Quantitative Organic Analysis via Functional Groups. New York, NY: Wiley Interscience; **1979**.
14. Pequot, C.; Standard Methods for the Analysis of Oils, Fats and Derivatives Part–1, 5th ed. Frankfurt, Germany: Pergamon Press; **1979**.
15. May, C. A.; Epoxy Resins: Chemistry and Technology. New York, NY: Marcel Dekker; **1973**.

CHAPTER 26

BIOSORPTION OF COPPER FROM AQUEOUS SOLUTION USING OSCILLATORIA SPLENDIDA

G. BABURAO[1*], P. SATYASAGAR[2], and M. KRISHNA PRASAD[3]

[1,2,3]Department of Chemical Engineering, GMRIT, Rajam, Srikakulam.
*E-mail: baburao01803@gmail.com

CONTENTS

26.1 Introduction 282
26.2 Materials and Methods 283
26.3 Results and Discussion 284
26.4 Conclusions 288
Keywords 290
References 290

26.1 INTRODUCTION

Water pollution due to toxic heavy metals has been a major cause of concern for environmental engineers. The industrial and domestic waste water is responsible for causing several damages to the environment and adversely affecting the health of the people. Several episodes due to heavy metal contamination in aquatic environment increased the awareness about the heavy metal toxicity. Metals can be distinguished from other toxic pollutants, since they are nonbiodegradable and can accumulate in living tissues, thus becoming concentrated throughout the food chain. A variety of industries are responsible for the release of heavy metals into the environment through their wastewater. These include iron and steel production, the nonferrous metal industry, mining, and mineral processing.

Copper is very common substance that occurs naturally. Copper metal and alloys have been used for thousands of years. In the Roman era, copper was principally mined on Cyprus, hence the origin of the name of the metal as Cyprium, "metal of Cyprus," later shortened to Cuprum. There may be insufficient reserves to sustain current high rates of copper consumption. Some countries, such as Chile and the United States, still have sizable reserves of the metal which are extracted through large open pit mines. Copper in our diet is necessary for good health. You eat and drink about 1,000 μg of copper per day. Drinking water normally contributes approximately 150 μg/day [1]. Immediate effects from drinking water which contains elevated levels of copper include vomiting, diarrhea, stomach cramps, and nausea.

In recent years, various removal methods have been developed for the treatment of heavy metal containing wastewaters. Chemical precipitation, solvent extraction, reverse osmosis, ion exchange, evaporation, fixation/solidification, filtration, adsorption, oxidation, reduction, dialysis/electro dialysis etc. are than most widely employed techniques for the removal of metal ions from industrial wastewaters [2]. In spite of their common usage for the removal of heavy metals from water and wastewater, they have some disadvantages, for example, they may be expensive, ineffective for low metal concentrations (around 1–100 ppm), produce toxic sludge, or other waste products that require the uses of high reagent and energy. Among these methods, adsorption has especially been applied as an efficient method for removal of present heavy metals at low concentrations [3]. Nowadays, activated carbon is also used as a common adsorbent for the metal removal from waste and wastewater. Although a variety of activated carbons are available commercially, very few of them are selective for heavy metals. Also, they are very costly. Therefore, the use of new and inexpensive adsorbents for the removal of metals from wastewaters seems necessary. In recent years, several low-cost and nonconventional natural materials, industrial byproducts, and agricultural wastes have been examined for the removal of metal ions from water systems [4]. The results obtained showed that these adsorbents were employed efficiently in the removal of copper ions from aqueous solutions [5]. In the present work, the biosorption of copper ions by oscillatoria splendida was studied by investigating the influence of different process variables on copper uptake, such as contact time, adsorbent dosage, adsorbent size and metal concentration. The adsorption isotherms constants were obtained and modeled according to both Langmuir and Freundlich models.

26.2 MATERIALS AND METHODS

26.2.1 MATERIALS AND METHODS

The Oscillatoria Splendida used in this study was collected from freshwater (rivers, wells, etc.). The collected material was washed for several times then it was kept drying in sunlight for nearly 15 days. The dried material was powdered using domestic mixer. In the present study the powdered material is separated into four different fractions using sieves were then directly used as adsorbent without pretreatment.

By taking 3.92 g of copper sulfate in 1,000 ml Volumetric flask and make up with the distilled water gives 1,000 ppm metal stock solution. We prepared 20, 40, 60, 80, and 100 ml from 1,000 ppm metal stock solution in five different 1,000 ml volumetric flasks and make up these flasks with distilled water.

26.2.2 ADSORBENT CHARACTERIZATION

They occur in island saline lakes, and a few species tolerate temperatures up to 56–60°C. Some species are mat-formers in streams and a number are planktonic in freshwater lake and warmer marine waters. A few *s*pecies occur in terrestrial habitats subjected severe drying or in a shallow ephemeral freshwater in Polar Regions where freeze-drying accompanies winter. *Oscillatoria Splendida s*pecies have been associated with toxic blooms. *Oscillatoria Splendidas*pecies have a worldwide distribution in fresh-water, marine, and brackish waters.

These genera are considered members of the order *Oscillatoriales* which all filamentous cyanobacteria that only produce vegetative cells. They do not produce Heterocyst's (nonphotosynthetic cells that fix nitrogen) or akinetes (resting cells that later reproduce). It is Gram-negative and its size is Trichome is 1.0 − 100+ μm in diameter. Production of hepatotoxins and neurotoxins has been observed in some species of these genera.

26.2.3 MECHANISM OF ADSORPTION

26.2.3.1 QUANTITATIVE ANALYSIS BY ATOMIC ABSORPTION

In the atomic absorption process light at the resonance wavelength of initial intensity I_o is focused on the flame cell containing ground state atoms. The initial light intensity is decreased by an amount determined by the atom concentration in the flame cell. The light is then directed on to the detector where the reduced intensity I is measured. The amount of light absorbed is determined by comparing I to I_o.

The atomic absorption spectrometer used for present study consists of back ground connection accessory. It is designed for the use with flame absorption, flame emission techniques. An efficiency burner system offers both spoiler and impact head for optimal performance with all sample types with built in gas controls.

Standards required for analysis of the metal required were prepared according to the data given under how spoiler conditions.

Before switching on the instrument the pressure reading of air and acetylene were checked. Then lamp of the required metal was placed in proper place and parameters viz., lamp current, integration time, number of replicates, liner or non liner mode, and number of standards with concentration were entered.

26.3 RESULTS AND DISCUSSION

26.3.1 THE EFFECT OF CONTACT TIME

Effect of contact time on percentage adsorption of copper onto Oscillatoria Splendida was studied over an agitation time of 1–60 min, using 0.1 g of Oscillatoria Splendida 50 ml of 10 ppm solution of individual metal concentration at pH 6, temperature 303°K and 180 rpm shaking speed. The data obtained from adsorption of copper ions on to the Oscillatoria Splendida showed that the contact time of 45 min was sufficient to achieve equilibrium and the adsorption did not change significantly with further increase in contact time it is shows in Figure 26.1 and Table 26.1. Hence, the uptake on adsorbed copper concentrations at the end 45 min is given as the equilibrium value.

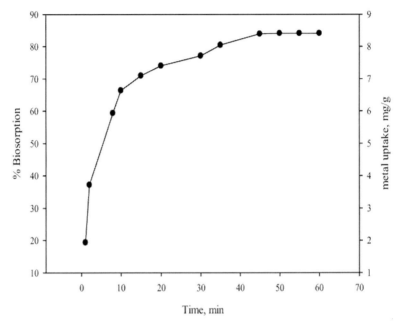

FIGURE 26.1 Effect of contact time on the percentage removal of copper.

TABLE 26.1 Range of variables covered in the present study

Variables	Minimum	Maximum	Max/Min
Time of contact (min)	1	60	60
Initial metal ion concentration (mg/L)	20	100	5
Average size of the adsorbent (µm)	85	200	2.35
Adsorbent dosage (g)	0.1	0.5	5

26.3.2 EFFECT OF METAL ION CONCENTRATION

Experiments were under taken to study the effect of initial copper concentration on the copper removal from the solution. From the data the copper metal uptake increases and percentage adsorption of the metals decreases with increase in the driving force, that is, concentration gradient. However, the percentage adsorptions of copper ions on Oscillatoria were decreased from 84 to 70 percent. Though an increase in metal uptake was observed, the decrease in percentage adsorption may be attributed to lack of percentage adsorption at higher concentration levels shows a decreasing trend whereas the equilibrium uptake of copper displays an opposite trend which is shows in Figure 26.2 and Table 26.1.

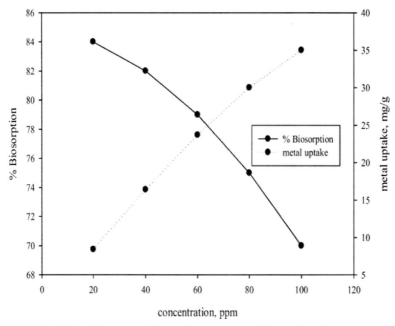

FIGURE 26.2 Effect of concentration on the percentage biosorption of copper.

At lower concentration, almost all the copper present in solution could interact with the binding sites and thus the percentage adsorption was higher than those at higher initial copper ion concentrations. At higher concentrations, lower adsorption yield is due to the saturation of adsorption sites. As a result, the purifications yield can be increased by diluting the wastewaters containing high metal ion concentrations.

26.3.3 EFFECT OF ADSORBENT SIZE

The effect of different adsorbent particle sizes (85–200 μm) on percentage removal of copper on Oscillatoria Splendida was investigated and showed in Figure 26.3 and Table 26.1. It reveals that the adsorption of copper on Oscillatoria Splendida decreases from 84 to 75 percent with the increases in particle size from 85 to 200 with 20 mg/L copper of concentration in solution. It is well known that decreasing the average particle size of the adsorbent increases the surface area, which in turn increases adsorption capacity. A marginal decline was also observed in uptake of copper with increases in size of the particle.

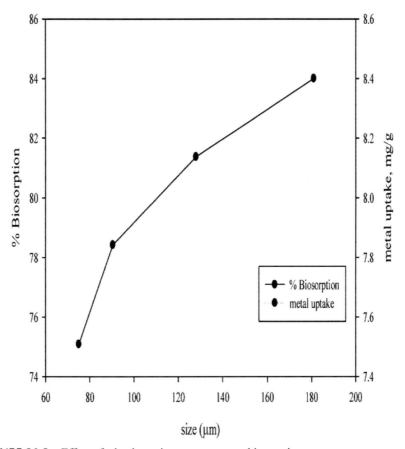

FIGURE 26.3 Effect of adsorbent size on percentage biosorption.

26.3.4 EFFECT OF ADSORBENT DOSAGE

The amount of copper adsorbed increased with an increase in adsorbent dosage from 0.1 to 0.5 at an initial concentration of 20 mg/L. The percentage copper removal was marginally increased from 84 to 94.66 for an increase in adsorbent concentration from 0.1 to 0.5 g. These shows that very small dosage of adsorbent, that is, 0.1 g is sufficient to treat about 50 ml of 20 mg/L copper concentration solution effectively which is shows in Figure 26.4 and Table 26.1.

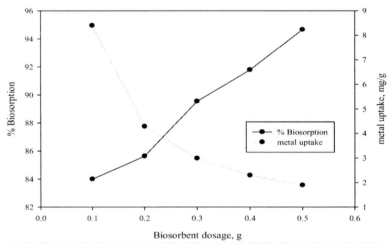

FIGURE 26.4 Effect of adsorbent dosage on percentage biosorption.

26.4 CONCLUSIONS

The biomass of the Oscillatoria Splendida demonstrated a good capacity of adsorption, highlighting its potential for efficient treatment processes. The data obtained from the aqueous solution of copper ions on the Oscillatoria Splendida showed that a constant time of 45 min was sufficient to achieve equilibrium and adsorption did not change with further increase in contact time. From the data the copper metal uptake increases and percentage adsorption of the metals decreases with increase in the driving force, that is, concentration gradient. However, the percentage adsorptions of copper ions on Oscillatoria were decreased from 84 to 70 percent. It was observed that the metal uptake increase and percentage adsorption of the metals decrease with increase in the initial metal ion concentration. The adsorption of copper on Oscillatoria Splendida decreases from 84 to 75 percent with the increases in particle size from 85 to 200 with 20 mg/L copper of concentration in solution. It reveals that the effect of different adsorption particle on the adsorption of copper is significant. The adsorption of the metal is decreased with increase in particle size for Oscillatoria Splendida.

The percentage copper removal was marginally increased from 84 to 94.66 for an increase in adsorbent concentration from 0.1 to 0.5 g The amount of copper adsorbed increases marginally in adsorbent dosage of Oscillatoria Splendida.

The experimental data gave good fit with Langmuir isotherm and the adsorption coefficient agreed well with the conditions of favorable adsorption which is shown in Figures 26.5 and 26.6 and also shown in Table 26.2.

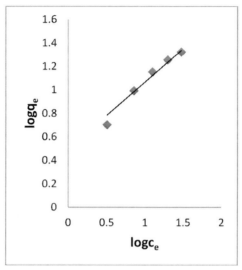

FIGURE 26.5 Freundlich isotherm.

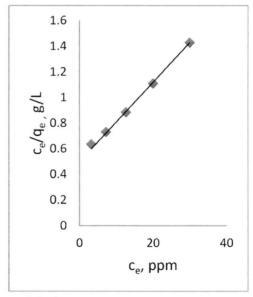

FIGURE 26.6 Langmuir isotherm.

TABLE 26.2 Equilibrium constants for copper onto Oscillatoria Splendida

Isotherm	Constants	Copper
Langmuir	Qmax (mg/g)	32.36
	b (L/mg)	0.0618
	R2	0.9964
Freundlich	Kf (mg/g)	3.162
	N	0.5663
	R2	0.9654

KEYWORDS

- Adsorbent
- Adsorption isotherms
- Biosorption
- Copper
- Oscillatoria Splendida

REFERENCES

1. Comparison of adsorption modeling of copper and zinc from aqueous solutions by Ulvafasciatasp. Y. prasannakumar, p. king and V.S.R.K. prasad environmental pollution control engineering laboratory, department of chemial engineering, a.u. college of engineering, Andhrauniversity, Visakhapatnam 530003, india.
2. Zinc removal from an aqueous solution using an industrial by-product phospogypsamhasancesur and nilgunbalkaya, ondokuz mays university,art and science faclty, department of chemistry, 55139 kurupelit, samsun, turkey. Isatanbul university, engineering faculty, department of environmental engineering, 34320, turkey.
3. Biosorptive removal of Cd, andzn from liquid streams with a rhodococcusopacus strain Tatiana gisset P. Vasquez, Ana Elisa C. Botero, Luciana maria S. de mesquite and Maurici Leonardo Torem. Catholic university of Rio de janerio, department of Materials Science and Metallurgy, Rua Marques de Vicente, 255 gavea, 22453-900 rio de janeiro, RJ, Brazil.
4. The simultaneous biosorption of Cu (II) and Zn on Rhizopusarrhizus; application of the adsorption modles Y. Sa, A. Kaya and T. Kutsal Department of chemical Endineering, Hacettepe university, 06532, Beytepe, Ankara, Turkey.,
5. Comparative study of adsorption properties of Turkish fly ashes I. The case of nickel (II), copper (II) and zinc (II) BelginBayat, Department of Environmental Engineering, Faculty of Engg and Architecture, Cukurova University, Balcali, Adana 01330, Turkey.
6. Adsorption of Zinc from aqueous solution using matine green algae – Ulvafasciatasp. Y. Prasannakumar, P. King and V.S.R.K. Prassad. Environment pollution control engg

laboratory, department of chemical engg, A.U. College of engg, Andhra University, VSP 530003, India.

7. Themodynamics and isotherm studies of the biosorption of Cu(II), Pb(II), and Zn(II), by leaves of saltbush (Atriplexcanescens) Maather F. Sawalhaa, Jose R. Peralta-Videa, Jamine Romero-Gonzalezc, Maria Duarte-Garded and Jorge L. Gardea-Torresdeya;b. environmental science and engg, university of Texas at EI Paso, Tx 79968, United states, university of Texas at EI Paso, Tx 79968, USA,university of Guanajuato, gtto. 36000, mexico, department of health promotion, college of health science, university of texas at EI Paso, EI Paso, TX 79968, USA.

8. Adsorption of zinc on colloidal (hydro) oxides of Si, Al and Fe in the presence of a fulvic acid A. Dukera, A. Ledina, S. Karlssona and B. Allarda a department of water and Environment Studies, Link6ping University, S-581 83 Linkoping, Sweden.

9. Removal of Pb(II), Cd(II), Cu(II), and Zn(II) from aqueous solutions by Adsorption on Bentonite Gozen Bereket, Ay e ZehraArog, 1 and Mustsfa Zafer Ozel a Faculty of art and sciebces, Chemistry department of Chemistry, University of stanbul, 34850, Avcilar, stanbul, Turkey.

10. Removal of Candium, Zinc, manganese and chromium cations from aqueous solutions by a clay mineral maria g. da Fonseca, a, Michelle M. de Oliveriraa and Luiza N.H. Arakakia a departmento de Quimica, CCEN, Universidade Federal da Paraiba, 58059-900 Joao Pessoa, Paraiba, Brazil.

CHAPTER 27

FERMENTATION OF STARCH AND STARCH-BASED PACKING PEANUTS FOR ABE PRODUCTION: KINETIC STUDY

AYUSHI VERMA[1], SHASHI KUMAR[2], and SURENDRA KUMAR[3*]

[1,2,3]Department of Chemical Engineering, Indian Institute of Technology Roorkee, Roorkee 247667, Uttarakhand, India; *E-mail: skumar@iitr.ac.in

CONTENTS

27.1 Introduction 294
27.2 Material and Methods 294
27.3 Results and Discussion 295
27.4 Conclusions 297
Keywords 297
References 297

27.1 INTRODUCTION

Second generation biofuels from lignocellulosic biomass or agricultural wastes have many advantages over the first generation biofuel produced from food crops. For instance, most second generation biofuels are considered to be able to deliver substantial greenhouse gas emissions reductions when compared with petrol [1, 2]. Acetone, butanol, and ethanol (ABE) are produced from several biomass wastes such as palm oil waste, domestic waste, abundant agriculture crops by fermentation process [3–5]. Considering the important effect of substrate cost on economic feasibility of ABE production, efficient utilization of lignocellulosic wastes instead of costly food-based substrates like corn and molasses has been suggested to make the ABE production economical [5–8]. In the literature, experimental studies have been reported on the production of ABE by using bacterial culture [4, 9, 10]. The mathematical modeling studies describing the kinetics of the production are very few [11]. In the present study, it is proposed to develop kinetic model for the production of ABE from fermentation of starch and starch-based packing peanuts by *Clostridium beijerinckii*.

27.2 MATERIAL AND METHODS

In order to develop the kinetic model the experimental studies of Jesse et al. [9] have been considered. Jesse et al. conducted the experiments in a batch reactor at pH 6.8 and temperature $36 \pm 1°C$. The pH was adjusted to approximately 6.8 by the addition of buffer solution. Two substrates pure starch and starch-based packing peanuts which contain 85 percent starch were used as substrates for fermentation. The concentrations of ABE products were measured at different time intervals using gas chromatograph and the starch concentrations of the samples were determined using a modified method of Holm et al. as described by Jesse et al. [9].

27.2.1 KINETIC MODEL

A simple reaction network containing four parallel fermentation reactions for the production of ABE is considered [12]. In the present work, power law model is proposed for each reaction to describe the rate of reaction by assuming the non inhibiting effect of substrates and products on the rate of fermentation. The proposed kinetic model equations and their solutions with boundary conditions are given in Table 27.1.

TABLE 27.1 Kinetic model and its solutions

Substrate/Products	Models	Solution
Starch	$\dfrac{dC_S}{dt} = -k_4 C_S^{n_4}$	$C_S = C_{S0} e^{-k_4 t}$
Acetone	$\dfrac{dC_A}{dt} = k_1 C_S^{n_1}$	$C_A = \dfrac{k_1 C_{S0}^{n_1}}{k_4 n_1}\left(1 - e^{-k_4 t n_1}\right)$
Butanol	$\dfrac{dC_B}{dt} = k_2 C_S^{n_2}$	$C_B = \dfrac{k_2 C_{S0}^{n_2}}{k_4 n_2}\left(1 - e^{-k_4 t n_2}\right)$
Ethanol	$\dfrac{dC_E}{dt} = k_3 C_S^{n_3}$	$C_E = \dfrac{k_3 C_{S0}^{n_3}}{k_4 n_3}\left(1 - e^{-k_4 t n_3}\right)$

Initial conditions at $t=0$; $C_S = C_{S0}$; $C_A = C_{A0} = 0$; $C_B = C_{B0} = 0$; $C_E = C_{E0} = 0$.

Here, C_S, C_A, C_B, and C_E are the concentrations of substrate, acetone, butanol, and ethanol at any time respectively. n_1, n_2, n_3, and n_4 are the order of reaction with respect to acetone, butanol, ethanol, and substrate respectively. k_1, k_2, k_3, and k_4 are the rate constants, for production of acetone, butanol, and ethanol, and consumption of substrate respectively.

27.3 RESULTS AND DISCUSSION

The values of kinetic parameters (k, n) are estimated by performing nonlinear regression analysis in MATLAB (R 2012 a) software [13] and are compiled in Table 27.2. It is clearly seen from the value of correlation coefficient R^2, that the kinetic model predictions are in good agreement with the experimental data. The reaction orders vary between 0 and 1 which support the fact discussed by Levenspiel [14] for non inhibited fermentation reactions.

TABLE 27.2 Estimated values of rate constants and order of reactions

Substrates	C_{S0} (g/l)	Acetone			Butanol			Ethanol			Starch		
		k_1	n_1	R^2	k_2	n_2	R^2	k_3	n_3	R^2	k_4	n_4	R^2
Starch (100%)	60	0.14	−0.08	0.99	0.072	0.37	0.98	0.001	0.87	0.98	0.03	1	0.98
Packing peanuts containing starch (85%)	48	0.07	−0.32	0.92	0.194	−0.28	0.94	–	–	–	0.02	1	0.95

The fermentation of pure starch is considered to be carried out with 60 g/l starch in feed. After 72 h of fermentation, 89 percent of starch is converted to acetone, butanol, and ethanol. The concentration of acetone, butanol, and ethanol are estimated by solving the proposed kinetic model. The model predictions for concentrations are 7.95 g/l of acetone, 16.13 g/l of butanol, and 1.11 g/l of ethanol. Likewise, the fermentation of 48 g/l of starch-based packing peanuts (85 percent starch) converts 90 percent starch in 115 h resulting in 3.44 g/l of acetone, 10.59 g/l of butanol, and negligibly small amount of ethanol.

Figures 27.1a,b and 27.2a,b demonstrate the concentration time profiles of products ABE and substrates starch and starch-based packing peanuts respectively as observed by experiments and predicted by models.

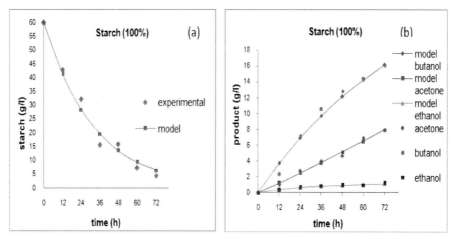

FIGURE 27.1 (a) Starch concentration versus time (b) concentration of product versus time of pure starch.

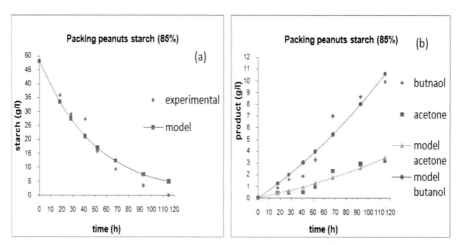

FIGURE 27.2 (a) Starch concentration versus time (b) concentration of product versus time of starch-based packing peanuts.

It is clearly seen from these figures that the model predictions agree well with the experimental results. Based on the above determined values and considering four parallel steps of fermentation reactions, it is reasonable to conclude that proposed kinetic model can be successfully used to describe the fermentation process.

27.4 CONCLUSIONS

The kinetic model scheme to represent the fermentation of starch and starch-based packing peanuts (85 percent starch) by *Clostridium beijerinckii* for the production of ABE was proposed. The kinetic parametric studies on the basis of experimental data were carried out by doing nonlinear regression analysis. In all cases the correlation coefficient R^2 was found to be close to unity and order of reaction was in between 0 and 1. Further, the model was used to predict the concentrations of products, and percent conversion of substrates. In case of pure starch, 89 percent conversion was achieved in 72 h yielding maximum production of butanol (16.13 g/l) and minimum production of ethanol (1.11 g/l). The waste starch-based packing peanut was 90 percent converted in 115 h to products giving negligibly small amount of ethanol and again maximum production of butanol (10.59 g/l). These computations agreed well with experimental results indicating correctness of the proposed model. Hence, the proposed kinetic model represents the kinetic behavior of fermentation of starch and waste starch-based packing peanuts quite well for the production of ABE. These models may be used for obtaining the optimal performance of other reactors and also for devising appropriate control strategies to control such processes.

KEYWORDS

- **Fermentation**
- **Kinetic model**
- **Starch**

REFERENCES

1. Sheehan, J.; Aden, A.; Paustian, K.; Killian, K.; Brenner, J.; and Wals, M.; *J. Ind. Ecol.* 2003, 7, 117–146.
2. Wu, M.; Wang, M.; and Huo, H.; Fuel-cycle assessment of selected bioethanol production pathways in the United States. Center for Transportation Research, Energy System

Division. Agronne National Laboratory. http://www.transportation.anl.gov/pdfs/TA/377.pdf (accessed November 7, 2006).
3. Pieternel, A.M.; Budde, A.W.; and Ana, M.; *J. Mol. Microbiol. Biotechnol.*2000,*2*(1), 39–44.
4. Zverlov, V. V.; Berezina, O.;Velikodvorskaya, G. A.;and Schwarz, W. H.;*Appl.Microbiol. Biotechnol.*2006,*71*, 587–597.
5. Ezeji, T.; Qureshi, N.; and Blaschek, H.P.; *Biotechnol. Bioeng.* **2007**,*97*(6), 1460–1469.
6. Claassen, P.A.M.; Van Lier, J. B.; Lopez Contreras, A.M.; Van Niel, E. W.; Sijtsma, J. L.; Stams, A.J.; and Weusthuis, R. A.; *Appl. Microbiol. Biotechnol.* **1999**, *52*,741–755.
7. Qureshi, N.; Saha, B. C.; and Cotta, M. A.; *Bioprocess Biosyst. Eng.* **2007**,*30*, 419–427.
8. Qureshi, N.; and Blaschek, H.P.; *J. Ind. Microbiol. Biotechnol.* **2001**, *27*(5), 292–297.
9. Jesse, T.W.; Ezeji, T.C.; Qureshi, N.; and Blaschek, H.P.; *J. Ind. Microbiol. Biotechnol.* **2002**, *29*, 117–123.
10. Maddox, I.S.; Qureshi, N.; and Roberts-Thomson, K.; *Process Biochem.* **1995**, *30*(3), 209–215.
11. Shinto, H.; Tashiro, Y.; Kobayashi, G.; Sekiguchi, T.; Hanai, T.; Kuriya, Y.; Okamoto, M.; and Sonomoto, K.; *Process Biochem.* **2008**, *43*(12), 1452–1461.
12. Ahmad, F.; Ahmad, T.J.; Kamarudin, M.H.; and Maizirwan, M.; *Afr. J. Biotechnol.* **2011**, *16*(81),18842–18846.
13. Phisalaphong, M.; Srirattana, N.; and Tanthapanichakoon, W.; *J. Biochem. Eng.* 2006,*28*, 36–43.
14. Levenspiel, O.; Chemical Reaction Engineering. New York: John Wiley; **1999**.

CHAPTER 28

PRODUCTION OF HOLOCELLULOLYTIC ENZYMES BY *CLADOSPORIUM CLADOSPORIOIDES* UNDER SUBMERGED AND SOLID STATE CONDITIONS USING VEGETABLE WASTE AS CARBON SOURCE

CHIRANJEEVI THULLURI, UMA ADDEPALLY*, and
BABY RANI GOLUGURI

Centre for Innovative Research, CBT, IST, Jawaharlal Nehru Technological University Hyderabad (JNTUH), Kukatpally-500085, Hyderabad, India
*E-mail: vedavathi1@jntuh.ac.in

CONTENTS

28.1 Introduction	300
28.2 Materials and Methods	300
28.3 Results and Discussion	303
28.4 Conclusions	306
Keywords	307
References	307

28.1 INTRODUCTION

Lignocellulosic biomass is the most profuse plant material available on the earth, which is a potential feed stock for the production of food, energy, and various biochemical products. Majority of lignocelluloses includeagriresidual, industrial and municipal solid wastes. In recent decades, due to the depletion offossil oil reserves, lignocellulosic biomass has been established into an alternative bioenergy resource [1]. Lignocellulose is a composite material comprising of cellulose, hemicellulose, and lignin among cellulose and hemicellulose contribute over 65 percent dry weight of the agriresidues [2]. Enzymatic hydrolysis has been considered as the most strategic method for generating fermentable sugars from holocellulose [3, 4].

Cellulases and xylanases are industrially important enzymes which can be employed in various fields including biofuels production, biobleaching of kraft pulp, cloths finishing, and clarification of fruit juices etc. [5]. In general, submerged fermentation is employed for industrial scale enzyme productions [6]. The sort of fermentation employed is chiefly reliant on the physiological adaptation of employed organism [7]. These days, solid-state mode of cultivation has been considered as an attractive and inexpensive route for enhanced productionof holocellulolytic enzymes [8–11].

Among a number of microbes, fugal species have more potential in the degradation of lignocellulosic materials [12]. India is an agricultural country and approximately 200 million tons of lignocellulosic biomass is generated annually from variousagro-processing industries, vegetable market places, and agricultural fields of urban areas etc. [13]. The aim of the present investigation is to study the holocellulolytic enzymes production by *Cladosporium cladosporioides* under submerged and solid-state cultivations using vegetable waste as carbon source.

28.2 MATERIALS AND METHODS

28.2.1 MATERIALS

Bovine serum albumin (BSA), D (+)-xylose, D (+)-glucose Di-nitro salicylic acid (DNS), Birch wood xylan, Carboxymethyl cellulose (CMC), Sodium potassium tartarate, Citric acid, Potato dextrose broth, Agar–Agar type I, Peptone, KH_2PO_4, $MgSO_4$, $FeSO_4.7H_2O$, $MnSO_4.2H_2O$, $CaCl_2.6H_2O$, NaOH, Na_2CO_3 were obtained from HiMedia, Mumbai, India.

28.2.2 VEGETABLE BIOMASS

Cabbage, spinachand cauliflower residual wastes were procured from local vegetable market and chopped into a size of 5–6 cm followed by washing with saline solution. Then the biomass was dried in an oven at 60°C to a constant weight. The dried material was pulverized using laboratory mixer and then sieved at a mesh size of 2mm. The resulted biomass was used as substrate for enzymes production.

28.2.3 MICROORGANISM MAINTENANCE AND INOCULUM PREPARATION

Cladosporium cladosporioides was isolated from agricultural fields of pea cultivation and identified by its morphological, cultural, and biochemical characteristics at Indian Type Culture Collection (ITCC), Indian Agricultural Research Institute (IARI), New Delhi and its ID No.7530. The culture was maintained on Potato Dextrose Agar (PDA) slants andstored at 4°C. These slants were used for inoculum preparation. To prepare the inoculum, the spores were scraped out from six days old slants and dispersed in sterilized Potatodextrose broth and incubated at 28°C and rpm of 180 for 24h to get optimal hyphae for inoculating the production media.

28.2.4 SUBMERGEDFERMENTATION(SMF)

Cladosporium cladosporioides was grown on production medium containing carbon source (vegetable waste) $40gL^{-1}$, KH_2PO_4 $1gL^{-1}$, Peptone $10gL^{-1}$, $MgSO_4.7H_2O$ $0.1gL^{-1}$, $FeSO_4.7H_2O$ $0.5mgL^{-1}$, $MnSO_4.2H_2O$ $0.15gL^{-1}$, $CaCl_2.6H_2O$ $2mgL^{-1}$, pH 7.0. 100 ml of this medium was placed in 250 ml Erlenmeyer flasks and sterilized at 121°C for 15 min. After sterilization the medium was allowed for cooling to room temperature and inoculated with 5 percent inoculum consisting of spore suspension at $2 \times 10^6 ml^{-1}$. Then the flasks were incubated in an orbital shaker (JSSI-300 C, JS Research Inc., MA, USA) at 28±1°C and 125 rpm for eight days.

28.2.5 SOLID STATE FERMENTATION (SSF)

Solid state fermentation was carried out using vegetable waste (carbon source 4 percent) mixed with other nutrient supplements mentioned above in 250 ml Erlenmeyer flasks. The flasks were autoclaved, cooled, and shaken thoroughly to split the solid mass. Five percent inoculum consisting of spore suspension at

2×10^6 ml^{-1} was added to the flask and the final moisture content was approximately 70 percent. Then the flasks were incubated in incubator at 28±1°C, for eight days.

28.2.6 ENZYME ACTIVITIES AND PROTEIN CONCENTRATION

The enzyme was extracted from the production flasks for every 24h. The solid-state flasks were extracted with sterile water (S/L 1:6ratios). The solid particles and fungal spores were removed by centrifugation at 4°C, 10,000 rpm for 20 min. The clarified supernatants were used for checking the enzyme activitiesas well as protein concentrations.

28.2.6.1 XYLANASE ASSAY

Xylanase activity was determinedby incubating 0.5ml of 2 percent (w/v) Xylan (substrate), 0.4ml of 50mm Citrate buffer (pH 5) and 0.1ml of culture filtrate at 50°C for 10 min. The amount of reducing sugar released was determined using DNS method of Miller [14]. One unit of xylanase activity is defined as the amount of enzyme (xylanase) that produces 1 μm of xylose/min.

28.2.6.2 ENDO CELLULASE (CMCASE) ASSAY

CMCase activity was determined by incubating the assay mixture comprising of 0.5mL of 1 percent (w/v) Carboxyl Methyl Cellulose (CMC) substrate, 0.4ml of 50mM citrate buffer (pH 3.5) and 0.1ml of culture filtrate at 50°C for 10 min. The amount of reducing sugar released was determined by DNS method of Miller [14]. One unit of CMCase activity is defined as the amount of enzyme (CMCase) that produces 1μm of glucose/min.

28.2.6.3 EXOCELLULASE (FPASE) ASSAY

FPase activity was determined by incubating the assay mixture consisting of 50mg of what man no.1 filter paper units, 0.5mL of 50mM citrate buffer (pH 3.5) and 0.5ml of culture filtrate at 50°C for 10 min. The released reducing sugars were estimated by DNS method of Miller [14]. One unit of Exocellulase activity is defined as the amount of enzyme (FPase) that produces 1μm of glucose/min.

28.2.6.4 PROTEIN CONCENTRATION

Protein concentration was determined by Bradford's method [15]. In brief, 1 ml of Bradford's reagent was added to appropriately diluted protein samples and to the protein blank. The absorbance was read at 595 nm within 5min. The protein concentration was calculated from the standard bovine serum albumin (BSA) plot.

28.2.7 PH AND TEMPERATURE OPTIMIZATION

The effect of temperature and pH on the enzymatic activity was determined by measuring the activities at varying pH 2.5–8.5 by employing buffers such as 50 mM sodium citrate buffer for pH 2.5, 3.5, 4.5, and 5.5; 50 mM sodium phosphate buffer for pH 6.5 and 7.5; and 50 mM Tris-HCL buffer for pH 8.5, and temperatures ranging from 20 to 80°C, respectively.

28.3 RESULTS AND DISCUSSION

28.3.1 COMPOSITIONAL ANALYSIS

The prime components of any lignocellulosic-based biomass are cellulose, hemicellulose, and lignin, and perhaps they account for 34–50, 19–34, and 11–30 percent, respectively [16]. Cellulose and hemicelluloses are polymeric carbohydrates which are targeted entities for biofuels generation by converting them into soluble fermentable sugars through various physicochemical treatments [2]. Whilst, lignin is the major obstacle for bring out the polymeric carbohydrates from lignocellulosic composite and its degraded products could be responsible for hindering fermentation process [17]. After procuring the vegetable biomass, it was analyzed for cellulose, hemicellulose (Table 28.1) and lignin contents and the same were found to be 41±1.7, 28±1.6, and 11.5±0.2 percent, respectively. The presence of significant amount of holocellulose 69±3.3 percent offers vegetable as a potential feedstock for the production of various bioproducts.

TABLE 28.1 Compositional analaysis of vegetable waste

Component	(% w/w)
Cellulose	41 ± 1.7
Hemicellulose	28 ± 1.6
Lignin	11.5 ± 0.2
Ash content	3.9 ± 1.1
Extractives	16.7 ± 1.0

28.3.2 PRODUCTION OF HOLOCELLULOLYTIC ENZYMES IN SUBMERGED AND SOLID STATE FERMENTATIONS

The production of enzymes and other industrially important bio-based products has long been carried out by submerged fermentation (SmF) [6].The submerged cultivation of *Cladosporium cladosporioides*exhibited 43.3, 5.4, and 7.1 IU/ml of xylanase and cellulase (CMCase and FPase) activities during sixth and fourth day of fermentation (Figure 28.1), respectively with maximum protein concentration of 0.47mg/ml (Figure 28.1). The enzymes excreted during the solid-state cultivation were extracted with sterile distilled water in 1:6 (solid to liquid) ratio and the activities of respective enzymes were expressed in international units per milliliter. The solid-state cultivation resulted in condensed holocellulolytic activities with high protein concentration (1.25 mg/ml) (Figure 28.2) when compared to submerged cultivation and they were found to be 22.7, 4.3, and 5.8IU/ml, for xylanase, CM Case, and FPase enzymes (Figure 28.2). However, while observed the activity of released enzymes per gram of carbon source taken, they were found to be in high levels as 340 IU of xylanase/g of biomass, 64.5 IU of CMCase/g of biomass and 87 IU of FPase/g of biomass.

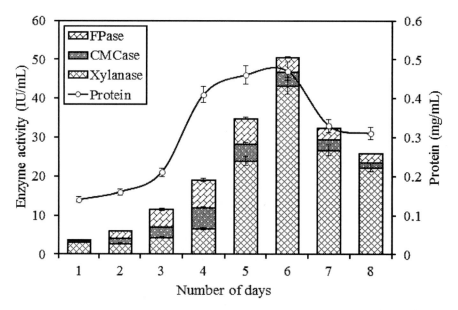

FIGURE 28.1 Holocellulolytic enzymes production by *Cladosporium cladosporioides* under submerged cultivation on vegetable waste.

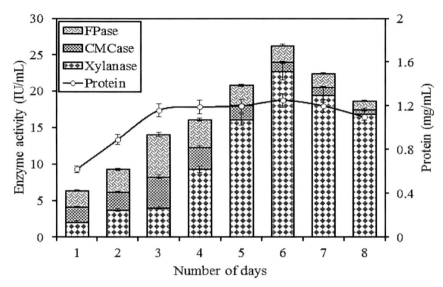

FIGURE 28.2 Holocellulolytic enzymes production by *Cladosporium cladosporioides* under solid-state cultivation on vegetable waste.

The growth of organism in solid-state fermentation is constrained to the surface of humidified insoluble solid media [18] and it can able to trigger the molecular machinery of fungi to produce more lignocellulolytic enzymes to degrade the targeted material effectively by providing the organism condition closer to the natural habitat like decayed organic matter, wood materials etc. [7]. But wherein the case submerged cultivation, the employed organism get exposed fully to hydrodynamic forces where it shows two distinct growth forms viz., filamentous and spores which can influence the enzymes production by means of mass transfer [19]. Moreover, the holocellulolytic enzymes production was also carried out independently with spinach waste (S), cabbage waste (C) and cauliflower residual wastes as carbonsubstrates, however considerable increment in the activities of enzymes was not observed.Previously, combination of wheat bran, wheat straw, beet pulp, and apple pomace was used for xylanases production using the organisms *Aspergillusniger*, *Chaetomiumglobosum* and obtained 20 and 45IU/ml of Xylanase activity [20].

28.3.3 PHAND TEMPERATURE OPTIMA

The pH optimization studies for cellulases (CMCase and FPase) and xylanases were carried out using substrates 1 percent Carboxy Methyl Cellulose, Whatman

No.1 filter paper units and 2 percent Birch wood xylan as substrates, respectively. The pH of the reaction mixtures was adjusted to 3.0–8.5 and the analyzed optimum pH points for all cellulases and xylanases were found to be 3.5 and 5.0 (Figure 28.3b). The optimum reaction temperature range for all three enzymes was found to be 50–60°C (Figure 28.3a). Most of the well-known microbial holocellulolytic enzymes exhibit optimum activity in the range of 50–65°C [21, 22]. However, thermo stable xylanases have also been found from various microbial organisms which exhibit optimal activity at 75°C or above [23].

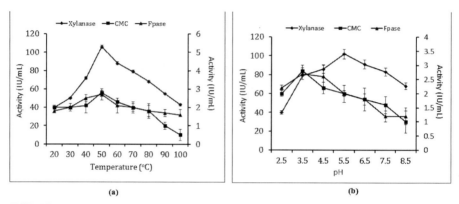

FIGURE 28.3 Temperature and pH optima of holocellulolytic enzymes produced from *Cladosporium cladosporioides*.

28.4 CONCLUSIONS

The holocellulolytic enzymes play a crucial role in the conversion of plant biomass to various value added bioproducts and biorefineries. The cocktail of these three enzymes is vital for the generation of soluble sugars from holocellulose which can be subjected to fermentation by specific microbes to generate biofuels including bioethanol, biobutanol, biohydrogen etc. In addition to this these enzymes are industrially pivotal especially xylanases can be employed in prebleaching of kraft pulp and cellulases are quite helpful in textile industries in cloths sizing and finishing. In the present investigation, novel isolated fungal organism, *Cladosporium cladosporioides* was grown in submerged and solid modes of cultivation for holocellulolytic enzymes production using vegetable market waste as carbon source. The organism utilized the employed substrate effectively by secreting the hydrolytic enzymes. The submerged fermentation contributed to the effective release of holocellulolytic enzymes with lower protein concentrations. The hydrolytic reaction operational conditions such as temperature and pH optima were optimized for effective usage of the produced enzymes.

KEYWORDS

- **Holocellulolytic enzymes**
- **SmF**
- **SSF**
- **Vegetable waste**

REFERENCES

1. Lucia, L. A.; *Bioresources.* **2008**, *3*(4), 981–982.
2. Yang, B.; and Wyman, C. E.; *Biofuels, Bioprod. Bioref.* **2008**,*2*, 26–40.
3. Ladisch, M. R.; Lin, K. W.; Voloch, M.; and Tsao, G. T.; *Enzyme Microb. Technol.* **1983**, *5*, 82–102.
4. Saddler, J. N.; and Mes-Hartree, M.; *Biotechnol. Adv.* **1984**, *2*, 161–181.
5. Bhat, M.K.; *Biotechnol. Adv.* **2000**, *18*, 355–383.
6. Harvey, L. M.; and McNeil, B.; Liquid fermentation systems and product recovery of *Aspergillus.* In: J. E. Smith (Ed.), Biotechnology Handbooks 7. New York: Plenum Press; **1993**, 141–176.
7. da Silva, R.; Lago, E. S.; Merheb, C. W.; Macchione, M. M.; Park, Y. K.; and Gomes, E.;*Braz. J. Microbiol.* **2005**, *36*, 235–241.
8. Dogaris, I.; Vakontios, G.; Kalogeris, E.; Mamma, D.; and Kekos, D.; *Ind. Crops Prod.* **2009**, *29*, 404–411.
9. Umikalsom, M. S.;Arrif, A. B.; Shamsuddin, Z. H.; Tong, C. C.; Hassan, M. A.; and Karim, M. I. A.; *Appl. Microbiol. Biotechnol.* **1997**, *47*, 590–595.
10. Jatinder, K.; Chadha, B. S.; and Saini, H. S.; *World J. Microbiol. Biotechnol.***2006**,*22*, 15–22.
11. Latifian, M.; Hamidin-Esfahani, Z.; and Barzegar, M.; *Bioresour. Technol.* **2007**,*98*, 1–4.
12. Lederberg, J.; Cellulases. In: Encyclopedia of Microbiology (Vol. 1; A-C). New York: Academic Press, Inc.; **1992**.
13. Das, H.; and Singh, S. K.; *Crit. Rev. Food Sci. Nutr.* **2004**, *44*, 77–89.
14. Miller, G. L.; *Anal.Chem.* **1959**,*31*, 426e8.
15. Bradford, M. M.; *Anal. Biochem.***1976**, *72*,248–254.
16. Bobleter, O.; *Prog. Polym. Sci.* **1994**, *19*, 797–841.
17. Palmqvist, E.; and Hahn-Hägerdal, B.; *Bioresour. Technol.* **2000**, *74*, 25e33.
18. Archana, A.; and Satyanarayana, T.; *Enzyme Microbiol. Technol.* **1997**, *21*, 12–17.
19. Mitchell, D. A.; and Lonsane, B. K.; Definition, characteristics and potential in solid substrate cultivation. In: H. W. Doelle; D. A. Mitchell; and C. E. Rolz (Eds.), Solid State Fermentation. New York: Elsevier Applied Science; **1992**, 455–467.

20. Wiącek-Żychlińska, A.; Czakaj, J.; and Sawicka-Żukowska, R.; *Bioresour. Technol.* **1994**,*49*(1), 13–16.
21. Martínez-Trujillo, A.; Pérez-Avalos, O.; and Ponce-Noyola, T.;*Enzyme Microbiol. Technol.* **2003**, *32*, 401–406.
22. Wejse, P. L.; Ingvorsen, K.;and Mortensen, K. K.; *Enzyme Microbiol. Technol.* **2003**,*32*, 721–727.
23. Gupta, S.; Bhushan, B.;and Hoondal, G. S.; *J. Appl. Microbiol.* **2000**, *88*, 325–334.

CHAPTER 29

REMOVAL OF AMMONIA FROM WASTEWATER USING BIOLOGICAL NITRIFICATION

P. B. N. LAKSHMI DEVI[1] and Y. PYDISETTY[2*]

[1,2]Department of Chemical Engineering, National Institute of Technology, Warangal-506004
*Email: psetty@nitw.ac.in,

CONTENTS

29.1 Introduction 310
29.2 Materials and Methods 310
29.3 Results and Discussion 311
29.4 Conclusion 319
Keywords 319
References 319

29.1 INTRODUCTION

Ammonia removal is one of the most important and commonly available elements in wastewater treatment and is normally carried out by biological nitrification process. In biological nitrification process, ammonia is first oxidized into nitrite (NO_2^-), and then oxidized to the much less toxic nitrate (NO_3^-) using *Nitrosomonas* and *Nitrobacter* as oxidizing bacteria. Equations (29.1) and (29.2) show the basic chemical conversions occurring in biological nitrification process [1, 2].

$$NH_4^+ + 1.5\, O_2 \qquad 2H^+ + H_2O + NO_2^- \qquad (29.1)$$

$$NO_2^- + 0.5\, O_2 \qquad NO_3^- \qquad (29.2)$$

Garrett [3] seems to be the first author who related microbial growth to the activated sludge process. A substantial step in understanding nitrification in the activated sludge process is due to a research group at the British Water Pollution Research Laboratory [4]. For the treatment of waste water, nitrification with autotrophic bacteria has received most attention. It has been shown that under these conditions autotrophic nitrification is favorable.

Biological nitrification can be accomplished in two types of systems: suspended and attached growth. Under a suspended growth environment, the microorganism is freely mobile in the liquid providing direct contact between the bacterial cells and the bulk water. In attached growth system, microorganisms had grown in a visco-elastic layer of biofilm that is attached on the surface of a solid support medium. Thus, this process is called a biofilm process in which the individual bacteria are immobilized. Attached growth on a fixed biofilm system offers several advantages when compared to suspended growth processes, such as handling convenience, increasing process stability in terms of withstanding shock loading and preventing the bacteria population from being washed off [5, 6]. The present work involves the removal of ammonia from wastewater with attached growth of microorganisms and optimal values of parameters were studied using Incubator Shaker.

29.2 MATERIALS AND METHODS

29.2.1 CULTURE PREPARATION

Nitrosomonas (NCIM No-5076) and *Nitrobacter* (NCIM No-5062) were obtained from National Chemical Laboratory, Pune, India. The subcultures were prepared according to their procedure. The culture was preserved in a refrigerator at a temperature of 4°C by periodic subcultures.

29.2.2 PREPARATION OF INOCULUM

All the chemicals used for the preparation of inoculums are LR grade only and with purity 98–99 percent. The liquid broth having the composition $(NH_4)_2CO_3$—0.303 g (Chemport Pvt Limited, Mumbai), $NaHCO_3$—25 g (Merck Limited, Mumbai), $MgSO_4.7H_2O$—0.257 g (Merck Limited, Mumbai), Na_2HPO_4—1.135 g (SD Fine-Chem Limited, Mumbai), KH_2PO_4—1.092 g (Merck Limited, Mumbai), $FeCl_3.6H_2O$—0.035 g (SD Fine-Chem Limited, Mumbai), and Sucrose—6.174 g per liter has been prepared [7]. The composition was adjusted to initial ammonium concentration of 100 ppm by varying the amount of ammonium carbonate proportionately. The pH of the solution was adjusted to a value of 7.5 using HCl and NaOH. In order to kill the undesirable microorganisms, the broth was sterilized in an autoclave for 15 min at 103 KPa pressure and 120°C. The broth was cooled to room temperature (28°C) and then medium of *Nitrosomonas* and *Nitrobacter* were introduced into the broth. Further, the medium was kept for the growth in an incubator for 24 hr for the formation of biofilm over polypropylene beads for attached growth of microorganisms where the temperature was maintained at 30°C. This is used as a medium in the bioreactor. 50 ml of the product was collected for an interval of 2 hr and was used for the analysis of final ammonium ion and nitrate ion concentration. The analysis for ammonium ion was carried out with Orion Ion Selective Electrode.

29.3 RESULTS AND DISCUSSION

29.3.1 EFFECT OF MECHANICAL AGITATION ON AMMONIUM ION CONCENTRATION

Kugaprasatham et al. [8] studied the effect of hydraulic conditions on nitrifying biofilm grown under a low ammonia nitrogen concentration (about 1 gm^{-3}) in a cylindrical reactor. In addition, Chen and Huang [9] reported higher nitrification rates in biofilters with high turbulence levels. These results are important for the design and optimal operation of biofilters, as they suggest that the nitrification rate may be significantly improved through increasing turbulence.

To assess the effect of mechanical agitation on the initial ammonium concentration studies at different agitations of 50, 75, 100, 125, and 150 rpm were made. Samples of fermentation media were prepared as per the composition discussed in Section 2.2. The conditions for all the samples were maintained at a temperature of 30°C, pH of 7.5 and initial concentration of ammonium ion at 100 ppm in incubator shaker with two-phase (liquid-solid bioreactor) system.

The concentration of ammonium ion was plotted against fermentation time and the optimal value has been observed for maximum removal of initial ammonium ion for different values of agitations. Figure 29.1 represents the removal of ammonium ion with liquid-solid bioreactor. It is observed that, from Figure 29.1, the optimal mechanical agitation has been found to be at 100 rpm in liquid-solid bioreactor, with maximum removal of ammonium ion concentration up to 16.2 ppm. As the agitation speed increased the removal percentage of ammonium ion increased and gets decreased at higher agitation due to less contact between biofilm and the substance. The optimal mechanical agitation of 100 rpm was used for further experiments.

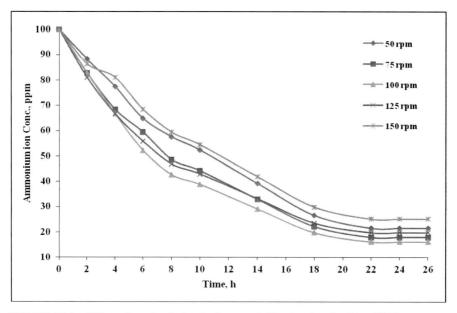

FIGURE 29.1 Effect of mechanical agitation on nitrification in a liquid-solid bioreactor.

29.3.2 EFFECT OF PH ON AMMONIUM ION CONCENTRATION

Several investigations conducted earlier, have been demonstrated, the pH effect on nitrification. However, poor agreement existed on how much, and what point, pH begins to effect nitrification rates [10]. Based on the review provided by Sharma and Ahlert [11], the optimal pH for the growth of nitrifying bacteria varies widely. The optimum pH for nitrification can range from 7.0 to 9.0 with the optimum pH ranging from 7.2 to 8.8 for *Nitrosomonas* and 7.2–9.0 for *Nitrobacter*.

However, Suzuki et al. [12] and Painter [13] suggested that free ammonia instead of ammonium is the substrate for ammonia—oxidizing bacteria (*Nitrosomonas*) based on the observation of a consistent Monod saturation constant under varying free ammonia concentration. Therefore, reduced nitrification activity at lower pH levels may result indirectly from substrate limitation since, the fraction of $NH_3 - N$ in the total ammonia nitrogen decreases with decrease of pH [14]. When a higher TAN (Total Ammonia Nitrogen) is used, a higher nonlimiting concentration of NH_3 may be maintained at lower pH values [10]. Interestingly, to evaluate the pH effect on ammonia oxidation activity, Groeneweg et al. [15] measured ammonia oxidation rates at a constant $NH_3 - N$ of 0.37 mg L^{-1} (varying TAN in accordance with pH) and a constant TAN of 5 mg L^{-1} ($NH_3 - N$ varies with pH) over a wide pH range (5–11). They found that the maximum ammonia oxidation rate was obtained between pH 6.7 and 7.0 (0.37 mg L^{-1} $NH_3 - N$) and pH 7.5 and 8.0 (5 mg L^{-1} TAN), while, the ammonia oxidation rate decreased sharply outside the optimum pH ranges.

The effect of pH on the initial ammonium concentration at different pH values varying from 6 to 8.5 were studied and observed in the present work. Samples of fermentation media were prepared as per the composition discussed in Section 2.2. The conditions for all the samples were maintained at a temperature of 30°C, initial concentration of ammonium ion at 100 ppm and agitation of 100 rpm in incubator shaker in a liquid-solid bioreactor.

The concentration of ammonium ion was plotted against fermentation time and the optimal value has been observed for maximum removal of initial ammonium ion for different values of pH. Figure 29.2 represent the removal of ammonium ion with liquid-solid bioreactor. It is observed from Figure 29.2 that the optimal pH among above considered values occurs at 7 in liquid-solid bioreactor, with maximum removal of ammonium ion concentration up to 10.8 ppm. It is also observed that the complete nitrification occurs at neutral pH and in acidic or basic media the percentage removal of ammonium ion decreased. The other parameters were optimized using the pH as 7.

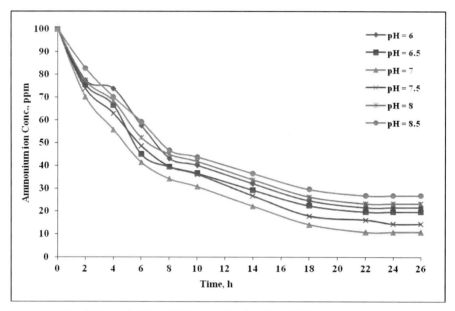

FIGURE 29.2 Effect of pH on nitrification in a liquid-solid bioreactor.

29.3.3 EFFECT OF TEMPERATURE ON AMMONIUM ION CONCENTRATION

It has been well accepted that a higher temperature enhances nitrification rate as the biochemical driven bacterial processes accelerate as temperature increases. This is true in a suspended growth system. For fixed film filters, however, the effects of temperature on nitrification kinetics are also influenced by other phenomena and parameters [16], especially substrate diffusion and transport. A general conclusion on the relationship between nitrification rate and temperature must also include the effect of mass transfer and bacteria. However, the impacts of change on nitrification rate in fixed film biofilters were understood [17]. Little information is available to quantify the effects of temperature on fixed film nitrification rate [18].

Zhu and Chen [19] studied the impact of temperature on nitrification rate through laboratory experiments, mathematical modeling and sensitivity analysis. They showed that in the case of oxygen limitation, temperatures from 14 to 27°C had no significant impact on nitrification rate. A lower nitrification rate was observed only at the lowest temperature they tested, 8°C. Temperature had a more significant effect on nitrification rate in the case of TAN limitation than in the case of DO limitation.

Experiments were performed for different temperatures varying from 24 to 30°C and the effect of temperature on initial ammonium concentration has been observed. Samples of fermentation media were prepared as per the composition as discussed in Section 2.2. The conditions for all the samples were maintained at a pH of 7, initial concentration of ammonium ion of 100 ppm and agitation of 100 rpm in incubator shaker in a liquid-solid bioreactor.

The concentration of ammonium ion was plotted against fermentation time and observed the optimal value for maximum removal of initial ammonium ion for different values of temperature. Figure 29.3 represent the removal of ammonium ion with liquid-solid bioreactor. It is observed that from Figure 29.3 the optimal temperature among above considered occurs at 30°C in liquid-solid bioreactor, with maximum removal of ammonium ion concentration up to 10.8 ppm. The percentage removal of ammonium ion is increased with increase in temperature, but at the higher temperature it gets decreased due to decreased activity of microorganisms. Further, experiments were carried out keeping the temperature at 30°C.

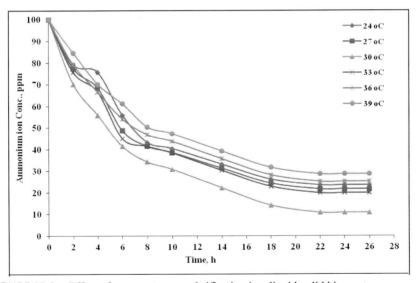

FIGURE 29.3 Effect of temperature on nitrification in a liquid-solid bioreactor.

29.3.4 EFFECT OF AMOUNT OF SOLIDS ON AMMONIUM ION CONCENTRATION

Sandu et al. [20] performed experimental work for three unique bed-height to diameter ratios for a fluidized bed nitrification filters which were charged with

2×4 ABS plastic beads with specific gravity 1.06, using column diameters of 12.7, 15.2, and 17.8 cm. They reported that total ammonia nitrogen (TAN) removal increased with column diameter at hydraulic loading rates (6, 8, 10, and 12 lpm) and ammonia concentration of 8.6 g per 100 g of synthetic substrate.

To observe the effect of amount of solids on the initial ammonium concentration at different concentrations of polypropylene beads (3 mm diameter) varied from 5 to 25 g/l were studied. Samples of fermentation media were prepared as per the composition as discussed in Section 2.2. The conditions for all the samples were maintained at a pH of 7, initial concentration of ammonium ion 100 ppm, 30°C of temperature and agitation of 100 rpm in incubator shaker in a liquid-solid bioreactor.

The concentration of ammonium ion was plotted against fermentation time and the optimal value for maximum removal of initial ammonium ion has been observed for different values of amount of solids. Figure 29.4 represent the removal of ammonium ion with liquid-solid bioreactor. It is observed from Figure 29.4 that the optimal amount of solids among above considered occurs at 10 g/l in liquid-solid bioreactor, with maximum removal of ammonium ion concentration up to 10.8ppm. The concentration of solids of 10 g/l was used for further experimental work.

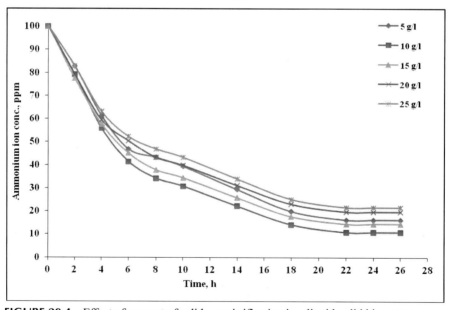

FIGURE 29.4 Effect of amount of solids on nitrification in a liquid-solid bioreactor.

29.3.5 EFFECT OF MAGNESIUM SULPHATE ON AMMONIUM ION CONCENTRATION

The effect of magnesium sulfate concentration on the initial ammonium concentration experiments at different concentrations varying from 0.1 to 0.5 g/l were studied and observed. Samples of fermentation media were prepared as per the composition as discussed in Section 2.2. The conditions for all the samples were maintained at a pH of 7, initial concentration of ammonium ion of 100 ppm, 30°C of temperature and agitation of 100 rpm in incubator shaker in a liquid-solid bioreactor.

The concentration of ammonium ion was plotted against fermentation time and observed the optimal value for maximum removal of initial ammonium ion for different values of magnesium sulfate concentrations. Figure 29.5 represent the removal of ammonium ion with liquid-solid bioreactor. It is observed that from Figure 29.5 the optimal concentration of magnesium sulfate among above values considered occurs at 0.3 g/l in liquid-solid bioreactor, with maximum removal of ammonium ion concentration up to 10.8 ppm. Further experiments were carried with the magnesium sulfate concentration as 0.3 g/l.

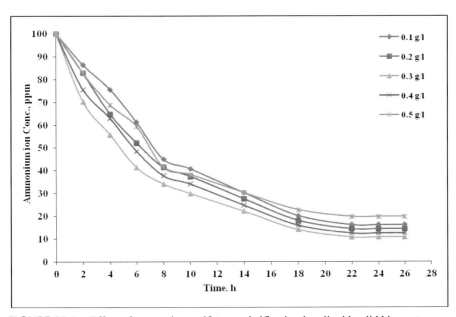

FIGURE 29.5 Effect of magnesium sulfate on nitrification in a liquid-solid bioreactor.

29.3.6 EFFECT OF POTASSIUM DI-HYDROGEN PHOSPHATE ON AMMONIUM ION CONCENTRATION [21]

To assess the effect of potassium di-hydrogen phosphate concentration on the initial ammonium concentration experiments at different concentrations varying from 1 to 5 g/l were studied. Samples of fermentation media were prepared as per the composition. The conditions for all the samples were maintained at a pH of 7, initial concentration of ammonium ion of 100 ppm, 30°C of temperature and agitation of 100 rpm in incubator shaker in a liquid-solid bioreactor.

The concentration of ammonium ion was plotted against fermentation time and the optimal value was observed for maximum removal of initial ammonium ion for different values of potassium di-hydrogen phosphate concentrations. Figure 29.6 represent the removal of ammonium ion with liquid-solid bioreactor. It is observed, from Figure 29.6 that the optimal concentration among above considered occurs at 1 g/l in a liquid-solid bioreactor, with maximum removal of ammonium ion concentration up to 10.8 ppm.

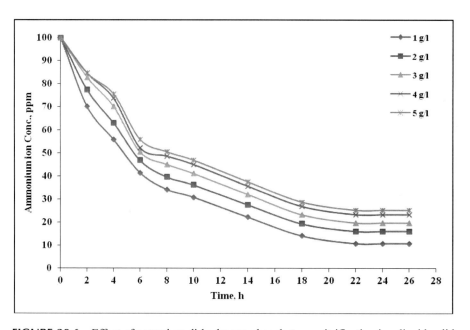

FIGURE 29.6 Effect of potassium di-hydrogen phosphate on nitrification in a liquid-solid bioreactor.

29.4 CONCLUSION

Biological reactions are commonly used in wastewater treatment plants to remove ammonia from water using *Nitrosomonas* and *Nitrobacter* as oxidizing bacteria. Experiments were performed for various parameters such as mechanical agitation, pH, temperature, amount of solids (polypropylene beads), magnesium sulfate and potassium di-hydrogen sulfate and the optimal values were determined and reported. It is observed, that the ammonium ion has been removed upto 89.2 percent with an initial concentration of 100 ppm as ammonium carbonate. Thus, these studies indicate that biological nitrification is a viable alternate process for removal of ammonia from wastewater.

KEYWORDS

- Incubator shaker
- Nitrification
- *Nitrobacter*
- *Nitrosomonas*
- Two-phase system
- Wastewater

REFERENCES

1. United States Environmental Protection Agency (US-EPA). Methods for the Chemical Analysis of Water and Wastewater. EPA-600/4-79-020. U.S. Environmental Protection Agency, Office of Research and Development, Environmental Monitoring and Support Laboratory, Cincinnati, OH; **1984**.
2. WPCF. Nutrient control, manual of practice. Publication Number FD-7 Water Pollution Control Federation. Washington, DC; **1983**.
3. Garrett, Jr, M. T.; Hydraulic control of activated sludge growth rate. *Sew. Ind. Wastes* **1958**, *30(3),* 253–261.
4. Water Pollution Research. Yearly Report. Ministry of Technology, Her Majesty's Stationary Office. London; **1964**.
5. Fitch, M. W.; Pearson, N.; Richards, G.; and Burken, J. G.; Biological fixed-film systems. *Water Environ. Res.* **1998**, *70,* 495–518.
6. Nogueira, R.; Lazarova, V.; Manem, J., and Melo, L. F.; Influence of dissolved oxygen on the nitrification kinetics in a circulating bed biofilm reactor. *Bioprocess Eng.* **1998**, *19,* 441–448.
7. Zhu, S.; and Chen, S.; Effects of organic carbon on nitrification rate in fixed film biofilters. *Aquacult. Eng.* **2001**, *25,* 1–11.

8. Kugaprasatham, S.; Nagaoka, H.; and Ohgaki, S.; Effect of short term and long-term changes in hydraulic conditions on nitrifying biofilm. *Water Sci. Technol.* **1991**, *23*, 1487–1494.
9. Chen, G. H.; Huang, J. C.; Determination of diffusion layer thickness on a biofilm. *J. Environ. Sci. Health.* **1996**, *31*, 367–386.
10. Biesterfeld, S.; Farmer, G.; Russell, P.; and Figueroa, L.; Effect of alkalinity type and concentration on nitrifying biofilm activity. In: Proceedings of Water Environment Federation Conference **2001**, Atlanta, GA, October, 12–17.
11. Sharma, B. and Ahlert, R.C. Nitrification and nitrogen removal. *Water Res.* **1977**, 11897–11925.
12. Suzuki, I.; Dular, U.; and Kwok, S. C.; Ammonia or ammonium ions as substrate for oxidation by Nitrosomonas europaea cells and extracts. *J. Bacteriol.* **1974**, *120*, 556–558.
13. Painter, H. A.; Nitrification in the treatment of sewage and wastewaters. In: Prosser, J. I.; Ed.Nitrification. Oxford: IRL Press;**1986**.
14. Allison, S.M.;and Prosser, J.I.; Ammonia oxidation at low pH by attached populations of nitrifying bacteria. *Soil Biol. Biochem.* **1993**, *25*, 935–941.
15. Groeneweg, J.; Sellner, B.; and Tappe, W.; Ammonia oxidation in *Nitrosomonas* at NH_3 concentrations near K_m: Effects of pH and temperature. *Water Res.* **1994**, *28*, 2561–2566.
16. Fdz-Polanco, F.; Villaverde, S.; and Garcia, P.A.; Temperature effect on nitrifying bacteria activity in biofilters: Activation and free ammonia inhibition. *Water Sci. Technol.* **1994**, *30*, 121–130.
17. Okey, R.W.; and Albertson, O.E.; Evidence for oxygen-limiting conditions during tertiary fixed-film nitrification. *J.Water Pollut. Control Fed.* **1989**, *61*, 510–519.
18. Wheaton, F. W.; Hochheimer, J. N.; Kaiser, G. E.; Krones, M. J.; Libey, G. S.; and Easter, C. C.; Nitrification principles. In: Timmons, M. B.; and Losordo, T. M.; Eds. Aquaculture Water Reuse Systems: Engineering Design and Management. Elsevier, Amsterdam;**1994**, 101–126.
19. Zhu, S.; and Chen, S.; The impact of temperature on nitrification rate in fixed film biofilters. *Aquacult. Eng.* **2002**, *26*, 221–237.
20. Sandu, S.I.; Boardman, G.B.; Watten, B.J.; and Brazil, B.L.; Factors influencing the nitrification efficiency of fluidized bed filter with a plastic bead medium. *Aquacult. Eng.***2002**, *26*, 41–59.
21. Devi, P.B. N. L.; and Setty, Y.P.; Effect of nutrients on biological nitrification. British Biotechnology J. 2015,4(12) 1283–1290.

CHAPTER 30

REMOVAL OF CHROMIUM FROM AQUEOUS SOLUTION USING LOW COST ADSORBENTS

P. AKKILA SWATHANTHRA[1*] and V. V. BASAVA RAO[2]

[1]Department of Chemical Engineering, S.V. University, Tirupati;
*Email: akhilasweety9@gmail.com

[2]College of Technology Osmania University, Hyderabad

CONTENTS

30.1 Introduction 322
30.2 Materials and Methods 322
30.3 Results and Discussion 324
30.4 Conclusion 328
Keywords 329
References 329

30.1 INTRODUCTION

Heavy metals are toxic to aquatic organisms even at very low concentrations. Environmental contamination by toxic metals is of great concern because of health risks on humans and animals like high blood pressure, irritation, depression, sleep disabilities, [2] etc., Among the toxic metal ions chromium is one of the common contaminants which gains importance due to its high toxic nature even at very low concentration. Water pollution by chromium is of considerable concern, as this metal has found widespread use in electroplating, leather tanning, metal finishing. The conventional methods used to remove Chromium from aqueous effluents, include chemical precipitation, ion exchange, electro flotation, membrane separation, reverse osmosis, electro dialysis, solvent extraction, etc. [1]. Many conventional methods for removing of metal ions from Industrial effluents suffer with high capital costs of material [4]. Therefore, these are not suitable for the small scale industries.

Thus, there is currently a need for new, innovative and cost effective methods for the removal of toxic substances from wastewater. However, these approaches have proved to be costlier and difficult to implement. Adsorption is one of the physico-chemical treatment process found to be effective in removing heavy metals from aqueous solutions. Adsorbents can be considered as cheap or low cost if it is abundant in nature, requires little processing and is by product of waste material from industrial or agricultural operations may have potential has inexpensive adsorbents [5]. Plant wastes are in expensive as they have no or low economic value and fly ash an inorganic residue from the combustion of powdered coal.

The aim of the present investigation is to detect the performance of fly ash, bagasse and sawdust on Chromium removal from aqueous solutions by varying Chromium concentration, P^H and adsorbent dosage. Langmuir and Freundlich isotherms were applied to fit the experimental data.

30.2 MATERIALS AND METHODS

All the chemicals used in this study were of analytical grade and were proceed from Sd. Fine Chem. Ltd.

The adsorbents are selected for removal of Chromium by fly ash, bagasse and sawdust. The adsorbent were grounded and wasted with deionised water. The adsorbents were dried at room temperature (30°C) till a constant weight of the adsorbents was achieved (after 20 hrs). Adsorption is an effective and versatile method for removing Chromium.

30.2.1 PREPARATION OF ADSORBENTS

Firstly the adsorbents are washed and dried at room temperature to avoid the release of color by adsorbent into the aqueous solution. The activation of adsorbents carried out by treating it with concentrated sulphuric acid (0.1N) and is kept in an oven maintained at a temperature range of 150°C for 24hr. Again is washed with distilled water to remove the free acid [6] and dried again at 100°C for 5 h.

30.2.2 BATCH EXPERIMENTS

A stock solution of Cr (VI) is prepared by dissolving 2.8287 g of 99.99% potassium dichromate ($K_2Cr_2O_7$) in distilled water [3] and solution made up to 1000 ml. This solution is diluted as required to obtain the standard solutions containing 5–500 mg/L of Cr (VI). The P^H is adjusted in the range of 2–10 by adding 0.1N H_2SO_4 and 0.1N NaOH solutions and measured by a P^H meter (ELICO, LI 613).

The batch experiments are carried out in 250 ml borosil conical flasks by shaking a pre-weighed amount of the bagasse with 100 ml of the aqueous chromium (VI) solutions of known concentration and P^H value. The metal solutions were agitated in a rotary shaken at 120 rpm for a desired time. The samples were with drawn from the shaken at the predetermined time intervals and adsorbent was separated by filtration. Chromium (VI) concentration in the filtrate was estimated by using AAS. The experiments were carried out by varying the chromium (VI) concentration in the solution (50–500mg/L), P^H (2–10). The adsorbent dosage 2–10 gr/lit. The samples were collected at different time intervals 15 min to 5 hrs and the adsorbent was separated by filtered using filter paper.

30.2.3 DETERMINATION OF CHROMIUM CONTENT

The chromium concentration of Cr (VI) ions in the effluent is determined by AAS. For this purpose, $K_2Cr_2O_7$ solutions of different Concentrations were prepared and their absorbance recorded by using AAS. A calibration plots for Cr (VI) were drawn between "%" absorbance and standard Cr (VI) solutions of various strengths [4, 6, 7]. Runs were made in triplicate. The percentage removal of chromium was calculated as follows:

% removal of chromium = $(C_{int} - C_{fin}) \times 100/C_{int}$

Where C_{int} and C_{fin} are the initial and final chromium concentrations, respectively.

30.3 RESULTS AND DISCUSSION

30.3.1 EFFECT OF PH

All Experiments were carried out in the pH range 2–10 (2, 4, 6, 8 and 10) Effect of solution pH on removal of Chromium was studied using bagasse, fly ash and sawdust as adsorbents.

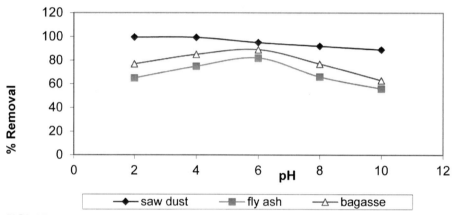

FIGURE 30.1 Effect of pH on % removal of Chromium by different adsorbents at 30°C.

Figure 30.1 shows the effect of pH on the batch adsorption studies of 100 mg/l Chromium at 30°C and adsorbent dosage 0.2 gr/100 ml. It is obvious that the increasing pH from 2 to 10 percentage removal is decreased for sawdust but in the case of bagasse and fly ash it is increased up to pH 6 then decrease up to 10. This may be attributed to the surface change development of the adsorbent and the concentration distribution of metal ions, which both are pH dependent.

30.3.2 EFFECT OF CONTACT TIME

The effect of contact time on adsorption of Chromium is investigated to study the percentage removal.

Figure 30.2 shows the percentage removal of Chromium for different initial concentration ranging from 50 mg/l to 500 mg/l at 30°C. The time is one of the most important factors for the adsorption of metals on adsorbent. It can be observed that, the adsorption rate is very rapid during the initial period of contact time and later it is very slow. However, the equilibrium is attained within 120 min fro fly ash and baagasse, but in the case of sawdust is nearly 240 min at temperature 30°C initial concentration 100 mg/L and adsorbent dosage 0.2 gr/100 ml.

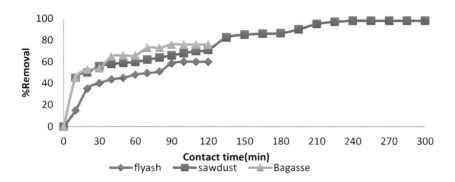

FIGURE 30.2 Effect of contact time on adsorption of Chromium by different adsorbents at 30°C.

30.3.3 EFFECT OF ADSORBENT DOSE

The effect of adsorbent dosage on the removal of Chromium metal by different adsorbents 2,4,6,8 and 10g/L at 30°C with an initial concentration of chromium metal ions at 100 mg/L at sawdust pH 2 and bagasse and fly ash pH is 6 at equilibrium time of chromium metal is shown in Figure 30.3

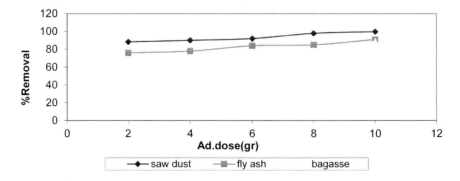

FIGURE 30.3 Effect of adsorbent amount on percentage removal of Chromium by different adsorbents at 30°C at different initial concentrations.

It is clearly indicates that the percentage removal of chromium metal increases with increasing in adsorbent dose. The adsorbent dose is increased percent removal is also increases due to the increasing surface area and adsorption sites available for adsorption. .

30.3.4 EFFECT OF INITIAL CONCENTRATION

The effect of initial concentration of stock solution on the adsorption of chromium metals shown in Figure 30.4.

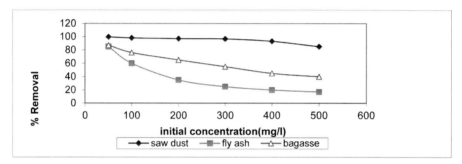

FIGURE 30.4 Effect of initial concentration on percentage removal of Chromium at 30°C by sawdust, bagasse and fly ash.

The percentage removal of chromium adsorption increased with decreasing the initial metal concentration because at lower concentration, the % removal is very high and it is lower at higher concentration [5]. This is due to the availability of same amount of adsorbent site for various amount of various amount of initial metal ion concentration. The effect of initial concentration on the removal of metals using different initial concentrations 100, 200, 300, 400 and 500 mg/L at 30°C while maintain dosage 10g/L at sawdust pH 2 and bagasse and fly ash pH is 6 at equilibrium time.

30.3.5 ADSORPTION ISOTHERMS

30.3.5.1 LANGMUIR AND FREUNDLICH ISOTHERMS

The adsorption isotherm plot between Ce/q_e verses Ce,

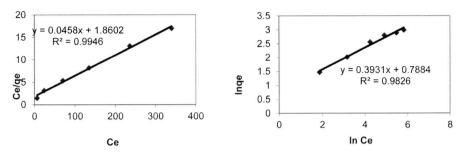

FIGURE 30.5 Langmuir and Freundlich isotherms for adsorption of Chromium by bagasse.

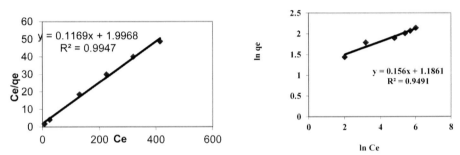

FIGURE 30.6 Langmuir and Freundlich isotherms for adsorption of Chromium by Fly ash.

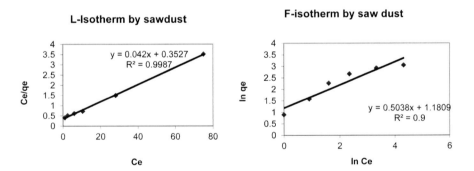

FIGURE 30.7 Langmuir and Freundlich isotherms for adsorption of Chromium by Sawdust.

Figures 30.5, 30.6, and 30.7 shows the Langmuir constant q_m, which is a measure of the monolayer adsorption capacity of adsorbents in mg/gr, the Langmuir constant b, and the value of regression correlation coefficient R^2 values are shown in Table 30.1

TABLE 30.1 Isotherm constants and regression data for various adsorption isotherms for adsorption of Chromium on bagasse, sawdust, and fly ash

Adsorbent	% Removal	pH	L-Constants				F-Constants			
			Q	b	R_L	R^2	1/n	n	k	R^2
Sawdust	99.5	2	23.8	0.11	0.08	0.99	0.5038	1.98	3.25	0.9
Fly Ash	91	6	8.5	0.05	0.29	0.9947	0.156	6.4	3.2	0.94
Bagasse	89	6	21.8	0.024	0.34	0.9946	0.7884	2.56	2.18	0.9826

Also the dimensionless parameter R_L, which is a measure of adsorption favorability is found to be 0.5(0< R_L <1) which confirms the favorable adsorption process for Chromium removal using bagasse, sawdust, and fly ash [5], R_L, also known as the separation factor, given by

$$R_L = 1/(1+bCe)$$

The value of R_L lies between 0 and 1 for a favorable adsorption, while R_L > 1 represents an unfavorable adsorption, and R_L = 1 represents the linear adsorption, while the adsorption operation is irreversible R_L = 0.

The Langmuir and Freundlich equations are given in equations

$$Ce/q_e = 1/q_m + (1/q_m) Ce$$

$$\ln q_e = \ln K_f + (1/n) \ln Ce$$

Freundlich isotherm is analyzed based on adsorption of chromium by using the same equilibrium data of adsorbents [5]. Freundlich constants, K_f and n are obtained by plotting the graph between log q_e versus log Ce. The values of K_f and n. shown in Table 30.1. It is found that, from the regression correlation coefficient of Freundlich Isotherm is more suitable than the Langmuir isotherm model.

30.4 CONCLUSION

A review of various agricultural adsorbents presented here shows a great potential for the elimination of chromium from waste water. The results obtained in this study for the removal of Chromium by sawdust, fly ash and bagasse. The following conclusions can be drawn based on the investigation: the percentage removal among these three adsorbents was observed as for sawdust (99.5%), fly ash (91%) and bagasse (89%). Langmuir Isotherm was better fitted for the experimental data since the correlation coefficients for Langmuir Isotherm was higher than the Freundlich Isotherm for three adsorbents. The constant value R_L in Langmuir Isotherm gives an indication of the favorable adsorption and when the temperature is increased from 30°C to 50°C, an increase of adsorption was indicating the adsorption process to be endothermic nature for all adsorbents. All these adsorbents are cheaper than commercial adsorbents and fly ash may be used as neutralizing agent in the treatment processes.

KEYWORDS

- Adsorption
- Atomic adsorption spectro photometer
- Chromium (VI)
- Low cost adsorbent (sawdust, bagasse, and fly ash)

REFERENCES

1. Ahsan Habib; Nazrul Islam; Anarul Islam; Shafiqual Alam, A. M. Removal of Chromium from Aqueous Solution Using Orange Peel, Sawdust and Bagasse. *Pak. J. Anal. Environ. Che.* **2007**, *1&2*, 21–25.
2. Amrit Kaur; Malik, A. K.; Neelam Verma; Rao, A. L. J. Removal of Chromium and Lead from Waste water by Adsorption on Bottom Ash, *Indian J. Environ. Protect.* **1991**, *11(6)*.
3. APHA. *Standard Methods for the Examination of Water and Waste Water*; (18th Ed.), APHA: Washington, DC, **1992**.
4. Himesh, S.; Mahadevaswamy, M. Sorption Potential of Biosorbent for Removal of Chromium. **1994** *36(3)*, 165–169.
5. Rajesh Kumar, P.; Akhila Swathanthra, P.; Basava Rao, V. V.; Rama Mohana Rao, S. Adsorption of Cadmium and Ainc Ions from Aqueous Solution Using Low Cost Adsorbents. *J. Appl. Sci.* **2013**.
6. Suresh Gupta; Babu, B.V. Removal of Toxic Metal Cr (VI) from Industrial Waste Water Using Sawdust as Adsorbent: Equilibrium, Kinetics and Regeneration Studies, *Conference Proceedings*.
7. Vikrant Sarin; Pant, K. K. Removal of Chromium from Industrial Waste by Sing Eucalyptus Bark, Elsevier. *Bioresour. Technol.* **2006**, *97*, 15–20.

CHAPTER 31

PREPARATION OF PANI/CALCIUM ZINC PHOSPHATE NANOCOMPOSITE USING ULTRASOUND ASSISTED *IN SITU* EMULSION POLYMERIZATION AND ITS APPLICATION IN ANTICORROSION COATINGS

B. A. BHANVASE[1,*], S. H. SONAWANE[2], and M. P. DEOSARKAR[1]

[1]Vishwakarma Institute of Technology, Pune, MS, India
[2]National Institute of Technology, Waragal, AP, India
*E-mail: bharatbhanvase@gmail.com

CONTENTS

31.1	Introduction	332
31.2	Materials and Methods	333
31.3	Results and Discussions	335
31.4	Conclusion	341
	Keywords	341
	References	342

31.1 INTRODUCTION

Polyaniline (PANI) has been paying substantial attention because of its exclusive electrical, optical, and opto-electrical properties as well as its simple preparation method and the most excellent environmental stability among the conducting polymer. There are several potential applications of PANI in electronics, energy storage, anticorrosion coatings, catalyst chemical sensing, biochemistry, electronics devices, electrostatic dissipation, etc. [1]. However, most of the conducting polymers are only slightly soluble in a few solvents and it becomes difficult to process it in a continuous film.Composite materials are the focus of interest of many research groups nowadays aiming to obtain the materials with enhanced mechanical and anticorrosive properties. Several reports on the synthesis of the nanocomposites of PANI with WO_3, ZnO, Fe_2O_3, PbO_2 or Au nanoparticles have been described [2–5].

Pigments like lead and chromates are not generally selected in the surface coating applications because of their toxicity and are now being replaced by the environment friendly pigments like phosphate based pigments. Zinc phosphate has been found to have wide range of application due to its nontoxic nature, excellent anticorrosive properties and can be readily used in the coatings [6–8]. Recently, in order to enhance the corrosion inhibition resistance the phosphate based pigments were modified with different ions like Ca, Si, Ca, Fe, K, etc. The additions of ions such as cation like calcium, potassium, aluminum, etc. during preparation of nanopigments, which are being used as anticorrosion pigments in the surface coating industry ranging in the micrometer size have been reported [7,9]. Ding et al. [10] have reported the synthesis of calcium zinc phosphate in nanosize range which will improve the properties like corrosion inhibition and mechanical properties significantly.

Further, simultaneous addition of inorganic nanoparticles in polymer matrix during synthesis of polymer nanocomposite improves the property profile of the resultant material because of uniform distribution of inorganic nanoparticles in the polymer matrix [11]. In order to have uniform distribution of inorganic nanoparticles in polymer matrix, surface modification of inorganic nanoparticles is essential to reduce the surface energy and increase the compatibility with organic polymer matrix [12,13].

The present work focuses on the preparation of PANI-calcium zinc phosphate nanocomposite by ultrasound assisted *in situ* emulsion polymerization process. The use of ultrasound irradiations during synthesis of PANI-calcium zinc phosphate nanocomposite is expected to improve the dispersion of the calcium zinc phosphate nanoparticles in PANI matrix. When ultrasonic waves pass through a liquid medium, large numbers of microbubbles form, grow, and collapse in very short time which leads to the cavitational effects of intense turbulence, and liquid circulation currents. In the present work, the collapse of cavitation bubbles near the interface of immiscible liquids will cause disruption of the phases because of the generated microjets and turbulence which results into the formation of very fine emulsions [11]. The use of ultrasonic irradiations during *in situ* emulsion polymerization can reduce the agglomeration of calcium zinc phosphate and can improve the dispersion of calcium zinc phosphate in PANI matrix. The prepared PANI-calcium zinc phosphate nanocomposite was characterized by XRD, TEM, and FTIR. Calcium zinc phosphate was modified with myristic acid, in order to enhance its compatibility with PANI.

31.2 MATERIALS AND METHODS

31.2.1 MATERIALS

Analytical grade chemicals such as zinc oxide, calcium hydroxide, phosphoric acid, ammonium Persulfate and sodium dodecyl sulfate were procured from S.D. Fine Chem. and used as received without further purification. Analytical grade chemicals such as sodium hydroxide, HCl, NaOH, NaCl, and ethanol were procured from Sigma Aldrich. The monomer aniline (AR, M/s Fluka) was distilled two times prior to the use. Deionized water from Millipore water purification system was used for all the experimental runs.

31.2.2 ULTRASOUND ASSISTED PREPARATION OF CALCIUM ZINC PHOSPHATE AND PANI-CALCIUM ZINC PHOSPHATE NANOCOMPOSITE

Synthesis of calcium zinc phosphate was performed by chemical precipitation method in presence of ultrasonic irradiations. In the beginning, an aqueous solution of calcium zincate was prepared in the presence ultrasonic irradiation (ultrasonic horn 10 mm diameter operating at frequency of 22 kHz and power as 240 W) by chemical reaction between zinc oxide (2.2 g) and calcium hydroxide (4.8 g) in 250 ml deionized water. The reaction mixture was then heated to 60°C and drop-wise addition of stoichiometric amount of dilute H_3PO_4 to the prepared mixture was carried within 30 min in the presence of ultrasound irradiation. The reaction was further continued for additional 30 min (total reaction time being 60 min). The obtained precipitate of calcium zinc phosphate nanoparticles was filtered and washed with water in order to remove unreacted reagents as well as impurities and then dried in oven at 80°C.

The synthesis of PANI-calcium zinc phosphate nanocomposite was carried out by sonochemical method at a constant temperature of 4°C in presence of myristic treated calcium zinc phosphate nanoparticles [1]. The functionalization of calcium zinc phosphate was carried out by adding 0.2 g myristic acid (MA) in 10 ml methanol and then adding it subsequently in the aqueous solution containing calcium zinc phosphate (1 g in 100 ml water) at 60°C. To begin with the surfactant solution was prepared by adding 3 g of SDS and 0.2 g of MA modified calcium zinc phosphate in 50 ml water and it was transferred to sonochemical reactor. An aqueous initiator solution was prepared separately with an addition of 3.5 g ammonium Persulfate in 20 ml of deionized water and then transferred to sonochemical reactor. The drop-wise addition of aniline monomer (5 g) was carried out with 30 min time period at 4°C in presence of ultrasonic irradiations. The total reaction time was 1.5 hr. Formed PANI encapsulated calcium zinc phosphate nanocomposite particles were separated by centrifugation and then washed with water. The separated product was dried at 60°C in oven for 4 hr.

31.2.3 CHARACTERIZATION AND PROPERTY TESTING

XRD diffraction patterns of calcium zinc phosphate and PANI-calcium zinc phosphate nanocomposite particles were recorded by using powder X-ray diffractometer (Philips PW 1800). The morphology of calcium zinc phosphate and PANI-calcium zinc phosphate nanocomposite was established using transmission electron microscopy (TEM) (Technai G20 working at 200 kV). FTIR analysis of calcium zinc phosphate and PANI-calcium zinc phosphate nanocomposite were carried out (SHIMADZU 8400S) in the region of 4,000–500 cm^{-1}. The particle size distribution measurements were carried out by Malvern Zetasizer Instrument (Malvern Instruments, Malvern, UK). Prepared PANI-calcium zinc phosphate nanocomposite/alkyd resin coatings (2.0 and 4.0 wt. % of freshly prepared PANI-calcium zinc phosphate nanocomposite dispersed in Soya alkyd resin using automatic pigment Muller and applied on mild steel panel) anticorrosion performance was tested using electrochemical corrosion analysis (Tafel plot (log $|I|$ vs. E)) which was carried out in 5 percent NaCl solution as an electrolyte at room temperature (25°C). All measurements were performed on computerized electrochemical analyzer (Autolab Instruments, Netherlands). Three different mild steel plates coated with neat alkyd resin, 2 and 4 wt. % PANI-calcium zinc phosphate nanocomposite dispersed in the same alkyd resin were used as working electrode, while Pt and Ag/AgCl were used as counter and reference electrodes respectively. The area of about 1 cm² was used for sample testing. The electrochemical window was −1.0 V to +1 V with 2 mV/s scanning rate.

31.3 RESULTS AND DISCUSSIONS

31.3.1 XRD ANALYSIS OF CALCIUM ZINC PHOSPHATE AND PANI-CALCIUM ZINC PHOSPHATE NANOCOMPOSITE PARTICLES

Figure 31.1 depicts the X-ray diffraction patterns of the calcium zinc phosphate nanoparticles and PANI-calcium zinc phosphate nanocomposite. As depicted in Figure 31.1, the X-ray diffraction peaks of calcium zinc phosphate has been compared with $CaZn_2(PO_4)_2$-$2H_2O$ in JCPDS file (Card No. 35-0495). The characteristics diffraction peaks at 25.9, 29.4, 31.8, 34.4, 36.4, 47.6, 56.6, 62.9, and 68° correspond to pure calcium zinc phosphate [14] without the presence of any impurities like unreacted species and calcium zincate media. For the characteristics diffraction peaks of calcium zinc phosphate, the formation of calcium zinc phosphate nanoparticles without any impurities can be confirmed. The estimated crystallite size of calcium zinc phosphate nanoparticles at 36.4° (2 Theta angle) using Scherrer's formula is about 10 nm. X-ray diffraction pattern of PANI-calcium zinc phosphate (4 wt. % loading) nanocomposite is shown in the Figure 31.1. The diffraction pattern at different 2θ values that is, 2θ = 18.3, 20.1, 21.2, 24, 25.8, 28.7, 29.2, 31, 31.8, 32.7, 34.2, 47.8° highlights the characteristic peak of PANI-calcium zinc phosphate nanocomposite [1]. The diffraction peaks at 2θ = 18.3, 20.1, 21.2, 24, 28.7, 31, 32.7° highlights the characteristic peaks of PANI and the presence of peaks at 2θ = 25.8, 29.2, 31.8, 34.2, 47.8° depicts the characteristics peaks of calcium zinc phosphate compound. Further it is observed that there is a slight shift of calcium zinc phosphate peaks when it is incorporated in PANI matrix. The presence of calcium zinc phosphate confirms the successful encapsulation of calcium zinc phosphate nanoparticles in PANI matrix to form PANI-calcium zinc phosphate nanocomposite.

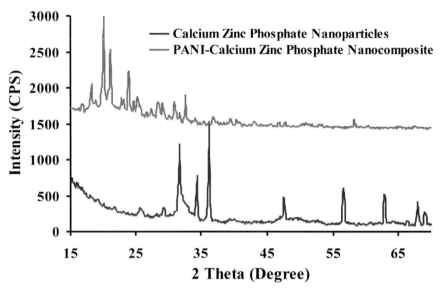

FIGURE 31.1 XRD pattern of calcium zinc phosphate nanoparticles and PANI-calcium zinc phosphate nanocomposite.

31.3.2 MORPHOLOGICAL ANALYSIS OF CALCIUM ZINC PHOSPHATE AND PANI-CALCIUM ZINC PHOSPHATE NANOCOMPOSITE PARTICLES

The TEM images of calcium zinc phosphate nanoparticles and PANI-calcium zinc phosphate nanocomposite particles synthesized by ultrasound assisted method are shown in Figure 31.2. As seen from Figure 31.2(a) the particle size of calcium zinc phosphate nanoparticle has been observed to be around 100 nm with spindle shape. The particles of calcium zinc phosphate nanoparticles with significantly smaller size and uniform shape was observed which is attributed to significant enhancement in micromixing, solute transfer rate, nucleation, and number of nuclei due to physical effects of ultrasound. A TEM image of PANI-calcium zinc phosphate nanocomposite is shown in Figure 31.2(b). The TEM image confirms the distinct formation of PANI-calcium zinc phosphate nanocomposite particles with slightly agglomerated morphology. It is also observed that the calcium zinc phosphate nanoparticles are finely dispersed in PANI matrix, which is attributed to cavitational effects generated through ultrasonic irradiations. In the initial stage of emulsion polymerization, because of hydrophobic nature myristic acid modified calcium zinc phosphate, it remains with aniline (organic phase). Further, micromixing and turbulence created by ultrasonic irradiations generates fine calcium zinc phosphate embedded emulsion droplets, which further leads to formation finely dispersed PANI-calcium zinc phosphate nanocomposite [1].

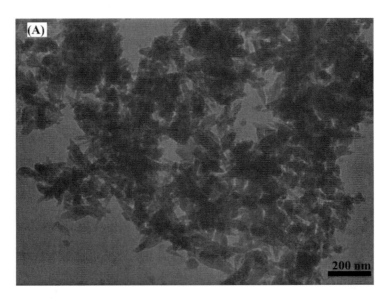

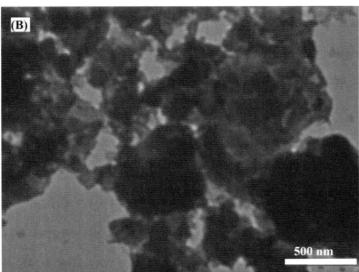

FIGURE 31.2 TEM image of (a) calcium zinc phosphate nanoparticles and (b) PANI-calcium zinc phosphate nanocomposite.

31.3.3 FTIR ANALYSIS OF CALCIUM ZINC PHOSPHATE AND PANI-CALCIUM ZINC PHOSPHATE NANOCOMPOSITE PARTICLES

Figure 31.3 shows the FTIR spectrum of calcium zinc phosphate, myristic acid modified calcium zinc phosphate and PANI-calcium zinc phosphate nanocomposite particles.

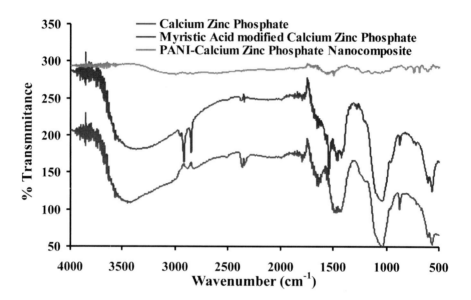

FIGURE 31.3 FTIR spectra of calcium zinc phosphate, myristic acid modified calcium zinc phosphate and PANI-calcium zinc phosphate nanocomposite.

The FTIR spectrum of calcium zinc phosphate shows the majorcharacteristics peaks at 574–615 and 1,070cm^{-1} which are assigned to the vibration modes of PO_4^{3-}[14]. The FTIR spectrum of the myristic acid modified calcium zinc phosphate shows the characteristic peaks at 2,920 and 2,848 cm^{-1} in addition to the peaks of calcium zinc phosphate which are attributed to stretching vibration of the C–H which came from the –CH$_3$ and –CH$_2$ groups in the myristic acid respectively. Further the characteristic peak at 1,543 cm^{-1} is attributed to bending of –OH. The FTIR spectrum of PANI-calcium zinc phosphate nanocomposite shows the characteristic peak at 1,159 cm^{-1} are due to the (C–N) stretching mode of the amine group of PANI. The characteristic peak at 1,500 cm^{-1} represents C=C stretching mode of the quinoid rings and C=C stretching of benzenoid rings respectively [14]. The characteristic peak at 1,579 cm^{-1} is attributed to the secondary =N–H bending. The presence of the characteristic peaks of PANI confirms the successful formation of PANI encapsulated calcium zinc phosphate nanocomposite.

31.3.4 PSD OF CALCIUM ZINC PHOSPHATE AND PANI-CALCIUM ZINC PHOSPHATE NANOCOMPOSITE PARTICLES

Figure 31.4 shows the particle size distribution of sonochemically prepared calcium zinc phosphate and PANI-calcium zinc phosphate nanocomposite particles.

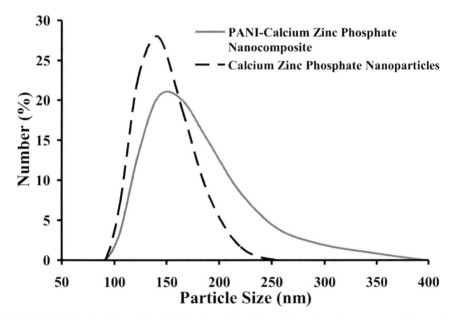

FIGURE 31.4 Particles size distribution of calcium zinc phosphate nanoparticles and PANI-calcium zinc phosphate nanocomposite prepared by ultrasound assisted method.

The particle size range of calcium zinc phosphate nanoparticles found is 91–255 nm with average particles size of 125 nm. The particles were formed with a fairly narrow size distribution and uniform shape could be observed from particle size analysis and TEM analysis. The reason for the considerable reduction in the particle size of calcium zinc phosphatenanoparticles is reported in earlier section. The particles size distribution of PANI-calcium zinc phosphate nanocomposite is depicted in Figure 31.4. The particle size of PANI-calcium zinc phosphate nanocomposite is found to be in the range of 105–342 nm with average particle size of 148 nm indicating consistency in the size distribution of PANI encapsulated calcium zinc phosphate nanocomposite particles. The smaller size of calcium zinc phosphate and PANI-calcium zinc phosphate nanocomposite particles are attributed to cavitational effects of ultrasound irradiations which form smaller particles size as well as create find emulsion of monomer containing myristic acid treated calcium zinc phosphate nanoparticles leading to smaller size and finely dispersed nanocomposite particles.

31.3.5 ELECTROCHEMICAL CHARACTERIZATION OF PANI-CALCIUM ZINC PHOSPHATE NANOCOMPOSITE/ALKYD COATINGS

Figure 31.5 reports the Tafel plot generated for pure alkyd resin and PANI-calcium zinc phosphate nanocomposite coatings on mild steel panels immersed in 5 wt. % aqueous NaCl solutions.

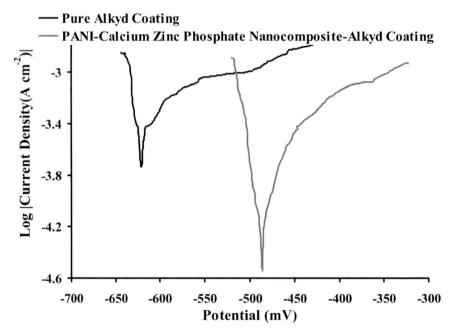

FIGURE 31.5 Tafel plots of mild steel samples coated with pure alkyd resin and PANI-calcium zinc phosphate nanocomposite coatings in 5 wt. % NaCl solution.

I_{corr} and E_{corr} values were estimated from intersection of coordinates of Tafel plot. It is found that electrochemical current was decreased from 0.041 to 0.02 A/cm^2, when neat alkyd resin coatings and PANI-calcium zinc phosphate nanocomposite coatings were tested in NaCl electrolyte solution. Further, E_{corr} value shows positive side shifts from −0.620 to −0.485 V by the incorporation of PANI-calcium zinc phosphate nanocomposite particles into alkyd coating. The decrease in the I_{corr} value and positive shift of E_{corr} value is an indication of improvement in the corrosion inhibition performance. The results of Tafel plot (polarization curve) indicates that with an addition 4 wt. % PANI-calcium zinc phosphate nanocomposite in alkyd resin shows a significant improvement in the corrosion inhibition performance compared to the neat alkyd resin coatings. The improvement in the anticorrosion performance is attributed to the presence of PANI and calcium zinc phosphate anticorrosion pigment and also formation of compact and homogeneous film.

31.4 CONCLUSION

The present work has clearly established that PANI-calcium zinc phosphate nanocomposites have been successfully synthesized using ultrasound assisted semibatch emulsion polymerization of aniline in presence of myristic acid modified calcium zinc phosphate nanoparticles. The dispersion process of calcium zinc phosphate nanoparticles has been improved significantly with the use of cavitation generated due to the ultrasonic irradiations during *in situ* emulsion polymerization. This method is efficient to increase the loadings of calcium zinc phosphate in the PANI composites, which can improve the corrosion inhibition performance of PANI-calcium zinc phosphate nanocomposites/alkyd coatings.

KEYWORDS

- **Functionalization**
- *In situ*
- **PANI/calcium zinc phosphate nanocomposite**
- **TEM**
- **Ultrasound**

REFERENCES

1. Bhanvase, B. A.; and Sonawane, S. H.;New approach for simultaneous enhancement of anticorrosive and mechanical properties of coatings: Application of water repellent nano $CaCO_3$-PANI emulsion nanocomposite in alkyd resin. *Chem. Eng. J.* **2010**, *156*, 177–183.
2. Zhang, H.;Huang,F.; Xu, S.;Xia,Y.; Huang, W.; and Li, Z.; Fabrication of nanoflower-like dendritic Au and polyaniline composite nanosheets at gas/liquid interface for electrocatalytic oxidation and sensing of ascorbic acid. *Electrochem. Commun.* **2013**, *30*, 46–50.
3. Phan, T. B.; Pham, T. T.; and Mai, T. T. T.; Characterization of nanostructured PbO_2–PANi composite materials synthesized by combining electrochemical and chemical methods.*Adv. Nat. Sci.: Nanosci. Nanotech.* **2013**, *4*, 015015.
4. Wu,C.; Xiao, H. M.; and Fu, S. Y.; Synthesis and simultaneous enhancements in electrical and magnetic properties of orientedγ-Fe_2O_3-nanoneedle/PANI nanocomposite films by cold stretching. *Polymer.* **2013**, *54*, 4578–4587.
5. Zhang, J.; et al. Ultra-thin WO_3 nanorod embedded polyaniline composite thin film: Synthesis and electrochromic characteristics.*Sol. Eng. Mater. Sol. Cells.* **2013**, *114*, 31–37.
6. Zeng, R.; Lan, Z.; Kong, L.; Huang, Y.; and Cui, H.; Characterization of calcium-modified zinc phosphate conversion coatings and their influences on corrosion resistance of AZ31 alloy.*Surf. Coat. Tech.* **2011**,*205*, 3347–3355.
7. Cheng, K.; et al. Composite calcium phosphate coatings with sustained Zn release. *Thin. Solid Films.* **2011**, *519*, 4647–4651.
8. Abd El-Ghaffar, M. A.; Youssef, E. A. M.; and Ahmed, N. M.; High performance anticorrosive paint formulations based on phosphate pigments. *Pigm. Resin Technol.* **2004**, *33*, 226–237.
9. Zin, I. M.; Lyon, S. B.; and Pokhmurskii, V. I.; Corrosion control of galvanized steel using a phospate/calcium ion inhibitor mixture. *Corros. Sci.* **2003**, *45*, 777–788.
10. Ding, S.; and Wang, M.; Studies on synthesis and mechanism of nano $CaZn_2(PO_4)_2$ by chemical precipitation. *Dyes Pigments.* **2008**, *76*, 94–96.
11. Bhanvase, B.A.; Pinjari, D.V.; Gogate, P.R.; Sonawane, S.H.; and Pandit, A.B.; Process intensification of encapsulation of functionalized $CaCO_3$ nanoparticles using ultrasound assisted emulsion polymerization. *Chem. Eng. Process.: Process Intensificat.* **2011**, *50*, 1160–1168.
12. Sonawane, S. H.; Khanna, P. K.; Meshram, S.; Mahajan, C.; Deosarkar, M.P.; and Gumfekar, S.; Combined effect of surfactant and ultrasound on nano calcium carbonate synthesized by crystallization process.*Int. J. Chem. Reactor Eng.* **2009**, *7*, A47.
13. Wang, C.; Sheng, Y.; Zhao, X.; Pan, Y.; Bala, H.; and Wang, Z.; Synthesis of hydrophobic $CaCO_3$ nanoparticles. *Mater. Lett.* **2006**, *60*, 854–857.
14. Bhanvase, B. A.; et al. Ultrasound assisted intensification of calcium zinc phosphate pigment synthesis and its nanocontainer for active anticorrosion coatings. *Chem. Eng. J.* **2013**,*231*, 345–354.

CHAPTER 32

SYNTHESIS OF 2K POLYURETHANE COATING CONTAINING COMBINATION OF NANO BENTONITE CLAY, NANO CACO3, AND POLYESTER POLYOL AND ITS PERFORMANCE EVALUATION ON ABS SUBSTRATE

S. A. KAPOLE[1], B. A. BHANVASE[1], R. D. KULKARNI[2], and S. H. SONAWANE[3*]

[1]Department of Chemical Engineering, Vishwakarma Institute of Technology, 666, Upper Indiranagar, Pune-411037, MS, India.

[2]University Department of Chemical Technology, North Maharashtra University, Jalgaon-402103, MS, India.

[3]Department of Chemical Engineering National Institute of Technology Warangal 506004 MS India; *E-mail: shirishsonawane@rediffmail.com

CONTENTS

32.1 Introduction 344
32.2 Materials and Methods 345
32.3 Results and Discussion 349
32.4 Conclusions 353
Keywords 354
References 354

32.1 INTRODUCTION

Polyurethanes (PU), which are available in two packs (2K) and one pack (1K), are extensively used in protective coatings due to its excellent properties like scratch resistance, impact hardness and weather resistance. Polyurethane coatings however, suffer from disadvantages such as, inferior thermal stability and abrasion resistance which can be overcome by using inorganic additives in the form of nanomaterials [1–6]. Corcione et al. [7] have studied the rheological properties of PU adhesive system containing of nanoclay improved the rheological and mechanical properties by intercalation and exfoliation mechanism. Somani et al. [8] have prepared the high solid PU coatings using castor oil based polyester polyol resin. Generally, the technique of simultaneous addition of the inorganic particles during the polymerization is not used [9] as it has been observed that the nano-size inorganic particles have a strong tendency for agglomeration because of their high surface energy [10]. The reduction in surface energy can be achieved by surface modification of inorganic particle with an organic molecule. Some studies demonstrated the use of surface modified nano-$CaCO_3$ [11], nanoclays [6, 7], and nanosilica for improvement in micro hardness, tensile strength, and abrasion resistance of the final film due to fine dispersion of surface modified inorganic nanoparticles in the polymer matrix. It is also reported that the use of nanosilica shows improved scratch resistance [11] in acrylic polyurethane and nano-ZnO imparts UV resistance to the matrix. Furthermore, it was also found that the use of small quantity of (2–5%) nano-$CaCO_3$ shows improvement in hiding power and limits the use of TiO_2 in the polymer coatings [12]. The aliphatic 2K PU finishes are used for exterior purpose due to superior UV stability. Plastic components like acrylonitrile butadiene styrene (ABS), poly propylene (PP), poly vinyl chloride (PVC), polyethylene (PE), fiber reinforced plastic (FRP) which are coated with 2K PU (top coat) have been commonly used in the automobile industry. Due to nucleation and isothermal crystallization mechanism, nanoparticles aredispersed in to the composite material which enhances performance of nanocomposite coatings. In the present study, synthesis of saturated polyester polyol was carried out. Phthalic anhydride, adipic acids were used along with neopentyl glycol (NPG) and trimethylol propane (TMP) as the saturated acid and alcohol respectively. Functionalized nanocalcium carbonate and nanobentonite clay particles were incorporated (1–5 wt. %) separately as well as in combination (4 wt. %) in to polyester polyol nanocomposite by *in situ* polymerization technique and subsequently 2K PU coating was prepared. The mechanical and barrier properties of 2K PU nanocomposite films on ABS substrate were evaluated as per ASTM standards.

32.2 MATERIALS AND METHODS

32.2.1 MATERIALS

For the preparation of 2K PU paint, pre polymeric isocyanate was procured from Bayer material science ltd, Thane, India (Dismodur N 75: % NCO 16.5, % NVM 74.62, % free moisture isocyanate as 0.3% confirmed by HPLC analysis). Neopentyl glycol (NPG) was procured from Ideal Chemi Plast, Badalapur, India; Trimethylol propane (TMP) was procured from Raybon Chemical and Allied products, Vadodara, India; Phthalic anhydride (M.P.1310 C) was procured from M/s Thrumalai Chemicals Private Limited, Ranipet , Tamilnadu, India; adipic acid (M.P. 1520C) was procured from M/s Triveni Aromatics And Perfumery Private Limited, Vapi Gujarat, India; Isopropyl alcohol, ABS panels, antistatic degreasing, commercial grade gasoline, tack-O-rag cloth were procured from local manufactures. Nano-$CaCO_3$ and nanobentonite clay particles were prepared and used according to procedure reported by Sonawane et al. [12]. Solvents like xylene (high b. p. 144°C), toluene (medium b. p. 115°C) and ethyl acetate (low b. pp. 76–78°C) were procured from Ramesh and Company, Pune, India.

32.2.2 SYNTHESIS OF POLYESTER-POLYOL RESIN AND ITS NANOCOMPOSITE USING IN SITU SOLVENT POLYMERIZATION

The detailed procedure for the synthesis of polyester polyol resin is schematically depicted in Figure 32.1 and formulation in Table 32.1.

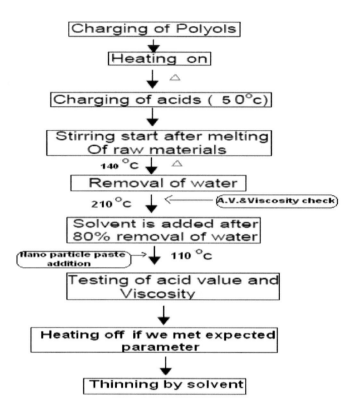

FIGURE 32.1 Flow sheet diagram of synthesis of nanocomposite based on polyester polyol, nano bentonite clay, and nano $CaCO_3$ preparation by *in situ* method.

TABLE 32.1 Composition of polyester polyol resin synthesis (Resin-nanocomposite composite—70% NVM)

Role of Raw Material	Name of the Raw Materials	Neat Polyester Polyol (wt. %)
	Neopentyl glycol	22.62
	Trimethylol propane	5.65
Monomers	Phthalic anhydride	22.62
	Adipic acid	11.3
	O-xylene and toluene	19
Solvent	Ethyl acetate	18.9
Total		100

Note: For the polyester polyol nanocomposite synthesis, paste of nano particles (4 wt. % $CaCO_3$ and nano bentonite clay is added separately during synthesis.

The detailed description of the reactor has been given in the earlier work [13]. Initially 45.24 g NPG and 11.30 g TMP were charged to the reactor and heated up to 80°C for 20 min. Then addition of phthalic anhydride and adipic acid was carried out and the temperature of reactor was increased and maintained at 140°C. Formed water during the course of the reaction was removed and collected in the measuring cylinder. Viscosity and acid value was measured with an interval of 10 min using Ford cup B-4 viscometer and acid base titration, respectively. The final hydroxyl value was measured at the end of the reaction. Further, temperature of reactor was raised to 210°C. After removal of 80 percent of water (based on the theoretical calculations) further solvent addition was carried out and temperature of reactor was maintained in the range 110–115°C. Addition of nanoparticles to the reactor was carried out in next 45 min. Predefined quantity (as reported in Table 32.1) of nano-$CaCO_3$ and nanobentonite clay were added in reaction mixture by making paste using o-xylene. During this addition, speed of agitation was maintained at 1,000 rpm to ensure uniform dispersion of nanoparticles into the resin sample.

32.2.3 PREPARATION, APPLICATION, AND TESTING OF 2K PU COATINGS (NEAT POLYESTER POLYOL AND ITS NANOCOMPOSITE SYSTEM)

Detail composition of base of 2K PU nanocomposite is reported in Table 32.2.

TABLE 32.2 Detailed composition for Base of 2K PU nanocomposite coating

Sr. No.	Component	Paint System (wt. %)
1.	Resin-nanocomposite composite—70% NVM, as reported in Table 32.1	70
2.	Coloring matter (rutile TiO_2)	05
3.	Solvent (Xylene: Toluene: EA 25:25:50% volume basis)	22
4.	Wetting and dispersing agent	03
	Total	100

Note: In above table for Sr. No. 1, the mentioned 70 percent quantity of composite consist of 62 wt. % resin and 8 wt. % nano particle.

Stoichiometric amount of isocyanate (calculated by using Eq. (32.1)) was mixed with polyester polyol nanocomposite as well as neat polyester polyol (in separate approach) by continuous agitation at room temperature under nitrogen environment. Required quantity of isocyanate hardeners is calculated using the following equation:

$$\% \text{ NCO content} = (4{,}200 \times \text{Hydroxyl value})/56{,}100 \tag{32.1}$$

Neat and nanocomposite 2K PU coating systems were applied by using spray gun (DeVilbiss, spray gun's air pressure was set to 0.35 MPa.) During the process of paint application, the dry film thickness was maintained about 30–35 µm. ABS Panels were then dried in an oven for 30 min at 70°C temperature. Pencil hardness of coating was tested according to ASTM D 3363. Flexibility was tested by using Conical Mandrel Apparatus as per ASTM D 522. Adhesion property analysis of the films was performed according to ASTM D 3359 method by cross cut adhesion tester. UV stability was tested by ASTM G 53 method .Other properties of polyester polyol nanocomposite were tested according to ASTM standards, as reported in Table 32.3.

TABLE 32.3 Properties of polyester polyol nanocomposite containing nano $CaCO_3$ and nano bentonite clay particles

Properties	Test Method	Polyester Polyol and Nanoparticles	Neat Polyester Polyol
Clarity	Visual	Turbid	Clear
% solid by weight	ASTM D 2369-10	62	60
Viscosity by ford cup B4	*ASTM D 1200-10*	70 sec	64 sec
Hydroxyl value (mg KOH/g)	ASTM D 1957	102	92
Acid value (mg KOH/g)	ASTM D 3644-06	6.2	9

32.2.4 CHARACTERIZATION

Fourier transform infrared (FTIR) spectroscopic measurement of the samples was performed using KBr pellets (2 mg/300 mg KBr) usinga spectrometer (Model 580, Perkin–Elmer) in the range of 400–4,000 cm^{-1}. Transmission electron microscopic (TEM) analysis of the samples was carried out using a CM200, PHILIPS Microscopy (20–200 KV, Resolution 2.4 Å and magnification 1,000,000X). The structural properties of pristine nanocalcium carbonate and nanoclay was carried out by using low angle X-ray diffraction (XRD) measurements using the CuKα radiation with λ= 1.5405 Å.

32.3 RESULTS AND DISCUSSION

32.3.1 CHARACTERIZATION

Figure 32.2 represents the X-ray diffraction pattern of neat Bentonite Clay, 2K PU-$CaCO_3$ Polyester Polyol Nanocomposite, 2K PU-Bentonite Clay Polyester Polyol Nanocomposite, 2K PU-$CaCO_3$-Bentonite Clay Polyester Polyol Nanocomposite and surface treated $CaCO_3$. The diffraction pattern at different 2θ values that is, at 2θ = 20.8, 21.3, 24.9, 26.6, 34.9, 36.6, 42.5, and 50.1° represents the characteristics peaks of bentonite clay (Figure 32.2, Pattern A). These peaks have been reflected in the X-ray diffraction pattern of 2K PU-bentonite clay polyester polyol nanocomposite (Figure 32.2, Pattern C), which confirms the intercalation of bentonite clay in polyester polyol polymer. Further it has been observed that the characteristics peaks are shifted toward lower 2θ values in the case of 2K PU-bentonite clay polyester polyol nanocomposite. This is an indication of increase in the d-spacing of the clay platelets which further confirms the intercalation of bentonite clay in polyester polyol polymer. X-ray diffraction (XRD) pattern of MA coated $CaCO_3$ has been shown in the Figure 32.2 (inset) in which the crystallite size of functionalized $CaCO_3$ nanoparticles was estimated as equal to 21 nm by Debye Scherrer's formula. Further, presence of the characteristics peaks of MA functionalized $CaCO_3$ and bentonite clay in 2K PU-$CaCO_3$-Bentonite Clay-polyester polyol nanocomposite confirms the formation of 2K PU-$CaCO_3$-Bentonite Clay-polyester polyol nanocomposite (Figure 32.2, Pattern D).

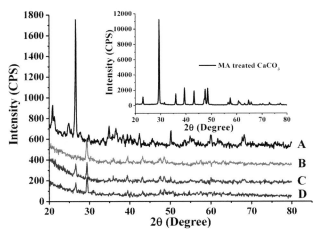

FIGURE 32.2 XRD pattern of (A) neat Bentonite clay, (B) 2K PU-$CaCO_3$ polyester polyol nanocomposite, (C) 2K PU-Bentonite clay polyester polyol nanocomposite, and (D) 2K PU-$CaCO_3$—Bentonite clay polyester polyol nanocomposite and MA coated $CaCO_3$.

Figure 32.3 shows the FTIR spectra of nanobentonite clay, 2K PU nanocomposite and nano-$CaCO_3$ in the range 4,000–500 cm^{-1}. Figure 32.3 (Pattern A) shows FT-IR spectra of $CaCO_3$ nanoparticles. Characteristic peaks of $CaCO_3$ nanoparticles are at 1,481; 875; and 712 cm^{-1}. In addition to these peaks, characteristics peaks at 2,920 and 2,850 cm^{-1} shows the stretching vibration of the C–H which came from –CH_3 and –CH_2 in the myristic acid respectively. This confirms the modification of the $CaCO_3$ by myristic acid. On comparing the spectra, it is found that the new characteristic peaks are observed in Figure 32.3 (Pattern B), which indicates the presence of nanobentonite clay and $CaCO_3$ particles in 2K PU nanocomposite. The IR spectrum of 2K PU nanocomposite coating shows N=C=O stretching at 2,254 to 2,285 cm^{-1}. It was also observed that –OH stretching is observed at 3,338–3,262 cm^{-1} which means hydroxylation has taken place in the film and it might be the reason to increase the mechanical property like flexibility of the final film. A C–H stretching peak was also seen for the stretching bands (2,887–2,982 cm^{-1}). This peak can be attributed to the resin composite or to the reacted resin material in the 2K PU-polyester polyol composites.

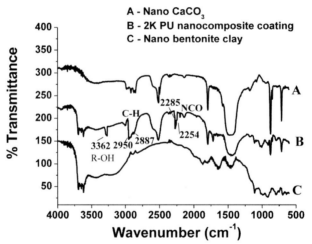

FIGURE 32.3 FTIR spectra for (A) nano $CaCO_3$, (B) 2K PU containing nanocomposite, and (C) nano bentonite clay.

TEM images of Figure 32.4(A) confirmed the spherical shape of nano-$CaCO_3$ finely dispersed into 2K PU polyester resin, which can help to enhance the barrier property like resistance to UV of the 2K PU-polyester composite paint film as compared with neat 2KPU polyester polyol paint film. The particle size of nano-$CaCO_3$ observed from TEM analysis is around 20 nm. Also, as shown in Figure 32.4(B), the nanobentonite clay particles are intercalated into polymer chains and its particle size is around 10 nm. This intercalated nanocomposite gives higher friction resistance, barrier to O_2 and moisture.

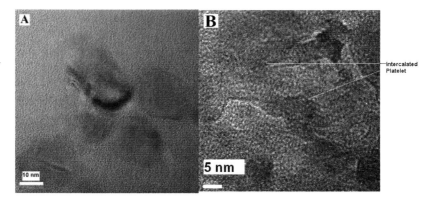

FIGURE 32.4 TEM images of (A) 2K PU-nano $CaCO_3$ nanocomposite and (B) 2K PU-nano bentonite clay nanocomposite.

32.3.2 EVALUATION OF FILM PROPERTIES

Table 32.4(a),(b) reports the data for different properties of coatings such as, pencil hardness, abrasion resistance, flexibility, and UV stability. As reported in Table 32.4(a), it was observed that 1 and 2 wt. % addition of nanobentonite clay and nano-$CaCO_3$ does not show any improvement in the properties significantly, while 3, 4, and 5 wt. % addition of nanobentonite clay shown improvement in the mechanical properties (pencil hardness, resistance to abrasion) and barrier properties (UV stability and resistance to moisture) and while, an improvement in flexibility and cross cut adhesion values found after an addition of nano-$CaCO_3$.

As reported in Table 32.4(a), it was also noted that optimum results were seen at 4 wt. % loading of $CaCO_3$ and nanoclay for mechanical and barrier properties; hence for further experimentation, loading of 4 wt. % of bentonite nanoclay and nano-$CaCO_3$ has been used.

TABLE 32.4 (A) Properties of coated panels using neat 2K PU and individual nano particles based 2K PU nanocomposites

Properties	Test Method	Neat 2K PU Film	Nano Bentonite Clay Based 2K PU Nanocomposite Film					Nano CaCO$_3$ Based 2K PU Nanocomposite Film				
			1wt. %	2wt. %	3wt. %	4wt. %	5wt. %	1wt. %	2wt. %	3wt. %	4wt. %	5wt. %
Pencil hardness	(ASTM D 3363)	H	H	H	3H	3H	2H	H	H	H	2H	H
Abrasion resistance (wt loss in g)	(ASTM D 4060)	130	130	124	104	89	98	130	130	129	128	128
Flexibility	(ASTM D522)	10.5 mm pass	10.5mm pass	10.5mm pass	10.5mm pass	10.0 mm pass	10.5mm pass	10.5mm pass	10.5mm pass	9.0 mm pass	8.0 mm pass	10.0 mm pass
Cross cut adhesion	(ASTM D 3359)	2B	2B	2B	3B	4B	3B	2B	2B	3B	5B	4B
UV stability (hr)	(ASTM G53)	1,600	1,600	1,600	1,600	1,600	1,600	1,600	1,600	1,630	1,680	1,632

TABLE 32.4(B) Properties of coated sample panels using Neat 2K PU-polyester polyol coating sample A and nanocomposite 2K PU-polyester polyol coating sample B

Properties	Test Method	Neat 2K PU-Polyester Polyol Coating (Sample A)	Nanocomposite 2K PU-Polyester Polyol Coating (Sample B)
Pencil harness	ASTM D 3363	2H fails	2H passes
Abrasion resistance (wt. g)	ASTM D 4060	130	89
Flexibility	ASTM D522	10.5 mm pass	9 mm pass
Cross cut adhesion	ASTM D 3359	2B	5B
UV stability (hr)	ASTM G53	1,600	2,000

As shown in Table 32.4(b) which was a comparative study between films of neat 2K polyester polyol polyurethane and nanocomposite of 2K polyester polyol polyurethane .It was noted that an improvement in pencil hardness from 2H fails to 2H passes, resistance toabrasion (by measuring weight loss by CS 17 wheel/1,000 revolution/1 Kg load) was observed which was attributed to intercalation of nanobentonite clay into polymer chain along with urethane backbone present in the 2K PU nanocomposite coating. NPG and TMP gives high cross linking polymer composite which improves the hardness. In addition, urethane backbone also gives network of secondary and reversible hydrogen bonding which absorbs the stress generated as a result of abrasion. Adhesive power of 2K PU film further improved due to the nano-$CaCO_3$ and nanobentonite clay which controls the flexibility and keeps the film intact. This was corroborated with cross cut adhesion test in which neat 2K PU coating (sample A) shows cross cut adhesion resultnear to 2B value (means affected cross cut area is greater than 15%), while nanocomposite of 2K PU coating (sample B) shows 4B value (means affected cross cut area is only about 5%). Further, UV stability results shows that gloss retention after first 500 hr was 96 percent for both the samples, but after 1,600 hr gloss retention value of sample A was 83 percent and for sample B it was 89 percent due to nano-sized calcite phase of the nano-$CaCO_3$.

32.4 CONCLUSIONS

Addition of 4 wt. % nano-$CaCO_3$ and nanobentonite clay in NPG based polyester polyol by *in situ* method plays an important role to augment the mechanical and barrier properties like flexibility, abrasion resistance, pencil hardness, cross cutadhesion, UV stability and resistance to humidity. Results from the TEM image confirms the intercalation of nanobentonite clay particles into polymer chain and shows relatively high resistance to UV stability. Due to a collective effect of urethane backbone and NPG, TMP based polyester polyol nanocomposite gives a network of secondary and reversible hydrogen bonding. The stress generated as a result of abrasion is absorbed by 2K PU nanocomposite coating. Overall, it can be said that, for exterior coating applications, such systems are highly recommended for ABS plastic due their stability against light, heat.

KEYWORDS

- 2K polyurethane
- ABS substrate
- Combination of nanoadditives
- Performance evaluation
- Polyester polyol

REFERENCES

1. Gao, X.; et al. Synthesis and Characterisation of well-dispersed polyurethane/$CaCO_3$ nanocomposite.*Colliods Surf. A.* **2010**, *371,* 1–7.
2. Li, J.; Hong, R.; Li, M.; Li, H.;Zheng, Y.; and Ding, J.; Effects of ZnO nanoparticles on the mechanical and antibacterial properties of polyurethane coatings. *Prog. Org. Coat.* **2009**, *64,* 504–509.
3. Chen, H.; Zheng, M.; Sun, H.; and Jia, Q.;Characterization and properties of sepiolite/polyurethane nanocomposites. *Mater. Sci. Eng. A.* **2007**, *445,* 725–730.
4. Zhou, S.;Wu, L.; Sun, J.;and Shen,W.; The change of the properties of acrylic- based polyurethane via addition of nano-silica.*Prog. Org. Coat.***2002**, *45,* 33–42.
5. Chen, Y.; Zhou, S.; Yang, H.;and Wu, L.; Structure and properties of polyurethane/nano silica composites.*J. Appl. Polym. Sci.***2005**, *95,* 1032–1039.
6. Jagtap, R.; Chimankar, Y.; and Jadhav, N.; Polyurethane hybrid nanocoatings containing organically modified MMT clay for wood applications.*Paint Ind.* **2008**,*58,* 77–88.
7. Corcione, C.; Prinari, P.; Cannoletta, D.; Mensitieri, G.;and Maffezzoli, A.; Synthesis and Characterization of clay- nanocomposite- solvent based polyurethane adhesive.*Int. J. Adhes. Adhes.***2008**, *28,* 91–100.
8. Somani, K.; Kansara, S.; Parmar, R.; and Patel, N.; High solids polyurethane coatings from castor –oil-based polyester-polyol.*Int. J. Polym. Mater.* **2004**, *53,* 283–293.
9. Qi, D.; Bao, M.; Weng,Y. Z.; and Huang, Z. X.; Preparation of acrylate polymer/silica nanocomposite particles with high silica encapsulation efficiency via miniemulsion polymerization. *Polymer.* **2006**,*47,*4622–4629.

10. Caris, C. H. M.;Louisa, P. M.;and Herk, A. M.;Polymerization of MMA at the surface of inorganic submicron particles.*Br. Polym. J.***1989**,*21,*133–140.
11. Yu, H.; Wang, L.; Shi, Q.;Jiang, G.; Zhao, Z.;and Dong; X.; Study on nano-$CaCO_3$ modified epoxy powder coatings.*Prog. Org. Coat.***2009**,*55,* 296–300.
12. Kulkarni, R.; Sonawane, S.;and Mishra, S.; Utilization of Nanoextenders for part replacement of TiO_2 for enhancement of hiding power of surface coatings. *Ind. Patent.* Application No: 11/mum/2004.
13. Kapole, S. A.;Kulkarni, R. D.;and Sonawane,S. H.;Performance properties of acrylic and acrylic-polyol polyurethane based hybrid system via addition of nano $CaCO_3$ and nanoclay.*Can. J. Chem. Eng.***2011**,*89,* 1590–1595.

CHAPTER 33

SIZE-CONTROLLED BIOSYNTHESIS OF AG METAL NANOPARTICLES USING CARROT EXTRACT

SHRIKAANT KULKARNI

Vishwakarma Institute of Technology, Pune (M.S.), India;
E-mail: srkulkarni21@gmail.com

CONTENTS

33.1 Introduction	358
33.2 Method and Materials	359
33.3 Results and Discussion	359
33.4 Conclusions	365
Keywords	365
References	365

33.1 INTRODUCTION

Noble metal nanoparticles are well known to have applications in the diverse fields of catalysis, biotechnology, bioengineering, textile engineering, water treatment, metal-based consumer products and other areas like biosensors, labels for cells and biomolecules and cancer therapeutics, electronic, magnetic, optoelectronics, and information storage[1, 2]. It is now understood that the intrinsic properties of a noble metal nanoparticles are influenced by its size, shape, composition, crystallinity, and structure (solid or hollow) [3–5].

The preparation of stable, uniform silver nanoparticles by reduction of silver ions using carrot extract is reported in the present paper. Silver metal nanoparticles is an important nanomaterial that can be synthesized using chemical routes, but so far it demanded eco-toxic solvents, harsh reaction conditions, etc. Therefore, there is a growing need to develop nontoxic, environmentally-benign "green" synthetic route for Ag nanoparticles synthesis accompanied by its stabilization in an aqueous environment. Synthesis of metal nanoparticles using bioreductants and biostabilizing agents or biosurfactants is an important area of research in nanobiotechnology which is an emerging eco-friendly science leading to well-defined sizes, shapes, and controlled monodispersity. In the recent past, research has been devoted to the synthesis of nanostructured materials under controlled conditions because of their unique chemical and physical properties which differ substantially from those of the bulk materials [6, 7]. In particular, metal nanoparticles have caught considerable attention because of their uniqueand distinct magnetic, optical, electrical, and catalytic properties and the promise and potential they hold in terms of their applications in nanoelectronics[2–6]. Recently biosynthetic methods employing either biological microorganisms or plant extracts have emerged as a simple and viable alternative to chemical synthetic procedures and physicalmethods. Following the initial report on intracellular Ag nanoparticles formation in Pseudomonas stutzeri, many reports on synthesis of nanoparticles using fungi or bacteria have also been reported in the literature [7–10]. Recently, excellent shape-selective formation of single crystalline triangular Au nanoparticles was observed using the extract of the lemongrass plant [11]. Sastry and coworkers reported a process for the rapid synthesis of stable Ag, Au, and their bimetallic nanoparticles at high concentration using Neem leaf broth and suggested that the flavanone and terpenoid constituents of the leaf broth are believed to be responsible for stabilization of nanoparticles [12, 13]. Using plants extract for nanoparticles synthesis can be advantageous over other biological processes because it eliminates the elaborate process of maintaining cell cultures and can also be suitably scaled up for large-scale nanoparticles synthesis [14–20]. Many physical and solid state chemical methods too have been developed over the time for synthesizing metal nanostructures with different morphology like nanowires, nanorods, nanobelts, and nanodots in addition to wet-chemical methods [21–25].

33.2 METHOD AND MATERIALS

The Chemicals used for the synthesis of Ag nanoparticles are 25 ml of $AgNO_3$ (0.001M), known but variable volume of carrot extract obtained by crushing weighed amount of fresh carrots fruits on washing thoroughly with Milli-Q water, as a reducing agent added upon by known amount of glucose as a stabilizing or capping agent. $AgNO_3$(Sigma Aldrich) was used as a precursor. Further, water used for preparation of solutions, extraction, and dilution purpose was Millipore. Solutions have been prepared meticulously by weighing precise amount of both extract and chemicals concerned and homogenization is brought about by sonication followed by dilution to the requisite strength in accordance with experimental requirement.

The experimental method followed consists of a semibatch reactor. A mixture of varied volume of carrot extract as a reducing agent with a given quantity of glucose as a capping agent is chilled to 0°C for 30 min. The chilled mixture is then taken in the semibatch reactor and $AgNO_3$ (0.001M) solution filled in burette is allowed to run down into the reaction mixture at the flow rate of drop/second. The reaction mixture is subjected to continuous stirring using magnetic stirrer during the progress of reaction.The color of the reaction mixture was found to change from red to pale yellow to golden yellow over a span of 40 min. After a regular interval of one minute from the start of the reaction samplewas collected for testing its absorbance using UV-Visible Spectroscopy (Model: UV-1650PC, Shimadzu make withUV Probe software) so as to check for its Surface Plasmon Resonance (SPR). The absorption spectrum was recorded in the form of overlay for all collected samples until the exhaustion of 25 ml of $AgNO_3$ in the burette and continued further until optimal results are obtained.

33.3 RESULTS AND DISCUSSION

UV-Visible spectroscopy is an important analytical technique to ascertain the first-hand information about formation and stability of metal nanoparticles in aqueous solution.

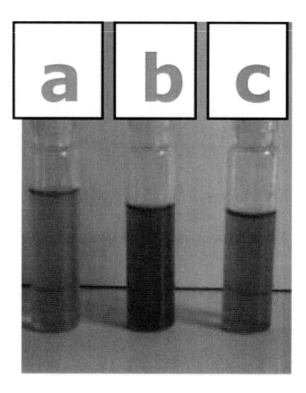

FIGURE 33.1 Shows vials containing (a) carrot extract, (b) reddish dispersion of Ag NP's synthesis, and (c) golden yellow sol (Colloidal solution) of Ag NP's.

Figure 33.1 shows vial marked as "A" which contains pure, fresh carrot extract, the bioactive compounds of which work as a reducing agents and glucose as a capping or stabilizing agent which facilitates Ag nanoparticles synthesis as well as preventing the NP's from getting agglomerated, vial "B" shows reddish colored colloidal solution (sol) indicating onset and furtherance of Ag NP synthesis with subsequent increase in color intensity, while vial "C" shows a golden yellow coloredcolloidal dispersion (sol) indicating conclusion of Ag NP synthesis.

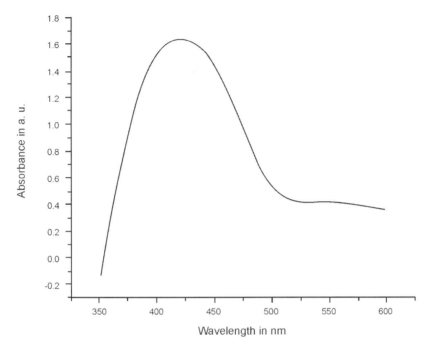

FIGURE 33.2 UV-Vis spectrum for Ag NP's (a) λmax at 415 nm with stabilizing agent(Glucose).

Figure 33.2(a) shows the UV-Visible spectrum of silver nanoparticles from precursor $AgNO_3$(0.001M;25ml) using carrot extract (30ml) as a reducing agent with glucose as a capping agent The rapid color change to the reddish solution and then finally to golden yellow indicate the formation of silver nanoparticles. Surface Plasmon Resonance (SPR) band appears at 415 nm which is in harmony with the earlier researchers reported data [16–19]. The Surface Plasmon Resonance produces a peak near 415 nm with peak width at half maximum (FWHM) of the order of 50–80 nm. This indicates approximate particle size of silver nanoparticles to be between 45–65 nm. This is further confirmed from particle size distribution graph which shows the mean size of nanoparticles which too lies in the same range [11–14] which is further confirmed by employing Debye-Scherrer formula.

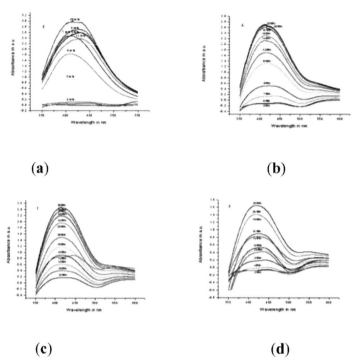

FIGURE 33.3 UV spectrum for the kinetics followed in the Ag NP's with (a) 30 ml, (b) 40 ml, (c) 50 ml, and (d) 60 ml of carrot extract with Glucose as a capping agent.

Figure 33.3(a), (b), (c), and (d) shows the UV-Visible spectrum of silver nanoparticles from precursor $AgNO_3$ (0.001M;25ml) using carrot extract (30, 40, 50, 60 ml) respectively. As a reducing agent with 1 gm glucose powder as a capping agent as received from the market. The sharpness in the SPR indicates the formation of spherical and anisotropic nanoparticles. The FWHM of all colloidal Ag specimens were almost identical (40–60 nm), which indicates that the samples were more of monodispersed type. Further, the Surface Plasmon Resonance (SPR) did not show any peak broadening but rather peaksharpness enhanced indicating not much change in size of nanoparticles but morphology of the NP's remain more of spherical in nature which is confirmed by further characterization like PSD, FTIR, SEM, etc.

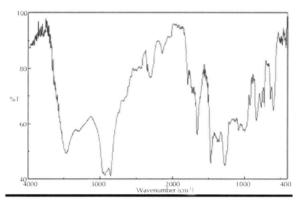

FIGURE 33.4 FTIR image for Ag NP's synthesized using carrot extract as a reducing agent.

Figure 33.4 shows the IR bands observed at 1,317 and 1,733 in carrot extract which are characteristic of the C–O and C=O stretching modes of the carboxylic acid group. The amide I band appears as very strong band at 1,619 cm^{-1} and amide II band as a medium broad in the form of shoulder at 1,546 cm^{-1}. These amide I and II bands arise due to carboxyl stretch and N–H deformation vibrations in the amide linkages of the proteins present in it. The medium broad band at 1,399cm^{-1} is assigned to the C–N stretching mode of aromatic amine group. The C–O–C and C–OH vibrations of the protein appear as a very strong IR band at1,022 cm^{-1}. Further, the band due to C–O stretching at1,314 cm^{-1} is intense in the spectrum of silver nanoparticles.

FIGURE 33.5 Scanning electron microscope (SEM) image of the spherical Ag nanoparticles.

Figure 33.5 shows scanning electron microscopy (SEM) image for Ag nanoparticles derived on formation of golden yellow sol by evaporation of the solvent followed by washing of nanoparticles with ethanol followed by water was used to study the morphology and size of the synthesized Ag NPs. The Ag NPs obtained with the help of Carrot extract as a reductant and glucose as a capping agent were predominantly spherical with diameters ranging from a few nanometers to well above 30 nm [26–32].

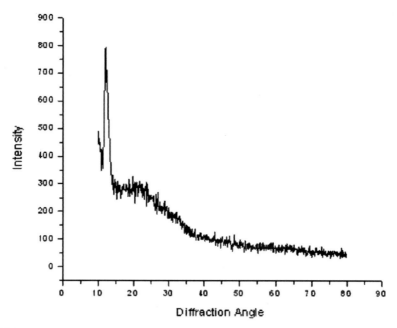

FIGURE 33.6 XRD image for Ag NP's.

Figure 33.6 shows XRD image with a prominent peak which is sharp enough which shows that Ag nanoparticles synthesized are crystalline and are more of monidispersed. The structure of prepared silver nanoparticles has been investigated by X-ray diffraction (XRD) analysis. Typical XRD patterns of the sample, prepared by the present green method are shown in the Figure 33.6. The XRD study indicates the formation of silver (Ag) nanoparticles average particle size has been estimated by using Debye-Scherrer formula. Ag nanoparticles derived on formation of golden yellow sol by evaporation of the solvent followed by washing of nanoparticles with ethanol followed by water obtained in powder form are used for characterization purpose.

$$D = \frac{0.9\lambda}{W\cos\theta}$$

Where "λ" is wave length of X-Ray (0.1541 nm), "W" is FWHM (full width at half maximum), "θ" is the diffraction angle, and "D" is particle diameter (size). The average particle size is calculated to be around 35 nm.

33.4 CONCLUSIONS

A synthetic route using carrot extract which is so far unreported, inexpensive, nontoxic, eco-friendly, abundantly available green reagent was adopted for the consistent and rapid synthesis of Silver nanoparticles. Here Glucose was used successfully as a green capping agent. Ag nanoparticles so synthesized have the SPRat 415 nm which is in good agreement with the reported wavelength in the earlier literature and research papers. The red shift in λ_{max} (Bathochromic effect) to certain degree than the earlier reported wavelength of the order of 398 nm is attributed to more particle size and distinct morphology. Overlay absorption spectrum too with time, shows no change in SPR and remains at about 415 nm. The use of pure Carrot extract is quite effective with glucose as a capping or stabilizing agent as the bioactive compounds in the carrot extract serve as reductants in particular when it is sufficiently concentrated and in this case 50 ml of carrot extract as reductant and 1 gm glucose as a capping agent for 25 ml $AgNO_3$ yields excellent results. More is it concentrated more is it reducing capacity which leads to formation of monodispersed, almost spherical and crystalline Ag Np's. The process can be further investigated and optimized for achieving a greater degree of efficiency, and yield of Ag NP's.

KEYWORDS

- **Bioreductant**
- **Polydispersity**
- **SPR**

REFERENCES

1. Okuda, M.; et al. *Nano Lett.* **2005**, *5*, 991.
2. Rafiuddin, Z. Z.; *Colloids Surf. B: Biointerf.* **2012**, *89*, 211.
3. Fendler, J.H.; and Meldrum, F.C.;*Adv. Mater.* **1995**, *7*, 607.
4. Klaus, T.; Joerger, R.; Olsson, E.; and Granqvist, C. G.; *Proc. Natl. Acad. Sci. U.S.A.* **1999**, *96,* 13611.

5. Eby, D. M.; Shaeublin, N. M.; Farrington, K. E.; Hussain, S. M.; and Johnson, G. R.; *ACS Nano.* **2009**, *3,*984.
6. Zhang, Y.; Peng, H.; Huang, W.; Zhou, Y.; and Yan, D.; *J. Colloid Interface Sci.* **2008,** *325,*371.
7. Panacek, A.; Kolar, M.; and Vecerova, R.; *Biomaterial.* **2009**, *30,*6333.
8. Chwalibog, A.; Sawosz, E.; Hotowy, A.; Szeliga, J.; Mitura, S.; Mitura, K.; Grodzik, M.; Orlowski, P.; and Sokolowska, A.; *Int. J. Nanomed.* **2010**, *5,*1085.
9. Shankar, S. S.; Rai, A.; Ankamwar, B.; Singh, A.; Ahmad, A.; and Sastry, M.; *Nat. Mater.* **2004**, *3,*482.
10. Ahmad, N.; et al. *Colloids Surf. B: Biointerf.* **2010,** *81,*81.
11. Shankar, S. S.; Rai, A.; Ahmad, A.; and Sastry, M.; *J. Colloid Interf. Sci.* **2004**, *275,*496.
12. Krishnaraj, C.; Jagan, E. G.; Rajasekar, S.; Selvakumar, P.; Kalaichelvan, P. T.; and Mohan, N.; *Colloids Surf. B: Biointerf.* **2010**, *76,* 50.
13. Heiz, U.; et al. *NanocatalysisSpringer.***2007**.
14. Anker, J. N.; et al. Biosensing with plasmonic nanosensors. *Nature Mater.* **2008**, *7,* 442–453.
15. Jin, R.; et al. Controlling anisotropic nanoparticle growth through Plasmon excitation. *Nature.* **2003**, *425,* 487–490.
16. Maier, S. A.; et al. Plasmonics—a route to nanoscale optical devices. *Adv. Mater.* **2001**, *13,* 1501–1505.
17. Heaven, M. W.; Dass, A.; White, P. S.; Holt, K. M.;andMurray, R. W.; Crystal structure of the gold nanoparticle [N(C8H17)4][Au25(SCH2CH2Ph)18]. *J. Am. Chem. Soc.***2008**, *130,* 3754–3755.
18. Atwater, H. A.;and Polman, A.; Plasmonics for improved photovoltaic devices. *Nature Mater.* **2010**, *9,* 205–213.
19. Arvizo, R. R.; et al. Intrinsic therapeutic applications of noble metal nanoparticles: Past, present and future. *Chem. Soc. Rev.* **2012**, *41,* 2943–2970.
20. Jadzinsky, P. D.; et al. Structure of a thiol monolayer-protected gold nanoparticle at 1.1A° resolution. *Science.* **2007**, *318,* 430–433.
21. Williams, R.; The Geometrical Foundation of Natural Structure. Dover, New York;**1979**.
22. Zhu, M.; et al. Correlating the crystal structure of a thiol-protected Au25 cluster and optical properties. *J.Am. Chem. Soc.* **2008**, *130,* 5883–5885.
23. Qian, H.; Eckenhoff, W. T.; Zhu, Y.; Pintauer, T.; and Jin, R.; Total structure determinationof thiolate-protected Au38 nanoparticles. *J. Am. Chem. Soc.* **2010**, *132,* 8280–8281.
24. Zeng, C.; et al. Total structure of the golden nanocrystal Au36(SR)24. *Angew. Chem. Int. Edn.***2012**, *51,* 13114–13118.
25. Noginov, M. A.; et al. Demonstration of a spaser-based nanolaser. *Nature.***2009**, *460,* 1110–1112.
26. Brust, M.; et al. Synthesis of thiolderivatised gold nanoparticles in a two-phase liquid-liquid system. *J. Chem. Soc.Chem. Commun.***1994**, 801–802.
27. Ebbesen, T. W.; et al. Large-scale synthesis of carbon nanotubes. *Nature.* **1992**, *358,* 20–222.
28. Bethune, D. S.; et al. Cobalt-catalysed growth of carbon nano-tubes with singleatomic-layer walls. *Nature.***1993**, *363,* 605–607.
29. Shichibu, Y.; et al. Extremely high stability of glutathione-protected Au25 clusters against core etching. *Small.* **2007**, *3,* 835–839.

30. Kratschmer, W.; Solid C60: Anew form of carbon. *Nature.* **1990**, *347,* 354–358.
31. Bakr, O. M.; et al. Silver nanoparticles with broad multiband linear optical absorption. Angew. *Chem. Int. Edu.* **2009**, *48,* 5921–5926.
32. Harkness, K. M.; et al. A silver–thiolate superatom complex. *Nanoscale.* **2012**, *4,* 4269–4274.

CHAPTER 34

A COMBINED EFFECT OF ULTRASOUND CAVITATION ON ADSORPTION KINETICS IN REMOVAL OF 4-[(4-DIMETHYL AMINO PHENYL) PHENYL-METHYL]-N, N-DIMETHYL ANILINE ALONG WITH BIO-ADSORBENT

G. H. SONAWANE*, B. S. BHADANE, A. D. MUDAWADKAR, and A. M. PATIL

[1]Department of Chemistry, Kisan Arts Commerce and Science College Parola, District Jalgaon, (M.S.) 425111 India; *E-mail: drgunvantsonawane@gmail.com

CONTENTS

34.1 Introduction	370
34.2 Materials and Methods	372
34.3 Results and Discussion	374
34.4 Adsorption Equilibrium	378
34.5 Kinetics of Adsorption	381
34.6 Adsorption Isotherms	386
34.7 Conclusions	387
Acknowledgement	387
Keywords	388
References	388

34.1 INTRODUCTION

Wastewater from fabric and yarn dyeing impose serious environmental problems, because of their color and their potential toxicity. The release of colored wastewaters in the ecosystem is a dramatic source of aesthetic pollution, eutrophication, and perturbations in aquatic life. Dye pollutants in wastewaters are the principal source of environmental aqueous contamination. Their removal from water is thus ecologically necessary to offer a cleaner environment as a requirement for human health and has attracted the most wanted attention of environmentalists, technologists, and entrepreneurs. Effluent streams generated by different industries such as textile, tannery, food, printing, pulp and paper, etc., contain dyes as one of the most commonly observed componentimparting obnoxious color to the effluent streams [1–3]. Discharge of effluent streams containing dyes into the natural streams is harmful to the aquatic life and also in long run to the human beings. Removal of dyes from wastewaters is a complex problem because of the difficulty in treating such wastewaters by conventional treatment methods. A variety of physical, chemical and biological methods are presently available for the treatment of wastewater discharged from various industries. Conventional methods of the treatment of dye wastewater include adsorption, [4] chlorination and ozonation, [5, 6] electrochemical methods, [7, 8], biological methods, [9–11] and chemical oxidation [12, 13]. Recently, photocatalysis has been successfully used to oxidize many organic pollutants and particularly to decolorize and mineralize dyes [14–18]. In the past, various attempts have been made to develop effective treatment technologies for dye bearing wastewaters, but no single solution has been found to be satisfactory [19]. Each technique has its own limitations such as generation of secondary effluent, hazardous intermediate products and slow rates of degradation. Thus, research into techniques for effective removal of dyes from the effluent stream holds promise and the present work deals with removal of dyes using adsorption and ultrasound cavitation as one of the approaches for the effective removal of dyes. Over the past two decades, sonochemical degradation of organic pollutants in water has been extensively investigated [20–24]. The chemical effects of sonication arise from acoustic cavitation, namely the formation, growth, and implosive collapse of bubbles in a liquid, which produces unusual chemical and physical environments. The collapse of bubbles generates localized "hot spots" with transient temperature of about 5,000 K, pressures of about 1,000 atm [25]. Under such extreme conditions water molecules dissociated into OH radical and H radical. The radical species can either recombine or react with other molecules to induce sonochemical degradations.

When a liquid is irradiated by ultrasound, microbubbles can appear, grow and oscillate extremely fast and even collapse violently, if the acoustic pressure is high enough. These collapses, occurring near a solid surface, will generate micro jets and shockwaves, resulting in cleaning and erosion of the surface, and fragmentation of the solid which enhances mass transfer rates and solid-liquid interface areas under ultrasound [32]. Cavitational reactors are a novel and promising form of multiphase reactors, based on the principle of release of large magnitude of energy due to the violent collapse of the cavities. Due to generation of hot spots and highly reactive free radicals in the system. H_2O_2 is generated during the collapse of cavities. These peroxide radicals help in reduction of large molecules into the smaller molecules. Cavitation has been looked upon as potential treatment method which is capable of converting chemical substrates like chlorinated hydrocarbons, aromatic compounds, textile dyes, phenolic compounds and esters into short chain organic acids, inorganic ions as final products [33–36].

It is expected that ultrasound will break the big size organic pollutants into smaller size. The purpose of introduction of ultrasound is multifold (i) effective micromixing, (ii) reduction in diffusion resistance, and (iii) liberation of H_2O_2 which further degrade large bulky molecules into smaller size. The chemical reactions that occur in the simplest sonochemical systems are exceedingly complicated with nineteen forward/backward reaction mechanisms occurring in a pure water system alone. All nineteen reaction mechanisms are accounted for in the present model. Of these mechanisms, seven account for most of the chemical kinetics occurring during sonocation. These seven primary reaction mechanisms are:

$$H_2 + OH\cdot \rightarrow OH\cdot \tag{34.1}$$

$$H + H \rightarrow \cdot H_2 \tag{34.2}$$

$$OH\cdot + OH\cdot \rightarrow H_2O_2 \tag{34.3}$$

$$H + O_2 \rightarrow HO_2\cdot \tag{34.4}$$

$$H\cdot + HO_2\cdot \rightarrow H_2O_2 \tag{34.5}$$

$$HO_2\cdot + HO_2\cdot \rightarrow H_2O_2 + O_2 \tag{34.6}$$

$$H_2O\cdot + OH\cdot H_2O_2\cdot + H\cdot \tag{34.7}$$

In present, the work degradation along with adsorption of azo dye Malachite green (MG) that is widely used in textile dyeing by means of ultrasonic cavitation is studied and the effects of contact time, dye concentration, pH, and adsorbent dose were investigated. A combined effect of ultrasound cavitation on adsorption kinetics in removal of dyes along with bio-adsorbent was studied.

34.2 MATERIALS AND METHODS

34.2.1 MATERIALS

Malachite green is anorganic compound that is used as a dyestuff and has emerged as a controversial agent in aquaculture. Malachite green is traditionally used as a dye for materials such as silk, leather, and paper.

34.2.1.1 STRUCTURES AND PROPERTIES OF ADSORBATE

Malachite green is classified in the dyestuff industry as atriarylmethane dye (Figure 34.1). Formally, Malachite green refers to the chloride salt $[C_6H_5C(C_6H_4N(CH_3)_2)_2]Cl$, Molar mass 364.911 g/mol $C_{23}H_{25}ClN_2$ although the term Malachite green is used loosely and often just refers to the colored cation. The oxalate salt is also marketed. The chloride and oxalate anions have no effect on the color. The intense green color of the cation results from a strong absorption band at 615nm (extinction coefficient of $10^5 M^{-1} cm^{-1}$). Malachite green is an industrial product and was obtained from Nice Chemicals Pvt. Ltd, Cochin. Other chemicals used were of analytical grade and used without further purification.

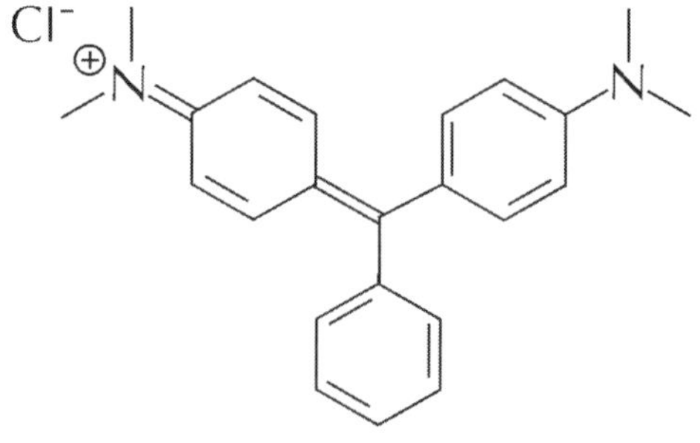

FIGURE 34.1 Malachite green, 4-[(4-dimethylaminophenyl)phenyl-methyl]-N,N-dimethylaniline.

34.2.1.2 ADSORBENT

Shells of groundnut (*Archishypogaea L.*) (GNS) were obtained nearby agricultural field, the collected shells were washed with double distilled water to remove adhering dirt and then were dried in sunlight, crushed and sieved through 60–250 μ size. This ground material was soaked in water for 18 hr and washed with hot and cold distilled water, till the wash water is colorless. This washed material was then dried in oven at 50°C for 12 hr and preserved in glass bottle for use as an adsorbent without any pretreatment.

34.2.2 APPARATUS

Equiptronics EQ-820, UV–Vis spectrometer was used to determine the concentration of Malachite green based on the absorption at the wavelength of 615 nm. Sonication was performed with a 22±3 kHz, Model DP-120 Ultrasonic Probe Sonicator at a power of 120 w (Dakshin). The experimental set up was shown in Figure 34.2. Hundred milliliter aqueous solution of Malachite green was sonicated. The aqueous solution was saturated with pure air before and during the sonication. The reactor was hermetically sealed and connected to a gas burette to ensure a constant pressure for 1 atm. The reaction temperature was controlled with the help of condensation water surrounding the reactor cell.

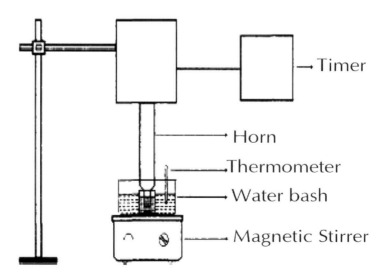

FIGURE 34.2 Ultrasonic probe sonicator.

34.2.3 ADSORPTION EXPERIMENT

Batch adsorption experiments were conducted in continuously stirred vessel. The experimental set up used is shown in Figure 34.2.The adsorption experiments were carried out in a batch process at constant temperature of 30°C by using aqueous solution of Malachite green. In each experiment an accurately weighed amount of GNS was added to 50 ml of the dye solution in 100 ml stopper bottles and mixture was agitated on a mechanical shaker for a given time at constant temperature of 30°C. The adsorbent was separated from solution by centrifugation. The absorbance of the supernant solution was estimated to determine the residual dye concentration. The residual dye concentration was determined at 615 nm with Equiptronics EQ-820, UV–Vis spectrometer. The experiments were carried out at initial pH values ranging from 2 to 11; initial pH was controlled by addition of 0.1 N HCl and 0.1 N NaOH solutions. Kinetics of adsorption was determined by analyzing adsorptive uptake of dye from aqueous solution at different time intervals. The effect of adsorbent dosage on percent removal was studied by varying with GNS dosage from 1 to 14 g/L, maintaining the dye concentration at 20, 40, and 60 mg/L. The effect of dye concentration from 5 to 100 mg/L at an initial pH of 8 was studied. Langmuir and Freundlich isotherms were obtained from dye concentration 20–60 mg/L and varying GNS dose from 1 to 14 g/L at pH = 8 and contact time 40 min.

34.3 RESULTS AND DISCUSSION

34.3.1 ADSORBENT CHARACTERIZATION

For structural and morphological characteristics FTIR, XRD, FE-SEM, and EDX of GNS were carried out.

34.3.1.1 SEM AND XRD ANALYSIS

As it is known, SEM (Scanning Electron Microscopy) is one of the most widely used surface diagnostic tools. The SEM micrographs of GNS and GNS dyed by Malachitegreen are shown in Figure 34.3. GNS has heterogeneous surface and micro pores as seen from its SEM micrographs. The XRD pattern of adsorbent GNS showed typical spectrum at 2θ of 22° and 22.8°, respectively.

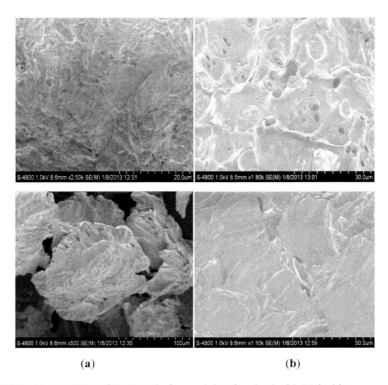

FIGURE 34.3 FE-SEM of GNS (a) before and (b) after dyed with Malachite green.

34.3.1.2 EDX AND ELEMENTAL ANALYSIS OF GNS

The EDX analysis of GNS before adsorption of Malachite green (Figure 34.4) shows the presence of 57.96 percent C, 40.64 percent O and mineral composition 0.83 percent Si, and irons 0.56 percent elements given in Table 34.1. The EDX analysis of GNS after adsorption of Malachite green (Figure 34.5).

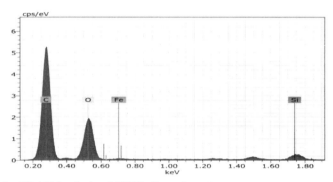

FIGURE 34.4 EDAX Analysis of GNS before adsorption.

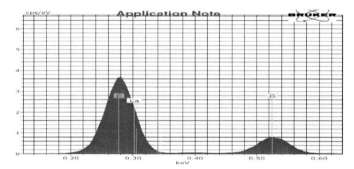

FIGURE 34.5 EDAX analysis of GNS dyed with Malachite green (after adsorption).

TABLE 34.1 EDAX Analysis in wt % of GNS before adsorption and afteradsorption.

Sr. No.	Element	Wt % of GNS Before Adsorption	Wt % of GNS After Adsorption
1.	Carbon	57.96	65.45
2.	Oxygen	40.64	34.36
3.	Silicon	00.83	00.00
4.	Iron Fe	00.56	00.00
5.	Calcium	00.00	00.20

34.3.1.3 FTIR ANALYSIS

The FTIR measurements of GNS showed the presence of peaks for large number of functional groups viz., 3845.8, 3755.1 free –OH group, 2922.0 alkyl C–H stretching, 2852.5 –C–CH$_3$ stretching, 2374.2 –OH stretching, 2275.8 –C–N, 1884.3–C–Cl, 1749.3 –C–OR stretching, 1556.4 > C = C < 1465.8 C–H bending of methylene group, 1037.6 –CHOH and 449.4 cm^{-1}O–Si–O bending. It is clear that the adsorbent displays a number of adsorption peaks, reflecting the complex nature of adsorbent (Table 34.2). The FTIR spectrum of dye adsorbed GNS shows peaks at 3647.3; 3435.8 –OH stretch; 2,920 CH stretch; 2,763 –C–CH$_3$stretch; 2,349 N–H stretch; 1,747 –C–OR stretching and 1,032 cm^{-1} C–O stretch. This indicates shifting of peaks of 3845.8–3647.3; 3755.1–3435.8; 2922.0 of alkyl C–H stretching shifted to 2,920; and 2852.5 –C–CH$_3$stretching shifted to 2,763; 2374.2 –OH stretching shifted to 2,349; 1749.3 –C–OR shifted to 1,747 and 1037.6 shifted to 1,032 cm^{-1}. The shifting of these peaks in the sample to lower frequency 3647.3; 3435.8 –OH stretch; 2,920 CH stretch; 2,763 –C–CH$_3$stretch; 2,349 N–H stretch; 1,747 –C–OR stretching; and 1,032 cm^{-1} C–O stretch. This indicates shifting of peaks of 3845.8–3647.3; 3755.1–3435.8; 2922.0 of alkyl C–H stretching shifted to 2,920; and 2852.5 –C–CH$_3$stretching shifted to 2,763; 2374.2 –OH stretching shifted to 2,349; 1749.3 –C–OR shifted to 1,747 and 1037.6 shifted to 1,032 cm^{-1}. The shifting of these peaks in the sample to lower frequency. Bending after adsorption, suggesting the participation of these functional groups in the adsorption of Malachite green by GNS.

TABLE 34.2 Comparison of FTIR peaks of GNS before and after adsorption of Malachite green

FTIR Frequency (cm^{-1})			
Before Adsorption			After Adsorption
3845.8	Free –OH group		3647.3
3755.1	3435.8		
2900.0	alkyl –C–H stretching		2,920
2852.5	–C–CH$_3$ stretching		2,763
2374.2	–OH stretching		2,349
2275.8		–C=N	
1884.3		–C–Cl	
1749.3	–C–OR stretching		1,747
1556.4		>C = C<	
1465.8	C–H bending of methylene group		
1037.6	–CH–OH		1,032
449.4	O–Si–O bending		

34.4 ADSORPTION EQUILIBRIUM

34.4.1 EFFECT OF SONICATION ON ADSORPTION OF MG ONTO GNS

The equilibrium adsorption of MG onto GNS before and after sonication was studied by measuring the isotherm of adsorption. Figure 34.6 represents the quantity of MG uptake against the equilibrium concentration of MG in the solution, and it corresponds to the equilibrium distribution of MG ions between the aqueous and solid phases when the concentration increases. From the graph of this isotherm, the maximum uptake of MG removal by the GNS and the sonicated GNS samples was determined as 13.03 and 12.85 mg/g, respectively Figures 34.13 and 34.14. The adsorption of dye decreases after the sonication. The decrease in the MG uptake can be explained by the destruction of adsorbing sites resulting from sonication. The percent removal of MG was 89.29 percent at 40 min contact time with GNS, while it decreases to 85.7 percent with sonication (Figures 34.6 and 34.7).

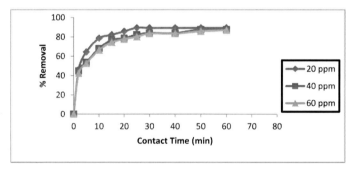

FIGURE 34.6 Effect of contact time on percent removal and concentration of Malachite green, pH 8, without ultrasound sonication.

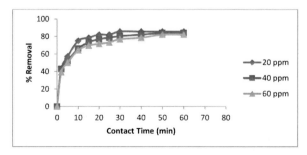

FIGURE 34.7 Effect of contact time on percent removal and concentration with sonication of Malachite green 25°C pH 8.

34.4.2 EFFECT OF PH ON ADSORPTION OF MG ONTO GNS

pH is one of the most important factors which controls the adsorption capacity of dye on clay surfaces. Change of pH affects the adsorptive process through dissociation of functional groups on the adsorbent surface active sites. Adsorption process usually depends on the electro kinetic properties of adsorbent which determines with H^+/OH^- amount. In order to study the effect of pH on the adsorption capacity of MG onto the sonicated GNS, experiments were performed using various initial solution pH values (pH 2–10). Figures 34.8 and 34.9 also indicates that adsorption of MG onto GNS shows the pH-dependent adsorption mechanism. The removal of MG increased from 46.42 to 94.6 percent with increase in pH from 2 to 10. GNS exhibited maximum removals of MG dye, at the highest pH tested (pH 10). This can be explained by considering the surface charge of GNS. The removal of MG increased from 4.64 to 9.46 mg/g with increase in pH from 2 to 10. The percent removal of MG was observed to be 60.4 to 95.3 percent and q_t was 6.04–9.53 mg/g with sonication (Figure 34.9).

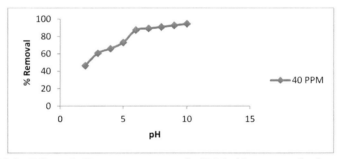

FIGURE 34.8 Effect of pH on percent removal of Malachite green, adsorbent dose 4g/L, contact time 40 min without ultrasound sonication.

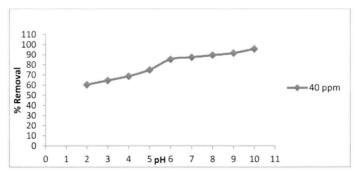

FIGURE 34.9 Effect of pH on percent removal of Malachite green, adsorbent dose 4g/L, contact time 40 min, with sonication of 2 min.

34.4.3 EFFECT OF CONTACT TIME AND INITIAL DYE CONCENTRATION ON ADSORPTION OF MG ONTO GNS

The initial concentration provides an important driving force to overcome all mass transfer resistance of ions between the aqueous and solid phases. A series of experiments were performed at different initial MG concentrations, viz., 5, 10, 20, 40, 60, 80, and 100 mg/L. Figures 34.10 and 34.11 show the effect of initial concentration on the adsorption kinetics of MG onto sonicated GNS at pH 8 and 25°C. The equilibrium adsorption capacity increased with increasing initial MG concentration, due to the increase in the number of ions competing for the available binding sites in the surface of GNS. The kinetics of adsorption represented a shape characterized by a strong increase of the capacity within the first few minutes of contact and gradually tailed off thereafter. This indicated that equilibrium time was independent of concentration. The figure also showed that the contact time to reach the equilibrium was about 40 min. As MG concentration was increased from 20 to 60 mg/L percent removal was decreases from 89.29 to 83.43 percent, when GNS is sonicated it decreases from 85.71 to 78.57 percent.

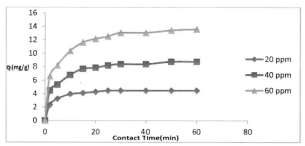

FIGURE 34.10 Effect of contact time on q_t and concentration of Malachite green, pH 8, without ultrasound sonication.

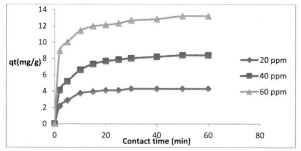

FIGURE 34.11 Effect of contact time on q_t and concentration of Malachite green, pH 8, with ultrasound sonication.

34.4.4 EFFECT OF ADSORBENT DOSE (GNS)

The effect of adsorbent dose was studied at different initial concentration of 20, 40, 60 mg/L with and without sonication from 1 to 14 g/L adsorbent dose(Figure 34.12). As dose increases percent removal also increases and reached to constant at 10g/L.

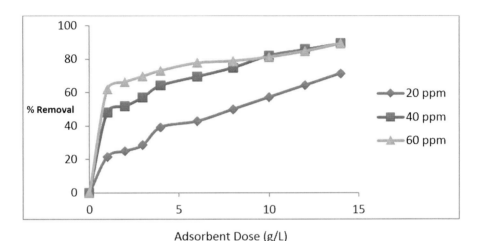

Adsorbent Dose (g/L)

FIGURE 34.12 Effect of adsorbent dose on percent removal and concentration of Malachite green, pH 8, without ultrasound sonication.

34.5 KINETICS OF ADSORPTION

Kinetics of adsorption is one of the most important characteristics to be responsible for the efficiency of adsorption. The adsorbate can be transferred from the solution phase to the surface of the adsorbent in several steps and one or any combination of which can be the rate-controlling mechanism: (i) mass transfer across the external boundary layer film of liquid surrounding the outside of the particle; (ii) diffusion of the adsorbate molecules to an adsorption site either by a pore diffusion process through the liquid filled pores or by a solid surface diffusion mechanism; and (iii) adsorption (physical or chemical) at a site on the surface (internal or external) and this step is often assumed to be extremely rapid. The overall adsorption can occur through one or more steps. In order to investigate the mechanism of process and potential rate controlling steps, the experimental kinetic data for the uptake of GNS at different initial concentration was modeled by the pseudo-first order by Lagergrenand the pseudo-second order by Ho and McKay and the intraparticle diffusion equations by Weber and Morris.

34.5.1 PSEUDO-FIRST-ORDER KINETICS

The adsorption data was analyzed in terms of pseudo-first-order and pseudo-second-order kinetic models. Adsorption rate constants were summarized in Table 34.3. The plots of log $(q_e - q_t)$ versus t gave straight lines with slope of $-k_1/2.303$ and intercept log q_e. For the pseudo-first-order adsorption rate expression Eq. (34.1), the linear plots of log $(q_e - q_t)$ against time were used to determine the rate constant, k_1, equilibrium adsorption q_e. Comparison of the rate constants, calculated and experimental q_e values with the correlation coefficients for different MG concentrations is shown in Table 34.3. The correlation coefficients for the first-order kinetic model obtained at all the studied concentrations were low (<0.5649). The q_e values computed from the Lagergren plots deviated considerably from the experimental q_e values. This indicates that pseudo-first-order equation may not be sufficient to describe the adsorption mechanism of MG onto the sonicated GNS. The values of regression coefficient for pseudo-first-order model (Table 34.3) indicated that the adsorption kinetics of MG on sonicated GNS was not diffusion controlled.

$$\log (q_e - q_t) = \log q_e - t \frac{K1}{2.303} \quad (34.1)$$

TABLE 34.3 Comparison of adsorption rate constants, calculated and experimental q_e values for different initial dye concentrations and with and without sonication for different kinetic models for MG-GNS

Adsorbent (g/L)	Dye Concentration (mg/L)	Pseudo-Second Order				
		$K_2 \times 10^{-2}$ (g/mg min)	q_e (cal) (mg/g)	t ½	h	r^2
4 (without US)	20	12.33	4.65	1.346	2.66	0.9999
	40	3.48	9.25	1.244	2.98	0.9999
	60	2.11	14.28	1.516	4.31	0.9999
4 (with US)	20	10.62	4.48	0.999	2.13	0.9999
	40	3.69	8.84	1.928	2.89	0.9999
	60	3.86	13.69	1.709	7.24	0.9999

34.5.2 PSEUDO-SECOND-ORDER KINETICS

The plots of t/q_t versus t (Figures 34.13 and 34.14) gave straight lines with slope of $1/q_e$ and intercept $1/k_2q_e^2$. The values of regression coefficient for pseudo-second-order model were nearly unity (>0.9999) for all initial dye concentrations studied. The calculated q_e values were very close to that of experimentally obtained q_e. Thus the adsorption of MG on sonicated GNS was explained by the pseudo-second-order kinetic model.

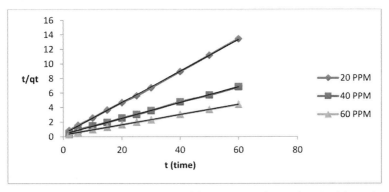

FIGURE 34.13 Pseudo-second-order plot of t/q_t vs. contact time (minute) without ultrasound sonication.

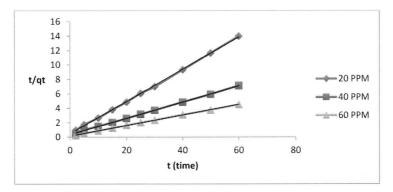

FIGURE 34.14 Pseudo-second-order plot of t/q_t vs. contact time (minute) with ultrasound sonication.

$$\frac{t}{q_t} = \frac{1}{K_2 q_e^2} + \frac{t}{q_e} \text{ and } h = K_2 q_e^2 \quad (34.2)$$

where K_2 is the rate constant of pseudo-second order.

The linear plots of t/q_t versus t for the pseudo-second-order model in Eq. (34.2) were obtained at different Malachite green concentrations (Figures 34.13 and 34.14). High correlation coefficients values (>0.9999) for the pseudo-second-order kinetic model in all conditions revealed that MG uptake process followed the pseudo-second-order rate expression. The amount of MG adsorbed "q_e" and rate constant "K_2" was calculated from the slope and intercept of the plot. A comparison of results of the kinetic constants with the correlation coefficients and calculated and experimental q_e values for different initial MG concentrations are given in Table 34.3. The calculated q_e values agree very well with the experimental data in the case of pseudo-second-order kinetics. According to the pseudo-second-order model, the adsorption rates become very fast and the equilibrium times are short, which are confirmed by the experimental results given in Figure 34.13 and 34.14.

34.5.3 INTRAPARTICLE MODEL

The intraparticular diffusion model was also tested to identify the diffusion mechanism and the results are tabulated in Table 34.4. The intra-particle diffusion model presents multilinearity (Figures 34.15 and 34.16), indicating that two steps take place. The sharper first-stage portion is attributed to the diffusion of adsorbate through the solution to the external surface of adsorbent or the boundary layer diffusion of solute molecules. The second portion describes the gradual adsorption stage, where intra-particle diffusion rate is rate-limiting. During these two stages, MG ions were slowly transported via intraparticle diffusion in the particles and were finally retained in the pores. From the figure, since the linear portion of the first initial stage does not pass through the origin, there is an initial boundary layer resistance between adsorbent and adsorbate. This deviation from the origin is proportional to the boundary layer thickness, which gives an insight into the tendency of the dye ions to adsorb to the adsorbent or remain in solution.

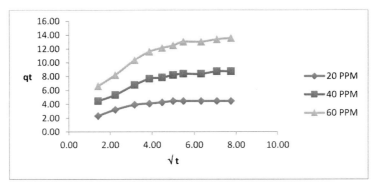

FIGURE 34.15 Plot of q_t vs. square root of t without sonication.

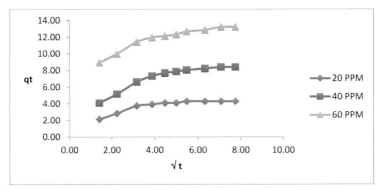

FIGURE 34.16 Plot of q_t vs. square root of t with sonication.

TABLE 34.4 Comparison of Intra particle diffusion adsorption rate constants, for different initial dye concentrations and with and without sonication for MG-GNS

Adsorbent (g/L)	Dye Concentration (mg/L)	K_p	r^2
4 (without US)	20	0.292	0.710
	40	0.661	0.847
	60	1.059	0.857
4 (with US)	20	0.640	0.876
	40	0.649	0.836
	60	0.735	0.876

34.6 ADSORPTION ISOTHERMS

Tables 34.5 and 34.6 indicates the results of Langmuir and Freundlich isotherm for adsorption of MG dye onto GNS from aqueous solutions before and after sonication at pH 8, for different contact time and adsorbent dose. The applicability of the Freundlich isotherm is analyzed by plotting log q_e versus log C_e, but data are not found in good agreement with the correlation coefficients less than 0.7435. The high correlation coefficients ($r^2>0.8288$) in all experimental conditions studied suggest that the adsorption of MG onto both natural and the GNS closely follow a Langmuir isotherm. The decreases in values of r^2 in with ultrasound suggest that this decreases in adsorption in sonication. The Langmuir monolayer capacity q_o of natural GNS and the sonicated GNS for MG dye at pH 8 and 25°C were found as 9.1743 and 20.00 mg/g, respectively. The values of R_L in all experimental conditions studied are presented in Tables 34.5 and 34.6. The R_L values calculated indicate that adsorption of MG dye on sonicated GNS is favorable ($0 < R_L < 1$) within the experimental conditions studied. The R_L values indicate that the adsorption is more favorable for the higher initial MG concentrations than for the lower ones.

TABLE 34.5 Freundlich and Langmuir isotherm constants for adsorption of MG on GNS for different dye concentration, and adsorbent dose of 1 to 14 g/L at pH=8, contact time 40 min

Dye Concentration (mg/L)	Freundlich Coefficient				Langmuir Coefficient			
	K_f (L/g)	n	1/n	r^2	a (mg/g)	b (g/L)	R_L	r^2
Without US								
20	9.4842	0.8576	1.166	0.7435	1.7793	0.2227	0.1833	0.9752
40	10.6712	1.1261	0.888	0.9148	8.0645	0.0993	0.2011	0.9078
60	13.7088	0.5144	1.944	0.9531	9.1743	0.1550	0.09708	0.8288
With US								
20	3.4119	1.0395	0.962	0.9274	11.9045	0.0262	0.6563	0.4873
40	3.0549	0.6978	1.433	0.8682	20.000	0.0402	0.3834	0.6182
60	25.0034	0.4192	2.385	0.9581	9.6153	0.1753	0.0868	0.9621

TABLE 34.6 Freundlich and Langmuir isotherm constants for adsorption of MG on GNS for different dye concentration and contact time, adsorbent dose of 4 g/L at pH=8.

Dye Concentration (mg/L)	Freundlich Coefficient				Langmuir Coefficient			
	K_f (L/g)	n	1/n	r^2	a (mg/g)	b (g/L)	R_L	r^2
Without US								
20	5.9979	2.849	0.351	0.9132	2.2026	0.7566	0.0619	0.9718
40	18.36532	2.369	0.422	0.9268	4.0983	0.2837	0.0809	0.9734
60	28.5101	2.652	0.377	0.9023	6.3291	0.2140	0.0722	0.9671
With US								
20	7.1285	2.217	0.451	0.9363	1.9531	0.5464	0.0838	0.9723
40	22.8559	1.9685	0.508	0.9348	3.6491	0.2235	0.1006	0.9689
60	24.7742	3.2894	0.304	0.9458	8.0646	0.2897	0.0544	0.9889

34.7 CONCLUSIONS

This study investigates the effect of ultrasonic treatment on the adsorption capacity of GNS for removing Malachite green. Sonication process resulted in a significant decrease in the removal of Malachite green. The ultrasonic pretreatment of GNS also improved the adsorption of MG so that maximum monolayer adsorption capacity increased from 1.7793 to 11.9047 mg/g and 8.0645 to 20.0 mg/g after the sonication. The experimental data fit well with Langmuir isotherm model. Adsorption process of MG onto GNS followed the pseudo-second-order rate expression and occurs by both physical adsorption and weak chemical interactions.

ACKNOWLEDGEMENT

The author (GHS) is thankful to UGC, WRO and Pune for financial assistance to carry out this research project. Authors are also thankful to Principal and Head Department of Chemistry, Kisan Arts Commerce and Science College, Parola, Dist. Jalgaon for providing laboratory facilities and Dr Shirish Sonawane, NIT, Warangal for his valuable suggestions and discussion.

KEYWORDS

- Adsorption kinetics
- Archis hypogaea
- Isotherms
- Malachite green
- Ultrasound

REFERENCES

1. Gupta, V. K.; Ali, I.; and Suhas, M. D.; Equilibrium uptake and sorption dynamics for the removal of a basic dye (basic red) using low cost adsorbents. *J. Colloid Interf. Sci.* **2003**, *265*, 257.
2. Kumar, K.V.; Ramamurthi, V.;and Sivanesan, S.; Biosorption of Malachite green, a cationic dye onto Pithophorasp a fresh water alga. *Dyes Pigments.* **2006**, *69*, 102.
3. Ahmad, R.;and Kumar, R.; Adsorptive removal of congo red dye from aqueous solution using bael shell carbon. *Appl. Surf. Sci.* **2010**, *257*, 1628.
4. Wang Y.;and Yu, J.; Adsorption and degradation of synthetic dyes on the mycelium of Trametesversi color, water quality international 98. Part 4, in: Wastewater: *Ind. Wastewater Treatment.* **1998**, *38*, 233–238.
5. Alaton, I. A.; Kornmuller, A.;and Jekel, M.R.; Ozonation of spent reactive dye-baths: Effect of alkalinity. *J. Environ. Eng.* **2002**, *128*, 689–696.
6. Brik, M.; Chamam, B.; Schoberl, P.; Braun, R.;and Fuchs, W.; Effect of ozone, chlorine and hydrogen peroxide on the elimination of colour in treated textile wastewater by MBR. *Water Sci. Technol.* **2004**, *49*, 299–303.
7. Guivarch, E.; Trevin, S.; Lahitte, C.;and Oturan, M.A.; Degradation of azo dyes in water by electro-Fenton process. *Environ. Chem. Lett.* **2003**, *1*, 38–44.
8. Vlyssides, A. G.; Loizidou, M.; Karlis, P. K.; Zorpas, A.;and Papaioannou, A. D.; Electrochemical oxidation of a textile dye wastewater using a Pt/Ti electrode.*J. Hazard Mater.* **1999**, *70*, 41–52.
9. Jo-Shu Chang;and Tai-Shin Kuo; Kinetics of bacterial decolorization of azo dye with Escherichia coli NO_3. *Bioresource Technol.* **2000**, *75*,107–111.
10. Novotny, C.; Svobodova, K.; Kasinath, A.; and Erbanova, P.; Biodegradation of synthetic dyes by irpexlacteus under various growth conditions. *Int. Biodeterior. Biodegrad.* **2004**, *54*, 215–223.
11. Chagas, P. E.; and Durrant, R. L.; Decolorization of azo dyes by phancerochaete chrysosporium and pleurotussajorcaju. *Enzym. Microb. Technol.* **2001**, *29*, 473–477.
12. Dutta, K.; Bhattacharjee, S.; Chaudhuri, B.;and Mukhopadhyay, S.; Chemical oxidation of C. I. Reactive Red 2 using Fenton-like reactions. *J. Environ. Monit.* **2002**, *4*, 754–760.
13. Donlagic, J.; and Levec, J.; Comparison of catalyzed and noncatalyzed oxidation of azo dye and effect on biodegradability. *Environ. Sci. Technol.* **1998**, *32*, 1294–1302.
14. Jae-Hong, P.; Hyoung, C.; and Young-Gyu, K.; Solar light induced degradation of reactive dye using photocatalysis. *J. Environ. Sci. Health A.***2004**, *39*, 159–171.

15. Nguyen Thi, D.; Nguyen, V.;and Jean-Marie, H.; Photocatalytic degradation of reactive dye RED-3BA in aqueous TiO_2 suspension under UV–visible light.*Int. J. Photoenerg.* **2005,** *7,* 11–15.
16. So, C. M.; Cheng, M. Y.; Yu, J. C.;and Wong, P. K.; Degradation of azo dye procion red MX-5B by photocatalytic oxidation.*Chemosphere.* **2002,** *46,* 905–912.
17. Liu, G.;and Zhao, J.; Photocatalytic degradation of dye sulforhodamine B: A comparative study of photocatalysis with photosensitization.*New J. Chem.* **2000,** *24,* 411–417.
18. Jae-Hong, P.; Euiso, C.;and Kyung, G.; Removal of reactive dye using UV/TiO_2 in circular type reactor.*J. Environ. Sci. Health A.* **2003,** *38,*1389–1399.
19. Gupta, V.K.;and Suhas; Application of low-cost adsorbents for dye removal-a review. *J. Environ. Manage.* **2009,** *90,* 2313.
20. Wang, X.K.; Chen, G.H.;and Guo, W.L.; Sonochemical degradation of kinetics of methyl violet in aqueous solutions. *Molecules.* **2003,** *8,* 40–44.
21. Hua,I.;and Hoffmann,R.M.; Optimization of ultrasonic irradiation as anadvanced oxidation technology.*Environ. Sci. Technol.* **1997,** *31,* 2237–2243.
22. Nagata, S.N.; Nakagawa, M.; Okuno, H.; Mizukoshi, Y.; Yim, B.;and Maeda, Y.; Sonochemical degradation of chloro- phenols in water.*Ultrasond Sonochem.* **2000,** *7,* 115–120.
23. Frim, J. A.;and Weavers,L.K.; Sonochemical destruction of free and metal-binding ethylene-diamine tetraacetic acid. *Water Res.* **2003,** *37,* 3155–3163.
24. Fındık, S.; Gundu, Z.E.;and Gundu, Z.; Direct sonication of acetic acid in aqueous solutions.*Ultrasond Sonochem.* **2006,** *13,* 203–207.
25. Didenko, Y. T.; McNamara, W.B.;and Suslick, K.S.; Hot spot conditions during cavitation in water. *J. Am. Soc.* **1999,** *121,* 5817–5818.
26. Yang, L.; Huan. H.; Liu, H.; Zou, Y.;and Chen, H.; Preparation and microwave absorption properties of FeNi/graphite nanocomposites. *Trans. Nonferrous Met. Soc. China.***2007,** *17,* 708–712.
27. Manocha,S.; Patel, N.; and Manocha, L.M.; Development and characterization of nanoclay from Indian clays. *Defense Sci. J.* **2008,** *58(4),* 517–524.
28. Gogate, P. R.; Treatment of wastewater streams containing phenolic compounds using hybrid techniques based on cavitation: A review of the current status and the way forward. *Ultrasonic Sonochem.* **2008,***15,*1–15.
29. Ji-Tai, L.; Mei, L.; Ji-Hui, L.; and Han-Wen, S.; Decolorization of azo dye direct scarlet 4BS solution using exfoliated graphite under ultrasonic irradiation.*Ultrasonic Sonochem.* **2007,** *14,* 241–245.
30. Hyun Jang, S.; Jeong, Y. G.; Gil Min, B.; Seok Lyoo, W.;and Cheol Lee, S.; Preparation and lead ion removal property of hydroxyl apatite/polyacrylamide composite hydro gels.*J. Hazardous Mater.* **2008,** *159,* 294–299.
31. Arrojo, S.; Nerín, C.; and Benito, Y.; Application of salicylic acid dosimetry to evaluate hydrodynamic cavitation as an advanced oxidation process. *Ultrasonics Sonochem.* **2007,** *14,* 343–349.
32. Ji-Tai Li; Mei Li; Ji-Hui Li;and Han-Wen Sun; Removal of disperse blue 2BLN from aqueous solution by combination of ultrasound and exfoliated graphite. *Ultrasonics Sonochem.* **2007,** *14,* 62–66.

33. Gültekin, I.; and Ince, N. H.; Degradation of aryl-azo-naphthol dyes by ultrasound, ozone and their combination: Effect of a-substituents.*Ultrason. Sonochem.* **2006,** *13,* 208–214.
34. Kenji, O.; Kazuya, I.; Yoshihiro, Y.; Hiroshi, B.; Rokuro, N.; and Yasuaki, M.; Sonochemical degradation of azo dyes in aqueous solution: a new heterogeneous kinetics model taking into account the local concentration of OH radicals and azo dyes.*Ultrason. Sonochem.* **2005,** *12,* 255–262.
35. Naomi Stock, L.; Julie Peller; Vinodgopal, K.; and Kamat, P. V.; Combinative sonolysis and photocatalysis for textile dye degradation.*Environ. Sci. Technol.***2000,** *34,* 1747–1750.
36. Xikui, W.; Zhongyan, Y.; Jingang, W.; Weilin, G.; and Guoliang, L.; Degradation of reactive brilliant red in aqueous solution by ultrasonic cavitation.*Ultrasonics Sonochem.* **2008,** *15,*43–48.

CHAPTER 35

FORMATION OF IRON OXIDE NANOPARTICLES IN CONTINUOUS FLOW MICROREACTOR SYSTEM

MAHENDRA L. BARI[2*], SHIRISH SONAWANE[1], and S. MISHRA[2]

[1]National Institute of Technology, Warangal, AP, India 506004

[2]Institute of Chemical Technology, North Maharashtra University Jalgaon, MS India - 42500; *E-mail: mlbari@gmail.com

CONTENTS

35.1 Introduction 392
35.2 Material 393
35.3 Results and Discussion 393
35.4 Conclusion 396
Keywords 397
References 397

35.1 INTRODUCTION

Unique properties of iron oxide, nanoparticles make it most versatile candidate for application in catalysis, medical diagnostics and cell imaging, as well as corrosion inhibitor, pigments and gas sensors applications, etc. [1–2]. The number of methods reported for production of iron oxide nanoparticles, some of the important methods are metal salt reduction, thermal decomposition of organometallic precursor, sol–gel and co precipitation, etc. [3–7]. Presences of impurities, use of organic solvents for production, wide particle size distribution are some of limitations of these methods. Particle aggregation is one of the important pronounced effects during synthesis as the particle size decreases, the surface charges on the particle increases, the particles come together to minimize the surface energy and they agglomerate [8]. Hence, to decrease the effect of surface charges and to keep the particles stable in solution, above said processes, utilizes surfactant for stabilization of nanoparticles. In wet chemical synthesis processes, the nanocolloids will form, which may find application in polymer coatings, conductive inks and biosensors etc. Metal salt reduction method is most widely used for synthesis of colloidal nanoparticles because of its ease in processing and precise control on process conditions and finally on the particle size. Reduction of metal salt aqueous solutions by sodium borohydride at room temperature produces both monometallic and bimetallic nanoparticles as amorphous powders [9]. The reduction process of metal salt for synthesis of colloidal iron oxide nanoparticles is simple but the stability and reproducibility of colloid is a great challenge. [10] Sania Pervaiz et al. [11] used vitamin C, sodium borohydrate for reduction of ferric chloride to yield the zero valent iron nanoparticles. Blum et al. [12] used Tiron for synthesis of monodisperse iron oxide nanoparticles. Hematite, maghemite, wustite are the main forms of iron oxide nanoparticles [13]. Maiyong et al. [14] synthesized hematite (μ-Fe_2O_3) form of iron oxide nanoparticles by hydrothermal route. Jing et al. [15] used anionic and cationic surfactant to modify the reaction conditions for preparation of μ-Fe_2O_3 nanoparticles through hydrothermal route. The most conventional method for obtaining maghemite Fe_2O_3 is by coprecipitation, Jeong et al. [16] fabricated γ-Fe_2O_3 nanoparticles by co precipitation. However other routes are also explored by researcher for synthesis of maghemite Fe_3O_4. Liu et al. [17] synthesized γ-Fe_2O_3 nanoparticles by the microemulsion method. Shafi et al. [18] have used the principles of sonochemistry for synthesis of γ-Fe_2O_3 nanoparticles and they reported that the size and morphology of nanoparticles are govern by the reaction parameter and reactor type. The batch and continuous reaction mode was intensively investigated to study effect of reactor type on nanoparticle size and morphology [19–21]. Microreactor was used for continuous synthesis of nanoparticles providing the fine control over the properties at nanoscale. Laminar flow conditions in microreactor due to low Reynolds number provide diffusion-based reaction condition [22–26]. Miniaturization of reaction volume provides the scope for easy and quick variation in process parameter to generate database of product properties with process parameters, which can be useful for invention of novel material [27, 28]. Microreactor was used for synthesis of metal, metal oxide, semiconductor, polymer nanoparticles by numerous researchers [29–31].

35.2 MATERIAL

Ferric Chloride (99% pure, Himedia Laboratories, Mumbai), Sodium Borohydride (99% pure, Merck PVT. LTD., Mumbai), Tween 80 (99% pure, SD Fine Chem LTD., Mumbai), were used as received. Deionized and distilled water was used in the experiments.

35.2.1 METHODOLOGY OF NANOPARTICLE SYNTHESIS IN MICROREACTOR

Synthesis of the iron oxide, nanoparticles was carried out in microreactor of 6 ml volume, (make Amar Equipment, Mumbai, India). The aqueous solutions of $FeCl_3$ and $NaBH_4$ were prepared of required concentrations and Tween 80 was added in both the solutions (1% W/V). Two peristaltic pumps were used to feed reactants to Y junction of reactor. The product was collected at outlet of the reactor.

35.3 RESULTS AND DISCUSSION

35.3.1 EFFECT OF PRECURSOR CONCENTRATION ON PARTICLE SIZE

Continuous synthesis of the iron oxide nanoparticles was carried out in microreactor. Reduction route was adopted for synthesis of iron oxide nanoparticles; $FeCl_3$ was reduced by $NaBH_4$ to create Fe^{3+} which was further oxidized with aqueous oxygen to form iron oxide nanoparticles. Tween 80 was used to stabiles the nanoparticles with 0.001 M, 0.002 M, and 0.005 M $FeCl_3$ concentration. The molar ratio of $FeCl_3$ to $NaBH_4$ was kept 1:4. At low concentration of $FeCl_3$ the nanoparticle size was small whereas, at higher concentration the nanoparticles size was larger as shown in Figure 35.1.

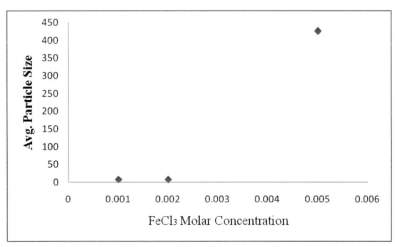

FIGURE 35.1 Effect of precursor concentration on size of nano particles.

Higher rate of reaction at higher precursor concentration yields large number of Fe^{3+} ion resulting in growth of nanoparticles so that larger nanoparticles were formed. The 8 nm sized particles were formed at 0.002 M $FeCl_3$ where as 425 nm particles were obtained at 0.005 M $FeCl_3$. The precursor concentration affects the morphology of iron oxide nanoparticles as shown in Figure 35.2. Iron oxide nanoparticles synthesized at 0.002 M $FeCl_3$ concentration were spherical and flower like structure of nanoparticles was obtained at 0.005 M $FeCl_3$.

(a)   (b)

FIGURE 35.2 FESEM images of nano particles at a.0.005 M $FeCl_3$ b.0.002 M $FeCl_3$.

35.3.2 EFFECT OF MOLAR RATIO OF PRECURSORS ON PARTICLES SIZE AND PARTICLES SIZE DISTRIBUTION

Experiments were carried at 0.002 M $FeCl_3$ with molar ratio ($FeCl_3$ to $NaBH_4$) of 1:1 and 1:2 to study the effect of molar ratio on nanoparticles size. The results in study of effect of precursor concentration show that the small size of nanoparticles was obtained at 0.002 M $FeCl_3$ with molar ratio of $FeCl_3$ to $NaBH_4$ 1:4. Figure 35.4 indicates that larger size of nanoparticles were obtained at molar ratio of 1:1 as compare to 1:4 for 0.002 M $FeCl_3$. Reduction in molar ratio of reactants causes the slowdown of reaction giving the low concentration of metal ion yielding few numbers of nuclei. Growth of these nuclei takes place along with the length of reactor as reaction mixture proceeds toward the out let of reactor and yields a larger size of iron oxide nanoparticles. Narrow size distribution of iron oxide nanoparticles was obtained at reactants molar ratio of 1:2 as shown in Figure 35.3.

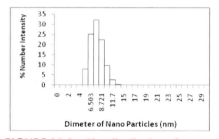
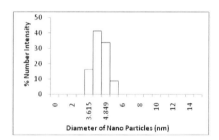

FIGURE 35.3 Size distribution of nano particles at molar ratio of (a) 1:4 and (b) 1:2.

35.3.3 EFFECT OF FLOW RATE OF PRECURSOR ON PARTICLE SIZE

The residence time of reaction mixture in microreactor is the function of flow rate; it decides the conversion of reactant in to product. Hence, the variation of flow rate can change the size of nanoparticles as well as size distribution. To study the effect of flow rate, the synthesis of nanoparticles was carried out with flow rate, 3 ml/min, 6 ml/min, 9 ml/min of $FeCl_3$. The size of nanoparticles increases with increase in flow rate in case of reactants molar ratio 1:2 as shown in Figure 35.4,

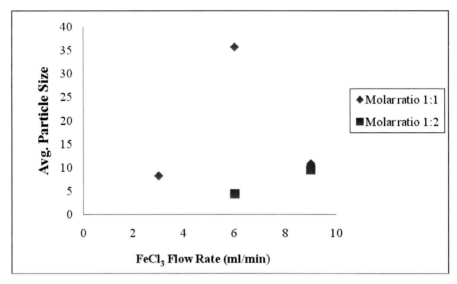

FIGURE 35.4 Effect of precursor flowrate on size of nano particles.

whereas, at reactants molar ratio of 1:1, the size of iron oxide nanoparticles increase with increase in flow rate from 3 to 6 mL/min Further, decrease in size was observed with increase in flow rate from 6 to 9 ml/min. The smallest size of nanoparticles with molar ratio ($FeCl_3$ to $NaBH_4$) of 1:1 and 1:2 were obtained at lower flow rate at higher flow rate, collision between the nanoparticles may occur resulting in aggregation of nanoparticles giving the higher size of nanoparticles.

35.4 CONCLUSION

In the present study the process parameters, precursor concentration, reactants stichometric ratio and the flow rate were investigated to find out their effect on size, morphology, and size distribution of iron oxide nanoparticles in the continuous microreactor process. The morphology of nanoparticles depends on concentration and it changes from spherical to flower like from a lower concentration to higher concentration respectively. The smallest size of nanoparticles with narrow size distribution in the range of 3–5 nm was obtained at molar ratio 1:2 for 0.002 M FeCl3 concentration. The flow rate of precursor affects the rate of reaction hence the size as well as the size distribution of nanoparticles get affected. The largest size of nanoparticles (90 nm) was obtained at a flow rate of 9 ml/min with molar ratio of 1:1. Whereas the smallest size (4 nm) was obtained at a flow rate of 9 ml/min with molar ratio of 1:2.

KEYWORDS

- **Iron oxide nanoparticles**
- **Microreactor**
- **Narrow size distribution**

REFERENCES

1. Jing, Z.; Wu, S.; Sn Zhang, and Huang, W.; "Hydrothermal fabrication of various morphological a-Fe_2O_3 nanoparticles modified by surfactants." *Mater. Res. Bull.* **2004**, *39*, 2057–2064.
2. Yan, A.; et al. "Solvothermal synthesis and characterization of size-controlled Fe_3O_4 nanoparticles." *J. Alloys Compounds.* **2007**, 487–491.
3. Shen, C. M.; et al. "Monodispersive CoPt nanoparticles synthesized using chemical reduction method." *Chin. Phys. Lett.* **2008**, *25*, 1479–1485.
4. Kulkarni, R. D.; Gosh, N.; Patil, U. D.; and Mishra, S.; "Surfactant assisted solution spray synthesis of stabilized Prussian blue and iron oxide for preparation of nano latex composite." *J. Vacuum Sci. Technol. B.* **2014**, *27(3)*, 1478–1483.
5. Veronica, S. M.; Miguel, A.; Michael, F.; Arturo, M. L.; Karl, S.; and Rodolfo, D.; "Synthesis and characterization of large colloidal cobalt particles." *Langmuir.* **2006**, *22*, 1455–1458.
6. Gubin, S. P.; Koksharov, Yu. A.; Khomutov, G. B.; and Yurkov, G. Y.; "Magnetic nanoparticles: Preparation, structure and properties." *Russ. Chem. Rev.* **2005**, *74*, 489–520.
7. Marie-Christine, D.; and Didier, A.; "Gold nanoparticles: Assembly, supramolecular chemistry, quantum-size-related properties, and applications toward biology, catalysis, and nanotechnology." *Chem. Rev.* **2004**, *104*, 293–346.
8. Wu, W.; He, Q.; and Jiang, C.; "Magnetic iron oxide nanoparticles: Synthesis and surface functionalization strategies." *Nanoscale Res. Lett.* **2008**, *3*, 397–415.
9. Ya, H. L.; Lien Loa, S.; Kuanb, W. H.; Lin, C-J.; and Weng, S. C.; "Effect of precursor concentration on the characteristics of nanoscale zerovalent iron and its reactivity of nitrate." *Water Res.* **2006**, *40*, 2485–2492.
10. Sileikaite, A.; Prosycevas, I.; Puiso, J.; Juraitis, A.; and Guobiene, A.; "Analysis of silver nanoparticles produced by chemical reduction of silver salt solution." *Mater. Sci.* **2006**, *12*, 4.
11. Pervaiz, S.; Farukh, M. A.; Adnan, R.; and Qureshi, F. A.; "Kinetic investigation of redox reaction between vitamin C and ferric chloride hexahydrate in acidic medium." *J. Saudi Chem. Soc.* **2012**, *16*, 63–67.
12. Korpany, K. V.; Habib, F.; Murugesu, M.; and Blum, A. S.; "Stable water-soluble iron oxide nanoparticles using Tiron." *Mater. Chem. Phys.* **2012**, 1–9.
13. Nidhin, M.; Indumathy, R.; Sreeram, K. J.; and Nair, B. U.; "Synthesis of iron oxide nanoparticles of narrow size distribution on polysaccharide templates." *Bull. Mater. Sci.* **2008**, *31*, 93–9.

14. Zhu, M.; Wang, Y.; Meng, D.; and Qin, X.; Diao, G.; "Hydrothermal synthesis of hematite nano particles and their electrochemical properties." *J. Phys. Chem. C.* **2012**, *116,* 16276—16276.
15. Jinga, Z.; and Wu, S.; "Synthesis and characterization of monodisperse hematite nanoparticles modified by surfactants via hydrothermal approach." *Mater. Lett.* **2004**, *58,* 3637–3640.
16. Jeong, J. R.; Lee, S. J.; Kim, J. D.; and Shin, S. C.; "Magnetic properties of γ-Fe_2O_3 nanoparticles made by coprecipitation method." *Phys. Status Solid B.* **2004**, *241,* 1593–1596.
17. Jing, Z.; Wu, S.; Zhang, S.; and Huang, W.; "Hydrothermal fabrication of various morphological Fe_2O_3." *Mater. Res. Bull.* **2004**, *39,* 2057–2064.
18. Shafi, K. V. P. M.; et al. "Magnetic enhancement of γ-Fe_2O_3 nanoparticles by sonochemical coating." *Chem. Mater.* **2002**, *14,* 1778–1787.
19. Kockmann, N.; and Roberge, D. M.; "Scale-up concept for modular micro structured reactors based on mixing, heat transfer, and reactor safety." *Chem. Eng. Process.* **2011**, *50,* 1017–1026.
20. Pohar, A.; and Plazl, I.; "Process intensification through microreactor application." *Chem. Biochem. Eng. Q.* **2009**, *23(4),* 537–544.
21. Mason, B. P.; Price, K. E.; Steinbacher, J. L.; Bogdan, A. R.; and McQuade, D. T.; Greener "Approaches to organic synthesis using microreactor technology." *Chem. Rev.* **2007**, *107(6),* 2300–2318.
22. Chang, C.-H.; Paul, B.K.; Remcho, V. T.; Atre, S.; and Hutchison, J. E.; "Synthesis and post-processing of nanomaterials using microreaction technology." *J. Nanopart. Res.* **2008**, *10,* 965–980.
23. Wagner, J.; and Kolhler, J. M.; "Continuous synthesis of gold nanoparticles in a microreactor." *Nano Lett.* **2005**, *5(4),* 685–691.
24. Lin, X. Z.; Terepka, A. D.; and Ya, H.; "Synthesis of silver nanoparticles in a continuous flow tubular microreactor." *Nano Lett.* **2004**, *4(11),* 2227–2232.
25. Chung, C. K.; Shih, T. R.; Chang, C. K.; Lai, C. W.; and Wu, B. H.; "Design and experiments of a short mixing-length baffled microreactor and its application to microfluidic synthesis of nanoparticles." *Chem. Eng. J.* **2011**, *168,* 790–798.
26. Hsin Hung, L.; Phillip Lee, A.; "Microfluidic devices for the synthesis of nanoparticles and biomaterials." *J. Med. Biol. Eng.* **2007**, *27,* 1–6.
27. Kanaris, A. G.; and Mouza, A. A.; "Numerical investigation of the effect of geometrical parameters on the performance of a micro-reactor." *Chem. Eng. Sci.* **2011**, *66,* 5366–5373.
28. Park, K.Y.; Ullmann, M.; Suh, Y. J.; and Friedlander, S. K.; "Nanoparticle microreactor: application to synthesis of titania by thermal decomposition of titanium tetraisopropoxide." *J. Nanopart. Res.* **2001**, *3,* 309–319.
29. Chang, C-H.; Paul, B. K.; Remcho, Vt T.; Atre, S.; and Hutchison, J. E.; "Synthesis and post-processing of nanomaterials using microreaction technology." *J. Nanopart. Res.* **2008**, *10(6),* 965–980.
30. Wu, C.; and Zeng, T.; " Size-tunable synthesis of metallic nanoparticles in a continuous and steady flow reactor." *Chem. Mater.* **2007**, *19(2),* 123–125.
31. Sun, L.; Luan, W.; Yuejin Shan; and Tu, S.-T.; One-step synthesis of monodisperse Au–Ag alloy nanoparticles in a microreaction system. *Chem. Eng. J.* **2012**, 451–455.

CHAPTER 36

COMPARATIVE STUDY OF PRODUCTION OF STABLE COLLOIDAL COPPER NANOPARTICLES USING MICROREACTOR AND ADVANCED-FLOW REACTORS®

M. SURESH KUMAR, S. NIRAJ, ANSHUL JAIN, SONAL GUPTA, VIKASH RANJAN, MAKARAND PIMPLAPURE, and SHIRISH SONAWANE*

Chemical Engineering Department, National Institute of Technology, Warangal, 506004;

*E-mail: shirishsonawane@rediffmail.com

CONTENTS

36.1 Introduction 400
36.2 Experimental 400
36.3 Results and Discussion 405
36.4 Conclusion 413
Acknowledgments 414
Keywords 414
References 414

36.1 INTRODUCTION

Nanomaterials research has been directed toward the synthesis of metal nanoparticles in order to explore their special properties and potential applications of metal nano-size particles [1–4]. Among various metal nanoparticles, copper nanoparticles have attracted considerable attention because of their superior electrical conducting properties when used in nanosize compared to the bulk size particles [5–8]. Colloid copper nanoparticles are found useful inpreparation of electronic circuits. Furthermore, because of their high electrical conductivity and chemical activity, Cu nanoparticles can replace Au and Ag in some potential applications such as conductive pastes and catalysts.Several methods have been developed for the preparation of copper nanoparticles, including thermal reduction [9,10], chemical reduction [11–13], sonochemical reduction [14], metal vapor synthesis [15], vacuum vapor deposition [16], radiation method [17,18], microemulsion techniques [19], and laser ablation [20]. Among these methods, the aqueous solution reduction method is the most widely employed, due to the advantages including high yield and quality of particles, simplicity of operation, limited equipment requirements, and ease of control.

In the present work, comparative study of synthesis of copper nanoparticles by chemical reduction method was attempted in continuous flow microreactor and Advanced-Flow reactor. Microreactor and Advanced-Flow reactor offers a variety of advantages over conventional technologies. The large surface area to volume ratio of microreactor and Advanced-Flow reactor offers enhanced heat and mass transfer in comparison with conventional reactors. Moreover, efficient mixing of is a key advantage for preparing monodispersed copper nanoparticle in microreactor and Advanced-Flow reactor. Scaled-up production of monodispersed colloidal nanoparticles has become an important research subject in recent years. In this regard, continuous flow reactors are generally favored over batch reactors. Advanced-Flow reactors can generate products on a continuous basis and are more appropriate for a large-scale production than the batch reactor. Furthermore, targeted reaction temperatures can be achieved in second or even millisecond time scales in a microreactor. We have controlled the size and size distribution of the particles by varying experimental parameters such as flow rate, temperature and concentration ratio of reducing agent to precursor.

36.2 EXPERIMENTAL

36.2.1 MATERIALS AND METHODS

Preparation of the copper nanoparticles was carried out by using copper (II) sulfate pentahydrate (Merck Specialities Private Limited, Mumbai) and Ascorbic acid (SD Fine Chemicals Ltd. Mumbai). Polyvinyl pyrrolidone (PVP, $(C_6H_9NO)_n$) (Sisco research laboratory Pvt. Ltd, Mumbai), Cetyltrimethyl Ammonium Bromide (Sisco research laboratory Pvt. Ltd, Mumbai) were used as surfactants. The purity of the all chemicals used was above 98 percent. Pure water (<1μS/cm conductivity) from Millipore was used for the preparation of aqueous solutions of the precursors.

36.2.2 PREPARATION OF STABLE COLLOIDAL COPPER NANOPARTICLES USING CONTINUOUS FLOW MICROREACTOR

As shown in Figure 36.1. Microreactor set up was fabricated using Low Density Polyethylene Tube (LLDPE), Syringe pumps were procured from M/S Universal medical instruments Indore MP India, length of microreactor was 150 cm and 0.8 mm inner diameter. Two syringes were used to insert the two precursors in the microreactor through a junction and a hot water bath was used to maintain the reaction temperature.

An aqueous solution of copper sulfate was prepared by adding 1.59 g of copper in 100 ml water. Aqueous solutions of ascorbic acid and polyvinylpyrrolidone (PVP) were prepared separately by taking 4.36 g ascorbic acid and 0.1 g of PVP in 100ml water. Ascorbic acid was used as a reducing agent whereas PVP was used as a surfactant. Aqueous solutions of copper sulfate, ascorbic acid, and PVP were passed through the microreactor tube and the reaction mixture was maintained between 75 and 80°C temperatures. Copper sulfate was reduced by ascorbic acid which results in the formation of copper nanoparticles. In general, copper nanoparticles are highly unstable as they tend to form copper oxide. However, due to reducing properties of ascorbic acid, formation of copper oxide is inhibited resulting in the formation of stabilized copper nanoparticles. As soon as the cooper nanoparticles were formed, PVP formed a layer of coating the copper particles, which prohibits further agglomeration. Formation of copper nanoparticles were confirmed when the color of reaction solution turned brick red. The formed copper nanoparticles were characterized using UV, PSD and FESEM analysis.

Initially, in water $CuSO_4$ dissociates to Cu^{2+} and SO_4^{2-} and Cu^{2+} ions are hydrolyzed into $Cu(OH)_2$ as a precursor. Further reduction of $Cu(OH)_2$ takes place in presence of ascorbic acid to form Cu_2O. Finally, Cu_2O was again reduced to form Cu nanoparticles. The overall reduction process can be represented as follows [21]:

$$CuSO_4 \rightarrow Cu^{2+} + SO_4^{2-} \quad (Rx\ 36.1)$$

$$Cu^{2+} + 2OH^- \rightarrow Cu(OH)_2 \qquad (Rx\ 36.2)$$

$$Cu_2O + C_6H_8O_6 \rightarrow Cu + C_6H_6O_6 + H_2O \qquad (Rx\ 36.3)$$

Molar ratio of copper sulfate to ascorbic acid was used as 1:4. PVP was used as surfactant. 50 ml of both solutions were filled in two syringes. Two syringes were used to pass two precursors in the microreactor through a junction and a hot water bath was used to maintain the reaction temperature. From the outlet of the microreactor colloidal Cu nanoparticles were collected. Temperature was maintained at 75–80°C. The reaction was carried out at different flow rates. Flow rate of precursors and size and shape of the microfluidic devices is one of the important hydrodynamic parameter which affects the particle size in the reactor. The space time should be less than the reaction time to complete the reaction, which will give a narrow particle size distribution. Further, the micromixing time τ_{mixing} should be less than the τ for instantaneous precipitation reactions [21]. Hence, following experiments (Table 36.1) were carried out to know the effect of space time on the particle size and data were analyzed using UV absorption spectra. A study was carried out to understand the effect of different flow rate (20, 50, 90, and 120 ml/h) on the particle size and yield of Cu nanoparticles.

Using AFR reactor is reported in Table 36.2

TABLE 36.1 The different conditions of precursors and flow rates and particle size obtained in microreactor

Run No.	$CuSO_4$ (Molar Solution)	Ascorbic Acid (Molar Solution)	Flow Rate-Total (in ml/h)	Grams of PVP (per 100 ml)	Temperature (in °C)	Particle Size in Microreactor by PSD (nm)
1.	0.01	0.04	20	0.1	75–80	591
2.	0.01	0.04	50	0.1	75–80	574
3.	0.01	0.04	90	0.1	75–80	357
4.	0.01	0.04	120	0.1	75–80	270

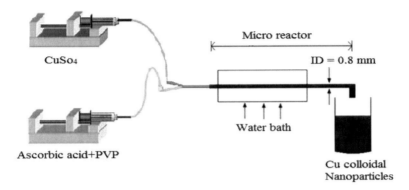

FIGURE 36.1 Schematic diagram of the experimental set up for synthesis of copper nanoparticles in microreactor.

As shown in Table 36.1 the details are given related to the various runs. The small quantity of the PVP (capping agent) was also added in order to restrict the growth of the particle size. As reported in the table the reactions were carried out at 75–80°C temperature while, the molar ratio of the precursors was kept constant. The reaction is pseudo-first order as the quantity of the precursor copper sulfate works as a limiting reactant.

36.2.3 PREPARATION OF STABLE COLLOIDAL COPPER NANOPARTICLES USING ADVANCED FLOW REACTOR

100 ml solution of copper sulfate of 0.02 molar concentrations prepared. 100 ml solution of ascorbic acid of 0.055 and 0.01 molar concentrations were prepared and 0.1 g of Cetyltrimethyl ammonium bromide (CTAB) added. Ascorbic acid was used as a reducing agent, while, CTAB was used as a surfactant. Aqueous solutions of copper sulfate, ascorbic acid, and CTAB were passed through the advance flow reactor and the reaction mixture was maintained at 85–90°C temperature. Copper sulfate was reduced by ascorbic acid which results in the formation of copper nanoparticles. The experimental conditions used for the copper nanoparticles synthesis

TABLE 36.2 The different conditions of precursors and flow rates and particle size obtained in Advanced-Flowreactor

Exp. No.	CuSO4 (Molar Solution)	Ascorbic Acid (Molar Solution)	Total Flow Rate (ml/h)	CTAB (per 100 ml Solution)	Temperature (in °C)	Particle Size (nm)
L1	0.02	0.055	50	0.1 g	85–90	320
L2	0.02	0.055	100	0.1 g	85–90	301
L3	0.02	0.055	150	0.1 g	85–90	218
L4	0.02	0.055	200	0.1 g	85–90	206
L5	0.02	0.11	100	0.1 g	85–90	224

the experiments were carried out at different flow rates from 50 to 200 mL/h. The change in the flow rates and precursor concentration was done as the Advanced-Flow reactor had problems of clogging of the particles. The specific heart shape has design which cannot handle the large formation of the precipitate as well as the long chain higher molecular weight surfactant molecules. Advanced-Flow reactor experimental setup is shown in Figure 36.2.

FIGURE 36.2 Corning® Advanced-Flow Reactors experimental setup.

36.2.4 CHARACTERIZATION

Absorbance measurements were made using UV-Visible spectrophotometer. Copper nanoparticles microstructure and morphology were studied by using field emission scanning electron microscopy (FESEM). Particle size distribution analysis were made using Malvern PSD instrument.

36.3 RESULTS AND DISCUSSION

36.3.1 COLLOIDAL COPPER NANOPARTICLES FORMATION IN THE CHEMICAL REDUCTION METHOD WAS USING MICROREACTOR/ADVANCED-FLOW REACTOR

In the chemical reduction method, various colloidal copper nanoparticles were prepared at different reaction conditions. In microreactor, copper nanoparticles were formed where the temperature was maintained at 75–80°C and the solution color became dark red when the molar ratio of the reducing agent to precursor was four. Figure 36.3 shows the absorption spectra of copper nanoparticles in the presence of PVP at different flow rates.

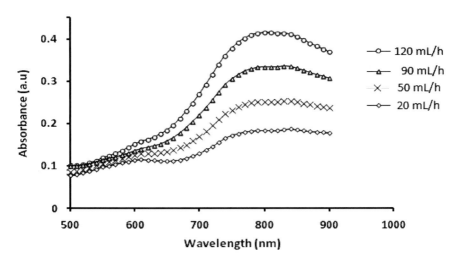

FIGURE 36.3 Effect of flow rate in the microreactor on formation of copper nanoparticles at 1:4 ratio of copper sulfate and ascorbic acid, at temperature 75°C–80°C.

The results showthat the increase in flow rate, the absorption increases by indicating the increase of number of copper nanoparticles. In the Advanced-Flow reactor, copper colloidal nanoparticles were formed at the temperature rangesfrom 85 to 90°C, the solution color became dark brown precipitate were formed due to the acceleration of the reduction rate. when the molar ratio of the reducing agent to precursor was 2.75. As shown in the Figures 36.4 and 36.5 it is found that the λ_{max} value obtained in between in 300–500 nm size for all the cases even though,the change in the flow rate conditions in Advanced-Flow reactor in the presence of CTAB. It is interesting to note that the nucleation growth and precipitation are three stages were involved. The heart shape design of the Advanced-Flow reactor gives the better mixing of precursor and generates the particles of very small size and having narrow distribution of the particles. From the UV absorbance the results indicates that the design of Advanced-Flow reactorsgive the lesser chance of agglomeration. Hence, the absorbance values and λ_{max} obtain is different than the microreactor which is based on production.

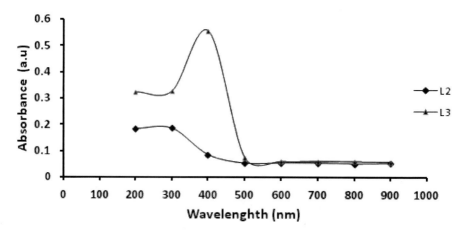

FIGURE 36.4 Effect of flow rate in the Advanced-Flow reactor on formation of copper nanoparticles at 2:5.5 molar ratio of copper sulfate and ascorbic acid, at temperature 85°C–90°C L2 at flow rate 100 ml/h and L3 at 150 ml/h.

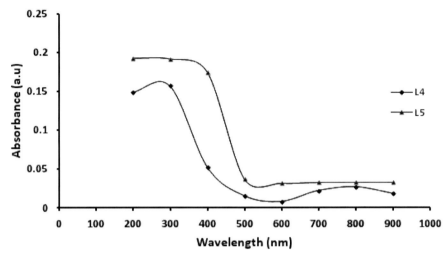

FIGURE 36.5 Effect of flow rate in the Advanced-Flow reactor on formation of copper nanoparticles at 2:5.5 molar ratio of copper sulfate and ascorbic acid, at temperature 85°C–90°C L4 at flow rate 200 ml/h and L5 at 100 ml/h with molar ratio of 0.2 to 0.11 molar ratio.

36.3.2 EFFECT OF FEED FLOW RATE ON FORMATION OF COLLOIDAL COPPER NANOPARTICLES USING MICROREACTOR/ADVANCED-FLOW REACTOR

In microreactor to determine the effect of feed flow rate (total flow rate of precursor and reducing agent) on the particle size, total flow rates from 20 to 120 ml/h were maintained at constant molar ratio (R/P) 4 and the temperature of 75–80°C in the presence of PVP surfactant. As shown in from Figure 36.6(a), (b), (c), and (d) the particle size decreases as the flow rate increased and the least particle size obtained 270nm at 120ml/h. The particle size analysis result clearly indicates that the higher residence time. Then there will be large particle size due to getting more time for agglomeration and the growth of particle.

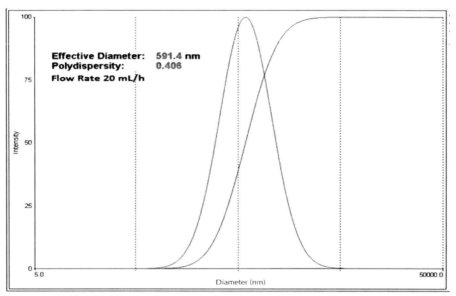

FIGURE 36.6(A) Particle size distribution of copper nanoparticles in microreactor at 20 mL/h flow rate.

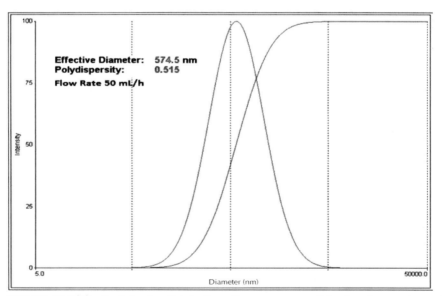

FIGURE 36.6(B) Particle size distribution of copper nanoparticles in microreactor at 50 ml/h flow rate.

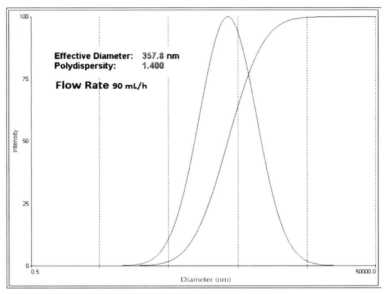

FIGURE 36.6(C) Particle size distribution of copper nanoparticles in microreactor at 90 ml/h flow rate.

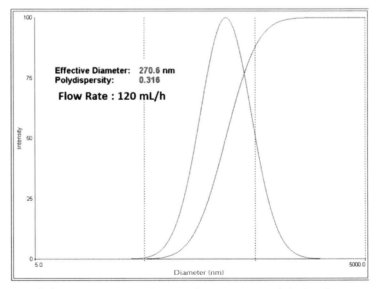

FIGURE 36.6(D) Particle size distribution of copper nanoparticles in microreactor at 120 ml/h flow rate.

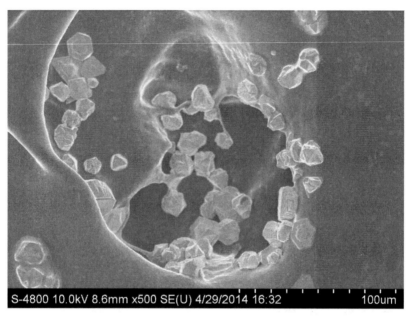

FIGURE 36.7(A) FESEM image of copper nanoparticles obtained in microreactor (CuSO4 and ascorbic acid the total flow rate of addition of the precursor was maintained at 20 ml/h).

FIGURE 36.7(B) FESEM image of copper nanoparticles obtained in microreactor (CuSO4 and ascorbic acid the total flow rate of addition of the precursor was maintained at 20 ml/h).

As Shown in Figure 36.7(a), (b) the FESEM images confirms the particle size obtained through microreactor is large in size near to 500 nm at 20 ml/h flow rate. Further, the particles obtained are not being well covered by the PVP coating. It is interesting to note that the shape of the particles is different rather than cubic and some of them are hexagonal.

In Advanced-Flow reactors, to determine the effect of feed flow rate (total flow rate of precursor and reducing agent) on the particle size, total flow rates from 50 to 200 ml/h were maintained at constant molar ratio (R/P) 2.75 and the temperature of 85–90°C in the presence of CTAB surfactant. As shown in Table 36.2 which increases in the ascorbic acid content, there is inverse effect on the particle size and it is found that, there is large impact on particle size distribution by the reactor configuration. As shown in Figure 36.8(a), (b), (c) the particle size will decrease with increasing flow rate. The mixing in Advanced-Flow reactors is also playing a role for reduction in the particle size. With increase in the flow rate, it is found that there is reduction in the particle size. At the higher flow rate 200 ml/h there is small size 206 is obtained and at the smaller flow rate 50 ml/h there is biggest particle size which is near to 320 nm. FESEM image of Figure 36.9 is also confirmed that at 200 mL/h flow rate the particle size is near to 200 nm size.

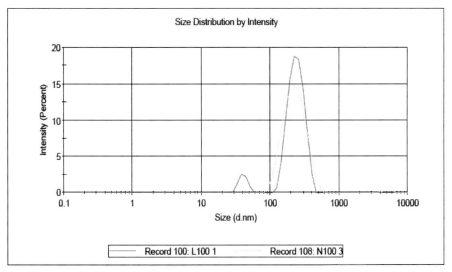

FIGURE 36.8(A) Particle size distribution of copper nanoparticles in Advanced-Flow reactor, L2 at 100 ml/h flow rate.

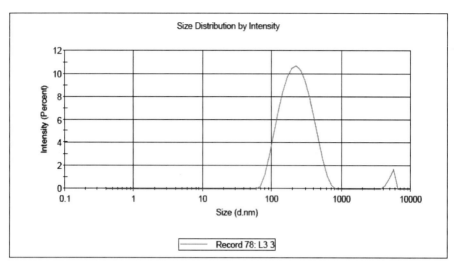

FIGURE 36.8(B) Particle size distribution of copper nanoparticles in Advanced-Flow reactor L3 at 150 ml/h flow rate.

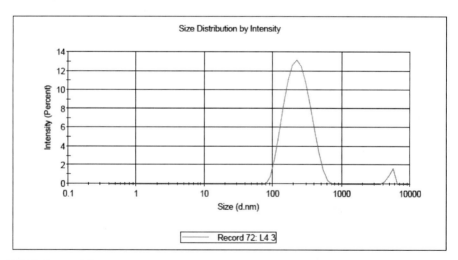

FIGURE 36.8(C) Particle size distribution of copper nanoparticles in Advanced-Flow reactor L4 at 200 ml/h flow rate.

FIGURE 36.9 FESEM image of the copper nanoparticles obtained in AFR at 2:5.5 molar ratio and flow rate at 200 ml/h, with temperature at 85°C–90°C.

Particle size formed at the flow rate of 50 ml/h from Tables 36.1 and 36.2 it is shown that 574 nm in the microreactor and 320 nm in the Advanced-Flow reactor, but molar ratio of R/P and stabilizers used are not same in both the cases. The dimension of the Advanced-Flow reactor is bigger than the microreactor but the Advanced-Flow reactor gives the lesser size particle due to the heart shape of Advanced-Flow reactors. Advanced-Flow reactors will provide the superior mixing of the precursor and reducing agent.

36.4 CONCLUSION

In this work, stable copper colloidal nanoparticle synthesized by chemical reduction method by using continuous flow microreactor and Advanced-Flow reactor. In microreactor, the copper colloidal nanoparticles formed in the temperature range of 75–80°C while, reducing agent to precursor ratio is 4 in the presence of PVP. Copper nanoparticle size was obtained 270nm at 120 ml/h flow rate. In Advanced-Flow reactor copper nanoparticle, particles in the temperature range of 85–90°C while, reducing agent to precursor ratio is 2.75 in the presence of

CTAB surfactant. At the feed flow rate of 200 ml/h copper nanoparticle size 206nm was obtained.

ACKNOWLEDGMENTS

Authors are thankful to Corning Incorporated for providing access to the Corning® Advanced-Flow Reactor facility and to the Chemical Engineering Department National Institute of Technology Warangal India.

KEYWORDS

- Advanced-Flow reactor
- Ascorbic acid
- Chemical reduction
- Colloidal copper nanoparticles
- Copper (II) sulfate
- Microreactor

REFERENCES

1. Macalek, B.; Krajczyk, L.; and Morawska-Kowal, T.; Colloidal copper in soda-lime silicate glasses characterized by optical and structural methods. *Phys. Status Solid (c)*.**2007**,*4*,761–764
2. Acharya, V.; Prabha, C. R.; and Narayanamurthy, C.; Synthesis of metal incorporated low molecular weight polyurethanes from novel aromatic diols, their characterization and bactericidal properties. *Biomaterials.***2004**, *25*,4555–4562.
3. Cioffi, N.;et al. Synthesis, analytical characterization and bioactivity of Ag and Cu nanoparticles embedded in poly-vinyl-methyl-ketone films. *Anal. Bioanal. Chem.***2005**,*382*,1912–1918.
4. Sun, S.; Murray, C. B.; Weller, D.; Folks, L.;and Moser, A.; MonodisperseFePt nanoparticles andferromagneticFePt. *Nanocryst. Superlattices. Sci.***2000**,*278*,1989–1992.
5. Dhas, N. A.; Raj, C. P.; and Gedanken, A.; Synthesis, characterization, and properties of metallic copper nanoparticles.*Chem. Mater.***1998**,*10*, 1446–52.
6. Vitulli, G.; Bernini, M.; Bertozzi, S.; Pitzalis, E.; Salvadori, P.; Coluccia, S.; and Martra, G.; Nanoscale copper particles derived from solvated Cu atoms in the activation of molecular oxygen.*Chem. Mater.***2002**,*14*, 1183–6.
7. Zhou, D. W.; Heat transfer enhancement of copper nanofluid with acoustic cavitation.*Int. J. Heat Mass Transfer.***2004**,*47*, 3109–17.

8. Quaranta, A.; Ceccato, R.; Menato, C.; Pederiva, L.; Capra, N.; and Dal Maschio, R.; Formation of copper nanocrystals in alkali-lime silica glass by means of different reducing agents. *J. Non-Cryst. Solids.* **2004,** *345/46,* 671–5.
9. Kapoor, S.; Palit, D. K.;and Mukherjee, T.; Preparation, characterization and surface modification of Cu metal nanoparticles. *Chem. Phys. Lett.* **2002,** *355,* 383–387.
10. Wang, J.; Huang, H.; Kesapragada, S. V.;and Gall, D.; Growth of Y-shaped nanorods through physical vapor deposition. *Nano Lett.* **2005,** *5,* 2505–2508.
11. Vijaya Kumar, R.; Mastai, Y.; Diamanta, Y.; and Gedanken, A.; Sonochemical synthesis of amorphous Cu and nanocrystalline Cu2O embedded in a polyaniline matrix. *J. Mater. Chem.* **2001,** *11,* 1209–1213.
12. Wu, S.;and Chen, D.; Synthesis of high concentration Cu nanoparticles in aqueous CTAB solutions. *J. Colloid Interf. Sci.* **2004,** *273,* 165–169.
13. Huang, H.; et al. Synthesis, characterization, and nonlinear optical properties of copper nanoparticles. *Langmuir.* **1997,** *13,* 172–175.
14. Wang, Y.; and Asefa, T.; Poly(allylamine)-stabilized colloidal copper nanoparticles: synthesis, morphology, and their surface-enhanced Raman scattering properties. *Langmuir.* **2010,** *26,* 7469–7474.
15. Engels, V.; Benaskar, F.; Jefferson, D. A.; Johnson, B. F. G.; and Wheatley, A. E. H.; Nanoparticulate copper routes towards oxidative stability. *Dalton Trans.* **2010,** *39,* 6496–6502.
16. Qi, L. M.; Ma, J. M.; and Shen, J. L.; Synthesis of copper nanoparticles in nonionic water-in-oil microemulsions.*J. Colloid Interf. Sci.* **1997,** *186,* 498–500.
17. Pileni, M. P.; and Lisiecki, I.; Nanometer metallic copper particle synthesis in reverse micelles.*Colloids Surf. A* **1993,** *80,* 63–8.
18. Lisiecki, I.; Sack-Kongehl, H.; Weiss, K.; Urban, J.; and Pileni, M. P.; Annealing process of anisotropic copper nanocrystals. *J. Cylinders Langmuir.* **2000,** *16,* 8802–6.
19. Pileni, M. P.; Ninham, B. W.; Gulik-Krzywicki, T.; Tanori, J.; Lisiecki, I.; and Filankembo, A.; Direct relationship between shape and size of template and synthesis of copper metal particles.*Adv. Mater.* **1999,** *11,* 1358–62.
20. Tilaki,R.M.; IrajiZad, A.;and Madhavi,S.M.; Size, composition and optical properties of copper nanoparticles prepared by laser ablation in liquids. *Appl. Phys. A: Mater. Sci. Eng.* *88,* 415–419.
21. Gaurav Prasad, et al. Stable colloidal copper nanoparticles for a nanofluid: Production and Application. *Colloids Surfaces A: Physicochem. Eng. Aspects.* **2014,** *441,* 589–597.

INDEX

A

Abscisic acid, 334
Acoustic grating, 74
Active packaging technology, 222, 225–226
Ajwan (Trachystermum ammi)
    harvesting and processing, 448
    spice essential oil, 448
Aloe Vera, 413
    agrotechniques, 414
    harvesting and postharvest technoligy, 414–415
Allspice (Pimenta dioica)
    harvesting and processing, 484–485
    quality issues, 485
    value added products, 485
Antimicrobial compounds
    cinnamon, 206
    EPS and EOs, 204
    in fresh-cut fruits, 207–208
    infusion of fruits, 205
    mechanisms of action, 205–206
    organic acids, 204
    Saccharomyces cerevisiae and Saccharomyces exiguous, 204
Anutritional requirements, vegetables, 146–149
Apples
    acidity, 107
    commercial cultivars, 109–110
    description, 106
    at different stations, 111–112
    eating quality, 108
    hybridization programs, 110
    ideal cider apple, 108
    polygenic control, 106
    polymorphism, 107
    scab breeding programme, 111
    texture, juiciness and flavor, 107
ATIS (Aconitum heterophyllum Linn.)
    agrotechniques, 415–416
    climatic and soil requirements, 415
    harvesting and yield, 416
    postharvest technolagy, 416
Atmosphere modification, 258–260
Auxins, 335

B

Banana, 104–106
Beet root, 166–167
Bio-switch
    "active packaging," 226
    antimicrobial packaging, 228
    antimicrobial starch-based film/food packaging, 229
    bioactive substances, 227
    concept, 227
    Escherichia coli and Bacillus subtilis, 230–231
    microbiological study, 230
    shelf-life, tomato (see Shelf-life)
Biotechnological approaches
    ABA biosynthesis, 498
    control oncytokininsbiosynthesis, 497–498
    control onethylene biosynthesis, 495–497
    fruit color and sweetness, 499–501
    maintaining fruit firmness, 494–495
    nutritionally essential food commodity, 492
    ripening and perishability, 493–494
    standards and expectations, 492
    vitamins and diet enrichment, 499
Blackberries, 113–114
Black cumin (Nigella sativa)
    distillation of oil, 449
    harvesting and processing, 449
    quality specifications, 449–450
Black pepper (Piper nigrum)
    harvesting, 476–477

processing of pepper, 477–478
Blue light method, 264–265
Brahmi/Thyme leaved gratiola
   agrotechniques, 418
   harvesting and postharvest technolagy, 418
Bud opening, 378
Bulb crops
   garlic, 163–164
   onion, 162–163
Burdock frutooligosaccharide (BFO), 263

C

Canopy position, 14–15
Cardamom
   curing, 452–453
   harvesting, 452
   packaging, 453
   quality issues, 453
   seeds, 454
Carrots, 165–166
Chebulic myrobalan fruit
   agrotechniques, 437–438
   harvesting postharvest technolagy, 438–439
Chemical postharvest treatments
   essential oils and plant extracts, 273–275
   fungicides, 270–273
   gasused, 268–270
   natural compounds, 267
Chilli (capsicum sp.)
   CFTRI method, 480
   drain excess emulsion, 480
   grading, 482
   harvesting, 479
   packing and storage, 482–483
   postharvest technology, 479
   solar drying, 481–482
   traditional sun drying, 479–480
   transportation, 484
Cinchona (Cinchona spp.)
   agrotechniques, 421
   description, 420–421
   harvesting and postharvest technology, 421
Cinnamon (Cinnamomum verum /Cinnamomum zeylanicum)
   harvesting, 486
   processing, grading and storage, 487
   scope of research, 488
Citronella (Cymbopogon winterianus Jowitt.)
   agrotechniques, 423
   harvesting and postharvest technoligy, 423–424
Citrus fruits, 114
Clove (Eugenia caryophyllata)
   harvesting and processing, 475
   quality, 475
Computed tomography (CT), 60
Concentration determination, 78
Continuous wave interferometer, 73
Controlled atmosphere (CA), 309–311
Coriander (Coriandrum sativum)
   harvesting and processing, 446
   postharvest treatment, 446
   quality issues, 446
Cucurbits
   cole crops, 171–172
   cucumber, 167–168
   melons, 169–171
   pumpkin and squashes, 168–169
Cumin (Cuminum cyminum)
   harvesting, 447
   processing and storage, 447
   quality specifications, 447–448
Curcuma (Curcuma spp.)
   agrotechniques, 419
   harvesting and postharvest technology, 419–420
Curry leaf (Murraya koenigii Linn), 485–486
Cut flowers (fresh)
   cooling, 385–386
   Indian scenario and trade, 350–352
   life-supporting processes, 348
   packing, 383–385
   quality control, 386–387
   quality loss
      flower senescence, 352
      growth, development and aging, 352
      leaf yellowing and senescence, 353

Index 419

shattering, 353
wilting, 352
transportation, 389–390
Cushioning materials, 32
Cytokinins, 334–335

D

Dimensional characterization, 64–66F

E

Edible coating
  alginate-based films, 255
  Arcitum lappa, 263
  atmosphere modification, 258–260
  b-aminobutyric acid (BABA), 263
  blue light, 264–265
  calcium and salicylic acid, 260–262
  chilgoza, 257
  chitosan, 256–257
  pectin, 255
  physical postharvest treatment, 266–267
  plant defence induction, 260
  semipermeable coatings, 254
  types, 254–255
Electromagnetic radiation, 56–57
Elicitors of induced disease resistance
  biologically induced resistance, 528–529
  induced resistance with chemical, 525
  inorganic elicitors, 526
  natural organic elicitors, 525–526
  physically induced resistance, 528
  synthetic organic elicitors, 526–527
Essential oils and plant extracts, 273–275

F

Fennel (Foeniculum vulgare), 450
Fenugreek (Trigonella foenum-graecum), 444–445
  harvesting, 445
  postharvest processing, 445
Flower senescence
  biochemical changes, 327–328
  description, 320
  down-regulated genes, 322–323
  ethylene-sensitive, 322
  factors, 324

inter-organ communication, 321
membrane permeability, 322
metabolic changes, 328
mRNAs, 323
nucleic acid synthesis and breakdown, 331–333
oxidative events, 329–331
physiological maturity, 320
pigments changes, 328–329
pollination, 320
remobilization process, 323
sexual reproduction, 321
ultrastructural changes, 324–327
Food safety indicators, 236–238
Forced air, 28
Forced-air ventilation, 39
Frictional characterization, 66
Fruit breeding
  approaches, 95
  characterization, 90
  description, 90
  quality
    appearance, 92
    control of ripening, 94
    softening, 93–94
    taste, 92–93
  seedless fruits, 114–120
  sensory analysis techniques, 91
Fungicides, 270–273, 511–512

G

Gibberelic acids, 335
Ginger
  candy, 460–461
  chutney, 462
  oil, 460
  oleoresin, 460
  pickle, 461
  shreds, 461
  soft drink, 461
Girdling, 12–13
Global market, 223–225
Grape, 121–125
Green chiretta (Andrographis paniculata)
  agrotechniques, 425

harvesting and postharvest technoligy, 425
Guava, 125–127

H

Harvesting
  forced air, 28
  fruits/vegetables, 25
  hydrocooling, 28–29
  mechanical method, 25–26
  precooling, 27–28
  sorting and grading, 30
  time, 27
  vacuum cooling, 29–30
Harvest factors, cut flowers
  description, 358
  Heliotropium arborescens and Lantana camara, 359
  length of flower stalk, 359
Heat treatment (HT), 252–254
Hormonal regulation, flower/petal senescence
  abscisic acid, 334
  auxins, gibberelic acids and jasmonic acid, 335
  cytokinins, 334–335
  ethylene, 333–334
Hot air treatment
  biological control and plant growth regulators, 308–309
  comparative study, 311–313
  controlled modified atmosphere treatments, 303–305
  grapefruit, 301–302
  High Temperature Forced Air (HTFA), 301
  irradiation, 306–307
  irradiation and heat, 309
  pesticide, 306
  postharvest quality, 301
  radiofrequency, 308, 311
  reducing/eliminating heat injury treatment, 302–303
Hot water dips, 300–301
Hot water immersion
  adult oriental fruit fly, 299

air treatment (see Hot air treatment)
Caribbean fruit fly, 300
fruit fly infested mango, 299
water dips, 300–301
Hydrocooling, 28–29

I

Indian ginsengor aswagandha (Withania somnifera)
  agrotechniques, 417
  description, 416–417
  harvesting and postharvest technolagy, 417–418
Indian scenario and trade, cut flowers, 350–352
Intelligent packaging, 223
Internal quality, 22, 68, 81, 114, 117, 387
Irradiation, 306–307
Irrigation, 11–12, 143, 294, 414–419, 421, 423
Induced mutations, 103

J

Jasmonic acid, 335

L

Leguminous vegetables, 173
Lemon grass (Cymbopogan flexuosus)
  agrotechniques, 425–426
  description, 425
  harvesting and postharvest technoligy, 426–427
Long pepper (Piper longum)
  agrotechniques, 427–428
  description, 427
  harvesting and postharvest technoligy, 428

M

Mace. See also Nutmeg (Myristica fragrans)
  essential oil, 474–475
  fixed oil, 473
  packaging and distribution, 472–473
Magnetic resonance imaging (MRI), 80–82
Mango
  "Amrapali" and "Imperial, 102

Index 421

breeding programme in Israel, 101
cross-pollination, 96
cultivar development, 96
hybrids, India, 96–101
Indian types, 95
Medicinal plants
 cultivation, 402–405
 farming and trade, 406–407
 organic farming, 405–406
 postharvest technoligy, 412–413
 quality control, 407–411
Medicinal yams (Dioscorea spp.)
 agrotechniques, 428–429
 harvesting and postharvest technoligy, 429–431
Microwave (MW) application, 251–252
Mineral nutrition, 9–11
Modified/control atmosphere storage, 202–204, 309–311

N

Natural microbial antagonists, 514–515
Natural ventilation, 39
Near infrared (NIR) spectroscopy, 67–69
Nondestructive quality measurement
 applications, 60–61
 assessment techniques, 54–55
 atom structure, 56
 computed tomography (CT), 60
 description, 53
 electromagnetic radiation, 56–57
 horticulture, 53
 image analysis, 62–64
 near infrared (NIR) spectroscopy, 67–69
 plantshigh-speed sorting, 54
 visual inspections, 62
 X-ray, 58–60
Nonhormonal regulation of flower senescence
 heterotrimeric G proteins, 335
 polyamines (PAs), 335
Nutmeg (Myristica fragrans)
 drying, 470–471
 fumigation, 472
 grading and flotation, 471
 harvesting and processing, 469
 inspection of grades, 471
 metal sieve grading, 471
 postharvest handling operations, 469
 sorting, 470
 yield, 469–470

O

Okra, 173–174
Oleoresin production, 459
Onion, 162–163
Optical characterization, 67
Organic production, 11–12

P

Packaging
 classification, 222–223
 definition, 218
 global active and smart packaging technology, 223–225
 historical development, 218–222
Pallatization, 31–32
Pallet construction, 389–390
Papaya, 127–130
Passive packaging, 222
Peel color, 16–18
Penicillium digitatum, 263–264
Penicillium italicum, 263–264
Peppers (hot and sweet), 156–159
Periwinkle (Catharanthus roseus)
 agrotechniques, 431–432
 harvesting and postharvest technoligy, 432–433
Pesticide, 306
Physical postharvest treatments. See also Edible coating
 gamma radiation, 249–250
 heat treatment (HT), 252–254
 microwave (MW) application, 251–252
 ultrasound, 250–251
 ultraviolet (UV) radiation, 247–249
Plant defence induction, 260
Plant growth regulators
 abscisic acid (ABA), 370
 cytokinins (CK), 369–370
 ethylene, 366–368
Postharvest disease
 antagonistic microorganisms, 247

biological control
  induced host resistance, 519–520
  microbial antagonists, 515–516
  natural microbial antagonists, 514–515
  nutrients and space, 516–518
  populations, microbial antagonist, 518
  production of antibiotics, 518–519
causes, 507–510
chemical control, 246
cold chain system, 245
consumption, 506
fungicides, 244
heat shock proteins (HSPs), 245
host resistance to infection, 512–513
hygiene practices, 513
mode of infection, 510–511
natural products' control
  acetic acid, 521
  chitosan, 522
  essential oils, 523
  flavour compounds, 520–521
  fusapyrone, 522
  glucosinolates, 522
  jasmonates, 521–522
  plant extracts, 523–524
  propolis, 522
physiological status of host, 510
preharvest factors, 513
prevention of injury, 513–514
Recommended Dietary Allowance/Reference Daily Intake (RDA/RDI), 244
thiabendazole (TBZ), 246
Postharvest factors affecting quality
  fruits breeding (see Fruits breeding)
  maturity stage
    abscission layer, 21
    acid ratio, 24
    aroma (flavor), 20
    firmness (flesh firmness), 21–22
    juice content, 22
    leaf/flowers/inflorescence condition, 20
    pallatization, 31–32
    peel color, 16–18
    shape, 18–19
    size, 19, 20
    specific gravity, 24
    starch pattern index (SPI), 23–24
    sugars (total soluble solids), 22
  packaging (see Packaging)
Postharvest factors, cut flowers
  bunching, 377
  grading, 375–377
  harvesting, 371–374
  microorganisms, 365–366
  pH, vase solution, 364–365
  pre cooling, 374–375
  respiration rate, 361–363
  storage temperature, 360–361
  treatment, floral preservatives
    biocides, 382–383
    bud opening, 378
    pulsing, 378
    vase solutions, 378–382
  water relation, 363–364
Postharvest pest management
  arthropods, 294
  horticulture, 294
  insects, 295
  nature and causes, 296–297
  organic treatments, 294
Postharvest treatments
  bacterial antagonists, 277–279
  eukaryotic microorganisms, 279–281
  heat and cold, 297–298
  hot water immersion, 298–300
  salts, 276
Preharvest factors, cut flowers
  environmental factors, 354
  genetic factors, 56–357
  light intensity, 354–355
  photo period, 355
  relative air humidity (RH), 355–356
  transgenic strategies, 357–358
  varieties, 354
Pruning, 12–13, 112–113
Pulsing, 378

Index

Q

Quality
  description, 2
  environmental factors
    cultural factors, 9–15
    temperature, 3–9
    fruits and vegetables, 78
    preharvest factors, 2–3

R

Radiofrequency, 311
Radish, 164–165
Root crops
  beet root, 166–167
  carrots, 165–166
  radish, 164–165
Rootstock, 13–14

S

Salts used in postharvest treatments, 275–276
Sandal wood (Santalum sp.)
  agrotechniques, 422
  harvesting and postharvest technoaigy, 422
Scavengers/emitters, 155, 224, 225
Serpentwood (Rauvolfia serpentina)
  agrotechniques, 435
  climatic and soil requirements, 435
  description, 434–435
Shelf-life, 14, 15, 27
  tomato
    colour change, 234
    microflora appearance, 235
    weight loss, 232–234
Small cardamom (Elettaria cardamomum), 451–452
Smart and intelligent packaging technology, 223, 226
Solanaceous vegetables
  eggplant, 153–156
  peppers, 156–159
  potato, 159–161
  tomato, 150–153
Solanums (Solanum spp.)
  agrotechniques, 436
  harvesting and postharvest technolagy, 436–437
Somatic mutations, 102–103
Spice crops
  damage and reproduction, 444
  microbiological safety, 443
  postharvest management, 443
Starch pattern index (SPI), 23–24
Storage
  description, 37–38
  paramount importance, 33
  storage temperature, 33–36
  types, 38–41
  fruits and vegetables (preservation)
    antimicrobial compounds, 204–208
    firmness, 195
    global production, 194
    modified/control atmosphere, 202–204
    persistent problems, 196–197
    quality attributes, 195
    relative humidity effects, 200–201
    temperature effects, 198–200
Sweet potato, 174

T

Temperature
  air pollutants, 8–9
  bioactive compounds, 5
  ethylene-treated exposed fruit, 3
  grapes and oranges, 4
  hail and frost damage, 8
  preharvest exposure, 3–4
  and relative humidity, 39–41
  sun exposure, field, 5
  sunlight, 5–7
  wind, 7–8
Thinning, 12–13
Tomato, 150–153
Transportation, 42–43
Tree age, 14
Tulsi (Ocimum sanctum)
  agrotechniques, 433
  climatic and soil requirements, 433
  harvesting and postharvest technoligy, 434

Turmeric (Curcuma longa)
   Agmark standards, 467
   grading, packing and storage, 465–467
   grinding and milling, 467
   harvest, 462–463
   oleoresin production, 467–468
   postharvest handling, 463–465

U

Ultrasonic methods
   characteristics, 70
   description, 69–70
   flow measurement, 79
   imaging techniques, 75–77
   level measurement, 79
   production, 70
   real-time application, 77–78
   temperature measurement, 79
   velocity and velocity measurements, 70–72

V

Vegetables breeding
   anutritional requirements, 146–149
   components of quality, 142
   cosmetic quality, 143–145
   nutritional requirements, 146
   taste qualities, 145